D0213009

Metamorphic petrology

Metamorphic petrology

Akiho Miyashiro
Professor Emeritus
State University of New York, Albany

New York
Oxford University Press
1994

First published in 1994 by UCL Press

UCL Press Limited
University College London
Gower Street
London WC1E 6BT

First published in the United States in 1994 by
Oxford University Press, Inc.
200 Madison Avenue
New York, NY 10016

Printed in Great Britain

Library of Congress Cataloging-in-Publication Data
Miyashiro, Akiho
Metamorphic petrology / Akiho Miyashiro
p. cm.
Includes bibliographical references and index
ISBN 0-19-521026-3
1. Rocks, Metamorphic. I. Title
QE475.A2M578 1994
552'.4—dc20 93-14016
CIP

Dedicated to the memory of
the late Professor Maurice Ewing

Contents

Preface

This book is intended to outline metamorphic petrology with a new conceptual framework based on its progress in the past 20 years. The main body of this book is divided into three parts.

Part I (Chs 1–6) describes the general principles of metamorphic petrogenesis. Since the 1950's, thermodynamic studies of metamorphic reactions and high-pressure synthetic experiments have come to play an increasingly important role in metamorphic petrology (Chs 3, 5, 6). Thermodynamic analyses made in direct connection with field petrological studies have greatly contributed to our better understanding of the actual physico-chemical conditions of metamorphism (Chs 2, 6). Although thermodynamic investigations shed new lights on the equilibrium relations involved in metamorphism, metamorphism is not merely static equilibrium states of rocks, but is a process of crustal changes through time. Petrologists, being fascinated by the advance in the study of equilibrium relations, tended to forget this trite fact, but were effectively reminded of the important significance of the fact by thermal model studies of regional metamorphism made in the 1970's. Equilibrium relations preserved in the mineral assemblages of metamorphic rocks usually record the state at or near the thermal peak of each rock, but it is important that rocks reach their thermal peaks at different times in different parts of a metamorphic region. The progress of research in these subfields has necessitated essential modifications of some basic concepts in metamorphic petrology such as progressive metamorphism, isograds, metamorphic reactions, buffering and chemical migration (Chs 2, 6).

Part II (Chs 7–10) deals with traditional subjects of metamorphic petrology such as metamorphic facies, a P/T ratio classification of regional metamorphism and mineralogical changes of metamorphic rocks in response to changing temperature and pressure, and further discusses the petrology and tectonics of metamorphic belts. The recent theoretical progress in the general principles outlined in Part I has greatly modified our ideas of metamorphic facies and mineralogical changes with temperature and pressure (Chs 7, 10). The advent of plate tectonics in the late 1960's opened a new viewpoint for understanding metamorphic processes in relation to tectonics. My former book, *Metamorphism and metamorphic belts* (1973), represented an early attempt to connect metamorphic petrology to tectonic studies. Since then, however, some regional-metamorphic terrains through the world have been intensively investigated by combinations of geologic, petrologic, thermobarometric, geochronologic, gravity and thermal model studies, and this has cast a new light on the nature and mechanisms of the tectonic processes relevant to regional metamorphism (Chs 8, 9).

Part III (Chs 11–13) gives a general survey of petrological and mineralogical data of metamorphic rocks. The data are classified and systematized according to the metamorphic facies and P/T ratio types of metamorphism to which they belong. Thus, Chapters 11 and 12 deal with data of metamorphic facies characteristic of the low- and medium-P/T ratio types and the high P/T ratio type respectively. Chapter 13 gives a very brief survey of metamorphic facies of contact metamorphism.

Appendix I gives a table of the symbols for mineral names used in this book. Appendix II is a glossary of metamorphic petrogenesis. It summarizes important technical terms in metamorphic petrogenesis which are discussed in scattered places through this book.

Appendix III describes a history of metamorphic petrology with special emphasis on the development of new ideas and viewpoints. The present state of metamorphic petrology represents a transient state in the course of historical development of the science, and so can be well understood only when it is seen in a historical perspective.

Since our science is ever advancing, all discussions and conclusions in this book are tentative. The examination and modification of basic concepts necessarily involve personal judgements and preference, and so are colored by my preoccupations. Moreover, no one can master the mushrooming literature in petrology in these days. There must be omissions of some important views and data in this book.

I am greatly indebted to Dr Fumiko Shido for her advice, encouragement and help in all stages of writing. Dr Roger Mason kindly read through an early manuscript of this book and gave me invaluable comments and advice. Drs W. G. Ernst, S. Banno, S. Maruyama, Jin-Han Ree, T. Hirajima, M. Hashimoto and R. Hall also read parts of an early version of the manuscript and gave helpful advice.

December 1992 — Akiho Miyashiro

PART I
GENERAL PRINCIPLES OF METAMORPHIC PETROGENESIS

1 Introduction: preliminary ideas of metamorphism

1.1 Metamorphism, temperature and local equilibrium

1.1.1 Metamorphism and metamorphic crystallization

Metamorphism is a collective name for the mineralogical, chemical and textural changes of rocks that take place in an *essentially solid state*, that is, without substantial melting, in the deeper parts of the Earth at various temperatures higher than those encountered on the Earth's surface. A small amount of fluid mainly composed of H_2O and/or CO_2 may be present between mineral grains. For the low-temperature limit of metamorphism, see Miyashiro (1973: 19–21).

These changes usually involve the formation of new minerals by various mechanisms, including nucleation and growth of metamorphic minerals at the expense of pre-existing amorphous, submicroscopic and/or clastic sedimentary materials, as well as of older metamorphic and igneous minerals. In this book, such changes are collectively called **metamorphic crystallization**. Metamorphic crystallization will lead to the formation of the appropriate equilibrium mineral assemblages under the pertinent conditions, if those conditions continue for a long time.

1.1.2 Temperature of metamorphism

New mineral grains form by chemical reactions, whose rates increase with temperature. When a rock is kept at a temperature higher than a certain value for a considerable length of time, most or all of the rock will be subjected to metamorphic crystallization, resulting in the formation of new mineral grains. This threshold temperature varies in accordance with several parameters: the composition and texture of the rock, the presence or absence of a fluid phase, the composition of the fluid phase, pressure and nonhydrostatic stress, and the length of time that the rock remains above the threshold temperature. It is as low as 200°C in some cases, but is more commonly around 350°C, and it can be even higher.

In the course of metamorphism, the temperature of rock changes with time. As discussed in the next chapter, it is believed that in the majority of cases each metamorphic rock preserves the mineralogical composition that it acquires at or near the **thermal peak**, which means the highest temperature that the rock experiences during metamorphism.

In ordinary regional metamorphism, granitic, pelitic and basaltic rocks begin to melt at temperatures between 600° and 750°C, if an aqueous fluid is present (Figs 1.1, 11.6). In the range of pressure up to about 13–17kbar, the temperature of onset of melting (solidus temperature) of such rocks decreases with increasing pressure. In some cases, melting may occur in the absence of an aqueous fluid. In such cases, the temperature of melting tends to be much higher than it is in the presence of an aqueous fluid. However, even in these cases, the presence of hydrous minerals such as muscovite and biotite considerably decreases the temperature of melting, as will be discussed in the next section. With increasing temperature, the degree of melting (i.e. the proportion of the rock that is melted) increases, and the rocks may pass into an igneous regime.

The highest value of the temperature of regional metamorphism (temperature of thermal peak) inferred from mineral assemblages of metamorphic rocks is about 1000°C (e.g. Ellis 1980, Barnicoat 1983, Sandiford et al. 1987).

1.1.3 Local equilibrium

While the Earth's crust is undergoing metamorphism, there is a temperature gradient, and so the crust as a whole cannot be in equilibrium. Moreover, mineral assemblages that are incompatible at the same pressure and temperature occur commonly in rocks only a few centimeters apart. So, a large volume in the crust is in neither thermal nor chemical equilibrium. It is generally assumed, however, that any part of the rock complex undergoing metamorphism is essentially in thermal and chemical equilibrium, if that part is sufficiently small. This is possible if no mutually incompatible phases are in direct contact, and if phases of variable composition are to show only continuous compositional variation from point to point. This is the concept of **local equilibrium** (J. B. Thompson 1959). It may also be assumed that spontaneous internal changes in rocks will be such as to establish local equilibrium. The application of equilibrium thermodynamics to the study of mineral assemblages in metamorphic rocks is based on these assumptions.

(a) Melting interval of muscovite granite

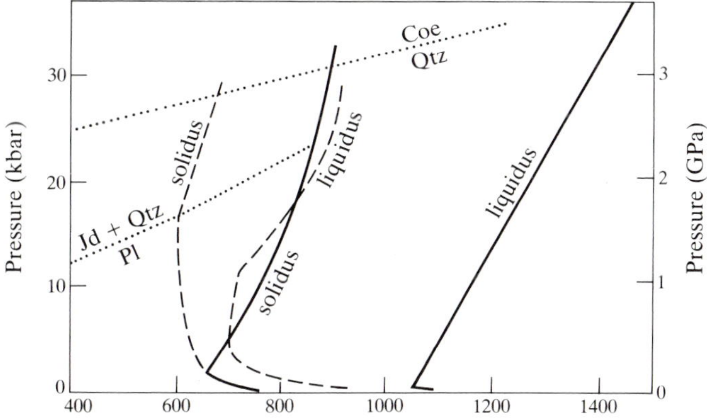

(b) Melting interval of tholeiitic basalt

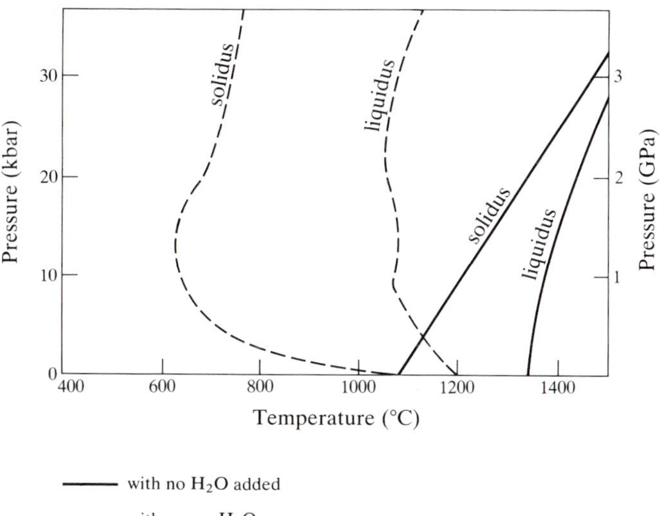

Figure 1.1 Experimentally determined melting intervals of (a) muscovite granite and (b) tholeiitic basalt, with no H_2O added (solid lines) and with excess H_2O (dashed lines). Rocks with excess H_2O are accompanied by an aqueous fluid. A detailed explanation of (a) is given in Section 1.2. The solidi of metapelites are close to that in (a). Modified (a) from Huang & Wyllie (1973), and (b) from Green & Ringwood (1967b) and Green (1982).

1.2 Metamorphism, intergranular fluid and melting

At the beginning of this chapter, metamorphism was defined as a process that occurs in an essentially solid state. The word "essentially" is used to allow for the fact that rocks undergoing metamorphism may contain a very small amount of aqueous fluid, or silicate melt, or both, between mineral grains. The following is a preliminary account of such aqueous fluids and silicate melts. More detailed discussions of intergranular aqueous fluids are given in Chapters 4 and 6, and the relation of dehydration to partial melting is discussed again in Chapter 11.

1.2.1 Distinction between intergranular fluid (vapor) and liquid (melt)

Rocks undergoing metamorphism at low or high temperatures commonly (but probably not always) contain a small amount of fluid mainly composed of H_2O or CO_2 or both between mineral grains. Metamorphism that takes place in the presence or the absence of such a fluid phase is called **fluid-present** and **fluid-absent** metamorphism, respectively.

When the temperature of metamorphism increases to a certain value above 600°C, some rocks begin to melt. The resultant melt phase (liquid phase) is composed of melted silicate substance with some H_2O dissolved in it. The temperature and mode of melting vary not only with pressure and bulk-rock composition but also with the presence or absence of intergranular fluid and, if it is present, the composition of that fluid. In discussions related to melting, it is important to distinguish clearly between the melt phase (liquid) and intergranular fluid (vapor or gas). We will use the terms "melt" and "vapor" to distinguish between the two in this section.

A vapor phase mainly composed of H_2O dissolves rock-forming substances (e.g. Fig. 4.1). The solubility of such substances in vapor is usually of the order of 10 wt % of the vapor at 10kbar, and it increases with pressure, as shown in Figure 1.2. On the other hand, a melt can dissolve H_2O. The solubility of H_2O in melt is of the order of 10–15 wt % of the melt at 10kbar, and increases with pressure (Fig. 1.2). There is a miscibility gap between vapor and melt at least to 35kbar, that is, over the whole possible range of pressure for crustal metamorphism, although the gap becomes smaller with increasing pressure (Stern & Wyllie 1973).

Melting of rocks usually advances over a range of temperature, and the proportion of melt in the rocks increases with increasing temperature, until melting is completed by the disappearance of all crystalline phases. Since melting is an endothermic reaction, it tends to impede the rise of temperature. The melt may be segregated into layers or veins to form migmatites. With an increasing proportion of melt, the rock enters a transitional state between metamorphic and igneous regimes. If the proportion of melt be-comes higher than a certain value (say, 20% or 30%), it may separate from the solid residues to form a discrete magma body. (For detailed discussions of melting of rocks, see Atherton & Gribble 1983 and Ashworth 1985).

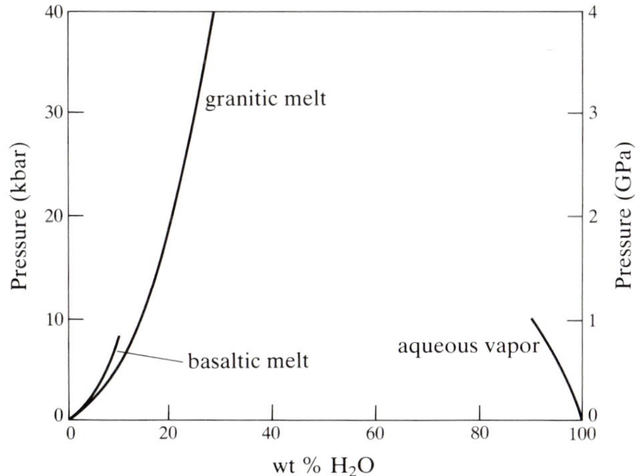

Figure 1.2 Solubility of H_2O in granitic and basaltic melts, and of solid in an aqueous vapor phase (Stern & Wyllie 1973).

1.2.2 Effects of H_2O and CO_2 on melting

1.2.2.1 Classification of metamorphic minerals according to their melting behaviors. From the viewpoint of melting, common minerals in rocks may be divided roughly into three classes as follows:

Class 1 includes quartz, alkali feldspar and the albite component of plagioclase. Mixtures of these minerals melt at a relatively low temperature, particularly where excess H_2O is present.

Class 2 includes hydrous silicates such as muscovite, biotite and hornblende. These minerals contain H_2O, Na_2O and K_2O, which have a strong effect of lowering the temperature of melting. In melting they react with associated minerals and commonly produce other minerals (i.e. they undergo incongruent melting).

Class 3 includes refractory minerals such as pyroxene, Al_2SiO_5 minerals, cordierite and garnet. These minerals may be formed by incongruent melting of hydrous minerals.

1.2.2.2 Change of melting behavior with the H_2O content of the system. Most metamorphic rocks contain some hydrous minerals when partial melting begins in higher amphibolite facies and granulite facies. This greatly complicates the processes of melting. For a better understanding, a review of a series of melting experiments on a muscovite granite by Huang & Wyllie (1973) is given below.

The granite used contained 13.8% muscovite, and hence had an H_2O content of 0.6% by weight bound to the crystal structure of muscovite. The H_2O content of the samples used for melting could be increased by the addition of varied amounts of H_2O to the rock. Figure 1.3 shows the results of the experiment on samples with varied H_2O contents under a pressure of 15 kbar. The results are extrapolated to 0% H_2O content; that is, a sample which was completely dehydrated beforehand. This extrapolation yields a solidus temperature of about 1100°C for a sample with 0% H_2O. For a rock sample with no addition of H_2O (i.e. H_2O 0.6%), the solidus temperature is 810°C. It should be noted that this small H_2O content causes such a great decrease of the solidus temperature. In this case, all the H_2O liberated by the melting of muscovite is dissolved in the melt without forming a vapor phase.

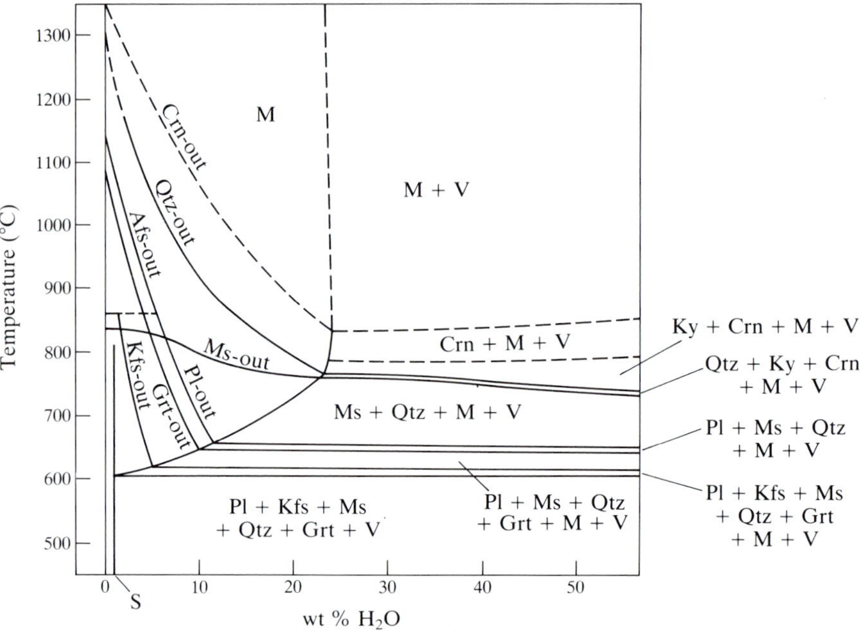

Figure 1.3 Melting behaviors of a muscovite granite at 15 kbar with varied amounts of H_2O added. S represents the H_2O content of the muscovite granite with no H_2O added (i.e. 0.6 wt % H_2O). V: vapor; M: melt. Corundum (Crn) is unstable. After Huang & Wyllie (1973).

When melting begins, any sample of muscovite granite with added H_2O has an aqueous vapor phase present, however small the amount of added H_2O may be. So the solidus temperature falls further to about 605 °C regardless of the amount of H_2O added.

If the amount of added H_2O is only a few per cent, a small increase of temperature beyond about 605 °C results in an increase of the amount of melt. The melt dissolves all the H_2O in the vapor phase, and the vapor phase disappears. With further increase of temperature, the melt becomes unsaturated with H_2O.

At temperatures above about 800 °C, the H_2O content of melt saturated with H_2O lies between 20% and 25% by weight. In this case, if a sample contains more than 25% H_2O, melting occurs always in the presence of a vapor phase. Not only the solidus but also the liquidus temperature is greatly decreased by the presence of vapor.

Figure 1.3 shows observed melting relations at 15 kbar; that is, at a higher pressure than the pressure range of ordinary regional metamorphism. The melting behavior of metamorphic rocks with hydrous minerals varies greatly not only with the H_2O content of the system but also with pressure. The complicated melting relations of metamorphic rocks with muscovite and biotite in the pressure range of ordinary regional metamorphism is reviewed in Chapter 11 (Figs 11.5, 11.6).

1.2.2.3 Effects of CO_2 on melting. The volatiles in rocks undergoing metamorphism usually include not only H_2O but also CO_2, CO, CH_4 and others. The proportion of CO_2 in particular may be very high in the metamorphism of calcareous rocks and in some high-temperature metamorphism (granulite facies).

In felsic melts, the solubility of CO_2 is one order of magnitude smaller than that of H_2O (Mysen 1977). When partial melting begins, most of the H_2O present in the system will be dissolved in the melt phase, and hence the coexisting gaseous fluid phase may be highly concentrated in CO_2.

The addition of volatile components such as CO_2 that are sparingly soluble in melt causes only a small change in the melting temperature and in the composition of the melt. If the fluid phase is composed only of H_2O, increasing pressure greatly decreases the melting point of ordinary silicate below several kbar. If the fluid phase is composed only of CO_2, on the other hand, increasing pressure increases the melting point. However, the melting point at high pressures in the presence of pure CO_2 fluid is still slightly lower than that in the absence of fluid (Mysen 1977).

1.3 Geologic causes of metamorphism in regions with a continental-type crust

1.3.1 Concept of regional metamorphism

Metamorphism occurs in regions not only with a continental-type crust but also with an oceanic-type crust, and further within the lithospheric mantle (e.g. Mason 1990, Chs 5, 9). However, this book deals with metamorphism only in regions with a continental-type crust.

Large-scale metamorphic complexes, commonly hundreds or thousands of square kilometers in extent, are exposed in many regions with a continental-type crust, including both continents and island arcs. In this book, the metamorphism that produced such a large-scale metamorphic complex is called **regional metamorphism**, if the *P–T* conditions of the metamorphism vary gradually on a regional scale (usually over a large distance), and are not controlled locally by individual igneous intrusions.

Regional metamorphism usually produces foliated metamorphic rocks. However, the formation of foliation is not regarded as an essential part of the definition of regional metamorphism. Regional-scale formation of unfoliated low-temperature metamorphic rocks, usually called **burial metamorphism**, is regarded as a variety of regional metamorphism when it occurs in a continental region.

It was known from the 19th century that regional-scale metamorphic rocks occur in folded mountain belts. As a result, regional metamorphism came to be regarded as genetically related to orogeny, and in the first half of the 20th century this relationship was interpreted from the viewpoint of geosynclinal theory. In the past 30 years, the advent of plate tectonics and the subsequent progress in the tectonic study of regional metamorphic complexes have made it clear that the tectonic environments of regional metamorphism are diverse. Regional metamorphism takes place in many cases within pre-existing continental-type crusts, but in other cases within accretionary complexes that are growing at the margin of pre-existing continents and island arcs, usually in genetic relation to subduction.

The regional metamorphic complexes now exposed on the surface of the Earth were crystallized at some depths in the crust and then uplifted so as to be exposed by erosion. The mechanism of uplift is, therefore, a crucial question. Where metamorphism has occurred in the middle or shallow depths of thickened crust, the uplift and exposure of the metamorphic rocks will have occurred subsequently by isostatic adjustment. In other cases, various processes involving faulting, thrusting and underplating will have caused such uplift.

1.3.2 Tectonic settings of regional metamorphism

As already stated, metamorphic crystallization occurs where the temperature of rocks becomes higher than a certain critical value. Hence, regional metamorphism occurs where the temperature of continental-type crust is raised on a regional scale, whatever the cause of the temperature rise may be. The temperature increase in the crust occurs in various tectonic settings for different reasons, and is controlled by many factors, which include the thickness and composition of the crust, the content and distribution of radioactive elements, the heat flux from the mantle to the crust, the thermal conductivity of rocks, the presence of magmatic intrusions, the movements of hydrothermal fluids, the thermal effects of other geologic bodies, and the rate of erosion.

The main tectonic settings of regional-scale heating and thus of regional metamorphism are outlined in the following, and more detailed discussions of each class are given in Chapter 9.

1.3.2.1 Arc–trench zones

ARC ZONES (ISLAND ARCS AND ACTIVE CONTINENTAL MARGINS). Some active continental margins have a very thick continental crust which leads to the accumulation of radiogenic heat. Other active continental margins and most mature island arcs have a continental crust of medium thickness. Island arcs and active continental margins are usually intensively and extensively intruded by magmas, which play an essential rôle in the temperature increase that causes regional metamorphism.

Intrusions may be broadly classified into two categories: gabbroic and granitoid. Gabbroic magmas are generated in the upper mantle, and may be intruded into the crust. The gabbroic intrusions transfer a large amount of heat from the mantle to the crust, contributing to the temperature rise of the crust. It is likely that intrusion of such gabbroic magmas occurs mainly into the lower crust, owing to their relatively high density, and this may be the reason why gabbros are relatively rare on the surface of such regions. In the lower crust, partial melting may generate granitoid magmas, which then rise to the middle and upper crust, and contribute greatly to the temperature increase that causes regional metamorphism at shallow depths.

TRENCH ZONES. Island arcs and active continental margins are accompanied by a trench zone on their oceanic side. There, an oceanic plate is subducted beneath the associated arc, and subduction may occur continuously over a long geologic time. Accretionary complexes form in the trench zone. Because of the subduction, the complexes may be carried to great depths to be metamorphosed at high pressures. The rises of temperature in such accretion masses are usually only moderate because the cold subducting oceanic plate underlies them. The result is high-*P/T* ratio metamorphic conditions. If an accretionary complex remains in a moderate or shallow depth, moderate- or low-*P/T* ratio metamorphic conditions may occur, or there may be no metamorphism at all.

1.3.2.2 Continental collision zones. Continental collision usually increases the thickness of the crust by folding, thrusting and other mechanical processes. In crust thickened in this way, the accumulation of radioactive heat may cause a marked increase of temperature, leading to metamorphic crystallization at depth. Moreover, continental collision commonly causes intrusion of plutonic masses, which also contributes to a large-scale increase of temperature inducing or promoting metamorphic crystallization.

Prior to a continental collision, the ocean basin between the two approaching continental masses must be subducted beneath one of the continents with the consequent formation of an arc–trench zone. High-*P/T* ratio metamorphism may occur in the trench zone as described earlier. Then, after the ocean basin has been completely consumed, the continental mass directly connected to the ocean basin may begin to be subducted beneath the other continent. This will cause high-*P/T* ratio metamorphism in the subducting continental mass and its overlying sedimentary piles. Owing to the buoyancy of the subducting continental mass, subduction stops after a short time, and an uplift may begin. The zone of continental plate subduction in the western Alps shows mineral assemblages characteristic of much higher pressures than in zones of oceanic plate subduction in circum-Pacific regions. This may mean that the greater buoyancy of subducted continental mass has caused uplift of metamorphic rocks from a greater depth.

1.3.2.3 Continental extension zones. Zones of extension and rifting in the continent such as the Basin and Range province in the western United States are characteristically accompanied by high heat flow, which suggests the existence of sufficiently high temperatures in the lower crust to cause metamorphic crystallization. In regions of this kind, heating of the crust occurs by thinning of the lithosphere (including the crust) with a resultant rise of high-temperature asthenospheric material and the consequent generation and rise of magmas in the mantle and crust.

Some regional metamorphic belts, such as the Pyrennean belt, have recently been claimed to represent an ancient extension zone within a continent.

Although these three tectonic settings probably cover almost all well crystallized Phanerozoic regional metamorphic complexes, they are not exclusive, and other types of setting may exist, particularly in the early Precambrian.

1.3.3 Igneous intrusion and the distinction between regional and contact metamorphism

There are many regional metamorphic complexes that are not accompanied by broadly coeval intrusions and so appear to have undergone a temperature increase by some cause other than magmatic heat supply. The most conspicuous example is metamorphism that produces glaucophane-schists.

On the other hand, other regional metamorphic complexes are accompanied by, and appear to have some genetic relationship to, broadly coeval igneous intrusions, commonly mainly granitoid with some accompanying gabbroic intrusions. Such igneous intrusions are even accompanied by the extrusion of coeval felsic volcanic rocks in some regions, e.g. the Mesozoic region of the western United States and the Ryoke belt of Japan. The relationship between this type of regional metamorphism and igneous intrusions has been interpreted in many different ways. Among the interpretations, we may recognize the existence of three different patterns. First, some geologists considered that both regional metamorphism and intrusions are concomitant processes of orogeny. One is neither the result nor the cause of the other. Harker (1932) was an early proponent of this attitude. Secondly, others regarded igneous intrusions as a result of regional metamorphism and associated partial melting (e.g. Wegmann 1935). Thirdly, still others considered that regional metamorphism is the result of regional-scale heating by a group of intrusions (e.g. Barrow 1893).

An isolated igneous intrusion commonly produces a relatively narrow metamorphic aureole around it. This is **contact metamorphism**. The question arises as to how we can distinguish contact and regional metamorphism, where the latter is closely associated with igneous intrusions. The most crucial criterion is the pattern of temperature distribution. The temperature of contact metamorphism is controlled by the relevant intrusion. Thus, isotherms show a concentric pattern around the intrusion. On the other hand, the temperature of regional metamorphism does not appear to be controlled by individual intrusions but shows a more regional-scale pattern of variation. The resulting isotherms tend to be nearly straight, irrespective of the shapes and positions of intrusions, as shown, for example, by Holdaway et al. (1982) and Shiba (1988). Some tracts of such regional metamorphism have a median line (axis) showing the highest metamorphic temperature, the position of which shows little or no relation to individual intrusive masses, although there may be many plutonic intrusions.

1.4 Pressure of metamorphism

1.4.1 Thermal-peak pressure

Pressure can vary during the course of metamorphism. Since metamorphic rocks usually preserve the mineral assemblage formed at, or near, their thermal peak, the pressure inferred from the mineral assemblage of a rock is the pressure at, or near, the thermal peak. Generally this does not coincide with the highest pressure that the rock experiences during metamorphism.

Pressure, which affects the phase equilibria of minerals, is probably usually close to the **lithostatic pressure** caused by the overlying column of rock. The value of lithostatic pressure can be calculated as $(dgh/10)$ bar or $(dgh \times 10^{-5})$ GPa at a depth of h km under a rock column with average density of d g cm^{-3} for the acceleration of gravity $g = 980$ cm sec^{-2}.

The pressure is about 10 kbar (i.e. 1 GPa) at the base of average stable continental crust, 35 km thick. The maximum observed thickness of the continental crust in present-day and Cenozoic orogenic belts is about 70 km, with a pressure of the order of 20 kbar (i.e. 2 GPa) at the base. According to determinations by geobarometers, most of the metamorphic rocks now exposed on the surface of the Earth crystallized in the pressure range 1–10 kbar, corresponding to depths of 3–35 km. At depths shallower than these, the prevailing temperature is probably usually too low to cause crystallization, whereas at greater depths metamorphism must be widespread, but the resultant metamorphic rocks will only rarely be uplifted to the extent that they are exposed at the surface.

Some crustal rocks metamorphosed in and around zones of subduction and continental collision appear to have been crystallized at depths of 100 km or greater, judging from the pressure inferred from their mineral assemblages. The most characteristic minerals indicating such high pressures are coesite and diamond, which are stable at pressures of about 30 kbar or higher (Fig. 12.1). Coesite has been found to occur in some metamorphic rocks in the western Alps (Chopin 1984), Norway and China. Diamond has been found in some crustal metamorphic rocks in Kazakhstan (Sobolev & Shatsky 1990) and China.

1.4.2 Stress and deformation

Rocks in the crust are subjected to forces caused partly by the weight of overlying rocks, and partly by tectonic activity. To express the force or stress at any point in a rock body, we can use three suitably chosen orthogonal planes on which the stresses are only normal to the planes. These planes and the stresses normal to them are called **principal planes** and **principal stresses** respectively.

The three principal stresses are denoted as σ_1, σ_2 and σ_3 in order of decreasing magnitude. The **mean stress**, $\bar{\sigma} = (\sigma_1 + \sigma_2 + \sigma_3)/3$, represents confining pressure. The ordinary thermodynamic pressure P, which affects phase equilibria of minerals and which is discussed in the preceding section, appears to be approximately equal to the mean stress.

The total state of stress at a point in a rock may be regarded as consisting of two components. One is the mean stress, which represents the hydrostatic component of the stress. The other component is **non-hydrostatic** or **deviatoric stress**. Stress in a rock causes **strain**, which is change in volume and/or shape. The hydrostatic component of stress causes a change in volume of the rock body, whereas the deviatoric stress causes a change in shape.

When strain is very small, rocks show **elastic behavior**, which means that strain occurs instantaneously upon application of stress and, when the stress is eliminated, the body returns almost instantaneously to the initial, undeformed state. When strain is increased beyond a certain value (called the **elastic limit**) at relatively low temperature and pressure and with a relatively high rate (temporal rate), the rock exhibits **brittle behavior**. This means that the rock breaks along fractures and shows cataclasis and faulting. At higher temperature and pressure, particularly when the rate of strain is low, rock shows **plastic behavior**, which involves pervasive permanent deformation that produces distorted mineral grains and folds without fracturing.

Under the P–T conditions of regional metamorphism, deformation of rocks occurs mainly by plastic flow. In this case, permanent deformations of rocks are caused dominantly by two mechanisms. One is **intracrystalline plastic deformation**, which includes slip (translation gliding), mechanical twinning and kinking in single crystals, combined with movements of crystal dislocation such as climb. The other is **intercrystalline plastic deformation**, which includes grain-boundary sliding and diffusive flow. A region of a mineral grain boundary under greater stress has greater Gibbs energy, and so is less stable, than other parts under less stress. Hence, chemical migration occurs from the former part of the crystal to the latter parts, where crystal growth results. This process is called the **diffusive flow** and it changes the shapes of crystals. It occurs in either the presence or the absence of an intergranular fluid phase. One type of diffusive flow through an intergranular fluid phase is the so-called **pressure solution**.

When a metamorphic rock is undergoing deformation, the mean stress may be considerably greater than the lithostatic pressure. The difference between the mean stress and lithostatic pressure has been called **tectonic overpressure**. Its magnitude depends largely on the strength of the rocks, which in turn varies with composition, temperature, rate of deformation and other factors (e.g. Rutland 1965). Since metamorphism takes place at high temperatures, the mechanical strength of rocks is not usually high. So, tectonic overpressure is usually relatively small, say, less than 1 kbar under normal metamorphic conditions (e.g. Brace et al. 1970).

Non-hydrostatic stress has an important effect on the kinetics of metamorphic reactions, and may conceivably have an effect on the phase relations in some cases, though this has not been demonstrated. Harker (1932) proposed the idea that certain minerals characteristic of crystalline schists, such as almandine and kyanite, are stable only in the presence of nonhydrostatic stress. Although this view was supported by the dominant majority of metamorphic petrologists in the 1930s and 1940s, it is no longer acceptable (Appendix 3.2).

1.5 Bulk-rock compositions

The chemical reactions that take place with changing P–T conditions are constrained by the chemical composition of the rocks.

In regional metamorphism, quartzite made up of quartz alone and limestone made up of calcite alone usually show a gradual increase of grain size with increase of temperature, but no mineral changes. This is because quartz by itself and calcite by itself are stable in all temperature ranges of regional metamorphism, and there is no other mineral to react with in the same rock. On the other hand, psammitic and quartzo-feldspathic rocks are more sensitive to temperature changes, and may undergo a few mineralogical changes with increasing temperature, such as the disappearance of chlorite and the appearance of garnet. Pelitic, mafic and impure calcareous rocks are most sensitive to temperature change.

1.5.1 Metapelites

Metapelites (metamorphosed pelitic rocks) often contain minerals that are too highly aluminous to occur in common igneous rocks, e.g. andalusite, kyanite, sillimanite, almandine garnet, chloritoid, staurolite and cordierite. In ordinary igneous rocks, most of the Al_2O_3 is combined with Na_2O, K_2O and CaO, making feldspars, in which the Al_2O_3/ (Na_2O + K_2O + CaO) ratio is 1.0. Many metapelites are comparable with common igneous rocks in their Al_2O_3 content, but are much lower in CaO and Na_2O (Table 1.1). Hence, metapelites usually show Al_2O_3/(Na_2O + K_2O + CaO) ratios considerably greater than 1.0. As a result, they contain highly aluminous minerals.

Metapelites contain chlorite, muscovite, biotite and other hydrous minerals.

Table 1.1 Average chemical compositions of common igneous and metamorphic rocks (wt %).

	AVERAGE					
	Granodiorite	Gabbro	Amphibolite	Low-grade metapelite	High-grade metapelite	Mica schist
SiO_2	65.01	48.24	50.3	59.93	63.51	64.3
TiO_2	0.57	0.97	1.6	0.85	0.79	1.0
Al_2O_3	15.94	17.88	15.7	16.62	17.35	17.5
Fe_2O_3	1.74	3.16	3.6	3.03	2.00	2.1
FeO	2.65	5.95	7.8	3.18	4.71	4.6
MnO	0.07	0.13	0.2	—	—	0.1
MgO	1.91	7.51	7.0	2.63	2.31	2.7
CaO	4.42	10.99	9.5	2.18	1.24	1.9
Na_2O	3.70	2.55	2.9	1.73	1.96	1.9
K_2O	2.75	0.89	1.1	3.54	3.35	3.7
H_2O	1.04	1.45	—	4.34	2.42	—
CO_2	—	—	—	2.31	0.22	—
A/NKC	0.93	0.71	0.68	1.56	1.91	1.65
F/FM	0.44	0.31	0.38	0.40	0.53	0.49

Note: A/NKC is the mole ratio $Al_2O_3/(Na_2O + K_2O + CaO)$, which in feldspars is unity. The value of A/NKC is near 1.0 in granodiorite, because the rock is mainly composed of feldspars and quartz. The value is much smaller than 1.0 in gabbro and amphibolite, which contains Ca-pyroxene and Ca-amphibole. The value is much higher than 1.0 in metapelite, which contains characteristic minerals with high A/NKC ratios. F/FM is the mole ratio FeO/(FeO+MgO).

Data sources: Daly (1933) for average granodiorite and gabbro; Poldevaart (1955) for average amphibolite and mica schist (H_2O-free); Shaw (1956) for low-grade and high-grade metapelites.

Pre-metamorphic pelitic sedimentary rocks usually contain some intergranular aqueous fluid. Most of the reactions that take place with increasing temperature in pelitic rocks release H_2O. Some or most pre-metamorphic pelitic rocks and low-temperature metapelites contain a small amount of carbonate minerals, which disappear by reaction with associated minerals, liberating CO_2, in relatively low-temperature stages of prograde metamorphism. Reactions that release H_2O and CO_2 are called **dehydration** and **decarbonation reactions**, respectively. Such reactions cause many mineralogical changes with increasing temperature that result in the appear-ance and disappearance of minerals.

Metapelites themselves show a considerable range of chemical composition. The sequence of metamorphic reactions that takes place with increasing temperature varies with small changes in bulk-rock chemical composition.

1.5.2 Metabasites

Metabasites (metamorphosed mafic rocks) are high in FeO, MgO and CaO, and so contain minerals which are high in these chemical components, e.g. chlorite, actinolite, hornblende and relatively calcic plagioclase. The original mafic rocks may have been, for example, anhydrous basaltic rocks, or strongly altered tuff breccias. To change anhydrous igneous rocks into low-temperature metabasites, which usually contain a large proportion of hydrous minerals, H_2O must be supplied from the surroundings, e.g. from associated pelitic rocks that are simultaneously undergoing metamorphism. With a further increase of temperature, they show a considerable number of mineralogical changes that release H_2O from minerals.

Usually, metabasites are made up of relatively few minerals, and hence have many degrees of freedom in the terminology of the phase rule (§3.1). For example, amphibolite, a typical metabasite, is essentially composed of only two minerals, hornblende and plagioclase. Commonly a major mineral in metabasites is amphibole, which is a very complex solid solution made up of many components. Hence, under given P–T conditions, the composition of amphibole varies in a complicated way with bulk-rock composition. Theoretical analysis of mineralogical changes is much more difficult in metabasites than in metapelites.

1.5.3 Impure calcareous rocks

Prior to metamorphism, impure calcareous rocks are mixtures of calcite and/or dolomite (or ankerite) with clayish and sandy substances. Metamorphic reactions in such rocks usually involve H_2O and/or CO_2, and are strongly influenced by the prevailing activities of H_2O and CO_2. This causes various complicated problems which will be discussed in later chapters.

1.6 The Scottish Highlands as a well investigated type regional metamorphic region

Metamorphic petrology in the proper sense began in the Caledonian metamorphic region of the Scottish Highlands with the work of George Barrow toward the end of the 19th century. Subsequently this metamorphic region was investigated by generations of British geologists, including such big names as Alfred Harker, C. E. Tilley, H. H. Read and W. Q. Kennedy. In the past 20 years, renewed active researches by young geologists have made it the best investigated regional metamorphic complex in the world.

This section offers a brief general description of the geology, petrology and tectonics of the Highlands, for two purposes. First, it will provide a preliminary idea about the regional metamorphic complex; secondly, it will offer a background knowledge for better understanding the advanced research findings on this region that are reviewed in many places in this book.

1.6.1 Geologic relations

Figure 1.4a shows the major tectonic lines and tectonic units as well as the distribution of Caledonian metamorphic rocks in Scotland.

There is a Precambrian craton on the northwestern side of the Moine Thrust. This craton is made up of metamorphic rocks of the Lewisian Complex which, before the Cretaceous–Tertiary opening of the northern North Atlantic, was continuous to the Precambrian of Greenland, repres-enting the northeastern part of the original Canadian Shield.

The Caledonian (early Paleozoic) metamorphic region of the Scottish Highlands, which is now under consideration, lies between the Moine Thrust Zone on the northwestern side and the Highland Boundary Fault on the southeastern side (Fig. 1.4a). The Moine Thrust Zone consists of a series of east-dipping thrusts. Movement along the Highland Boundary Fault continued throughout the Caledonian orogeny and subsequently. On the southeast side of the Highland Boundary Fault lies the Midland Valley, which is a graben that originated in late Caledonian time with considerable subsequent faulting, and is partly filled by the Devonian Old Red Sandstone and Carboniferous coalfields.

The Great Glen Faults cut the metamorphic region into two parts: the **Northern Highlands** on the northwestern side and the **Grampian Highlands** on the southeastern side. This fault is believed to have a lateral displacement of 100 km or more (Kennedy 1948).

The Caledonian metamorphic region is composed of the **Moine** and the **Dalradian** Supergroups. The Moine was deposited in the late Precambrian, and is now exposed in the Northern Highlands and a northwestern part of the Grampian Highlands. It is mainly composed of quartzo-feldspathic metamorphic rocks derived from psammitic sediments. On the other hand, the Dalradian was deposited in late Precambrian to Cambrian times, and possibly also in the early Ordovician. It is exposed in the eastern, southern and southwestern parts of the Grampian Highlands, and is mainly composed of pelitic, calcareous, quartzitic and mafic metamorphic rocks.

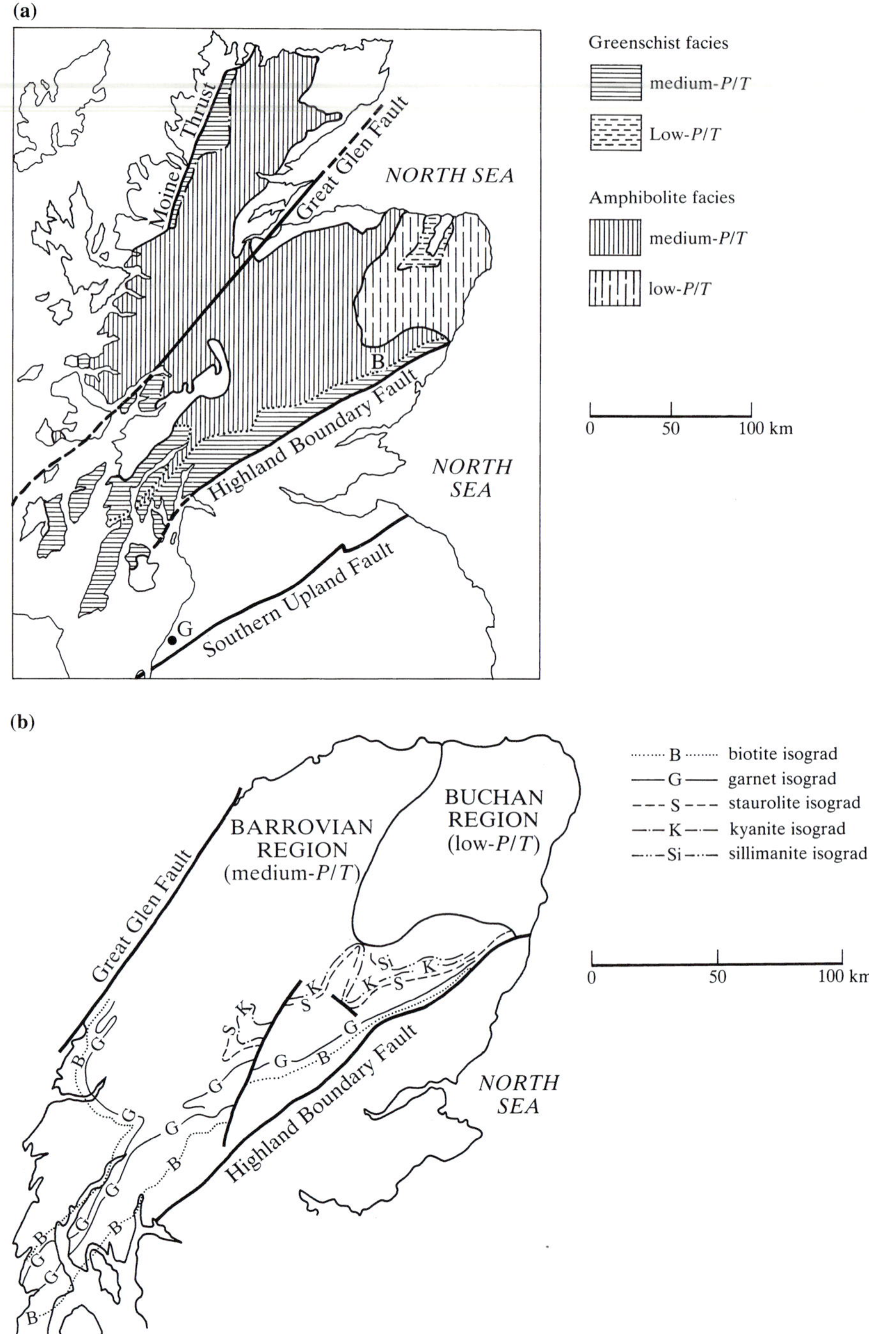

Figure 1.4 Metamorphic maps of the Caledonides in the Scottish Highlands, based on Fettes (1979), Chinner & Heseltine (1979), and Harte & Hudson (1979). (a) The Caledonian metamorphic region is indicated by hatching. Barrow's (1912) study area is indicated by letter B. The glaucophane schist mass of the Girvan area is indicated by G. (b) Isograds of Barrovian zones for pelitic metamorphic rocks of the Dalradian in the south-eastern Highlands (Grampian Highlands).

1.6.2 Zonal mapping of the Dalradian

In a small area (marked B in Fig. 1.4a) in the southeastern part of the Grampian Highlands, George Barrow (1893, 1912) discovered that pelitic rocks of the Dalradian showed regular mineralogical changes with increasing temperature. This area is about 40–70 km southwest of the City of Aberdeen and it lies directly on the northwest side of the Highland Boundary Fault between the rivers North Esk and South Esk. The temperature of metamorphism increases northward away from the Highland Boundary Fault. Barrow found that the area could be divided into a sequence of zones, characterized by the entrance of biotite, garnet, staurolite, kyanite and sillimanite, successively. This sequence of zones, composed of the lowest-grade zone (called the chlorite zone), succeeded by the biotite, garnet, staurolite, kyanite and sillimanite zones, is now widely called the **Barrovian sequence**. Tilley (1925) and many other workers extended similar mapping toward the southwest, reaching the west coast of Scotland.

The lines on a map which show the first appearance of biotite, garnet and so on, are called **isograds**. (Detailed discussions of isograds are given in §2.2–2.5) The distribution of the isograds delimiting Barrovian zones is shown in Figure 1.4b.

In Figure 1.4, the Caledonian metamorphic region of the Grampian Highlands is divided into two regions, which are characterized by medium-*P/T* ratio and low-*P/T* ratio metamorphic conditions, respectively (The meaning of these terms will be described in detail in Chapter 8.). In the formations where pelitic rocks are widespread, metamorphism of the medium-*P/T* ratio condition produces the above-mentioned Barrovian sequence of zones. On the other hand, metamorphism of low-*P/T* ratio produces different sequences of zones with cordierite, andalusite, and sillimanite in a pelitic region (e.g. Harte & Hudson 1979). The Dalradian region showing the Barrovian sequence of zones is called the **Barrovian region**, whereas that showing the low-*P/T* ratio metamorphic zones is called the **Buchan region** (Fig. 1.4b). The Barrovian region is the tract where Barrovian isograds are shown in Figure 1.4b. The stratigraphy and structure are continuous between the two regions. The two regions were subjected to the same metamorphism, but differed in *P–T* conditions, which changed continuously across the Grampian Highlands and between the two regions.

Although the petrographic characteristics of the Barrovian region were fairly well known, even before the Second World War, those of the Buchan region have been clarified only in recent years.

The distribution of actual *P–T* values in the Dalradian is shown in Figures 2.4 and 9.5, and its tectonic significance is discussed in § 9.2.2.

1.6.3 The Barrovian sequence of zones for pelitic rocks

The upper half of Figure 1.5 shows the progressive mineral changes of metapelites in the Barrovian sequence. Pelitic rocks of each zone show a range of chemical variation, and so of mineralogical variation. The diagram is intended to show the temperature ranges of occurrence of minerals in pelitic rocks now exposed in the region. All minerals shown for each zone do not necessarily coexist in the same rocks. The mineral assemblages that actually occur are described and discussed in §10.1 (Fig. 10.1).

A cursory description of progressive mineral changes in metapelites is given below in order of increasing temperature of metamorphism.

Chlorite zone. Common pelitic rocks in the low-temperature part of the chlorite zone are phyllites mainly composed of muscovite, chlorite and quartz in varying proportions. Albite and graphite-like matter are also widespread. Quartz may preserve the outlines of original detrital grains or may have lost them due to metamorphic crystallization. The muscovite is usually phengite. Paragonite has been found to occur only sporadically in metapelites. Pyrophyllite has not been found.

	Progessive metamorphic zone	Chlorite zone	Biotite zone	Garnet zone	Staurolite zone	Kyanite zone	Sillimanite zone
Metapelites	chlorite						
	muscovite						
	biotite						
	garnet						
	staurolite						
	kyanite						
	sillimanite						
	sodic plagioclase						
	quartz						
	Rock type	Greenschist			Amphibolite		
Metabasites	albite						
	oligoclase and more calcic plagioclase						
	epidote						
	actinolite						
	hornblende						
	chlorite						
	garnet						
	biotite						

Figure 1.5 Diagram showing the ranges of occurrence of minerals in metapelites and associated metabasites (so-called epidiorites) in the region of the Barrovian sequence in the Scottish Highlands. After Harker (1932), Wiseman (1934), McLellan (1985) and Baker (1985).

Biotite zone. Typical pelitic rocks in this zone are biotite–chlorite–muscovite–albite–quartz schist. The amount of biotite is usually very small and it tends to increase with increasing temperature. There are still many pelitic rocks that are rich in chlorite and devoid of biotite.

Garnet zone (almandine zone). With an increase of temperature, almandine garnet begins to occur. Almandine on the almandine isograd contains considerable amounts of MnO and CaO, but the contents of these components in the mineral decrease rapidly with increasing temperature within the zone. Typical pelitic rocks of this zone are almandine–biotite–muscovite–albite–quartz schists, commonly with compositional layering. Chlorite may still be present.

Staurolite zone. With a further increase of temperature, staurolite begins to occur in some rocks. Staurolite persists into the kyanite and the sillimanite zone. Typical pelitic rocks of the staurolite zone are staurolite–almandine–biotite–muscovite–plagioclase–quartz schists.

Kyanite zone. With a further increase of metamorphic grade, kyanite begins to join some of the staurolite-bearing schists.

Sillimanite zone. The sillimanite isograd represents the boundary between the stability fields of kyanite and sillimanite. Many rocks in the sillimanite zone contain kyanite as an unstable relict mineral, because kyanite is very refractory (non-reactive). The sillimanite occurs both as large porphyroblasts and as small fibrolite needles commonly growing at the grain boundaries between biotite and quartz.

Generally speaking, the temperature of metamorphism in the Barrovian region was not high enough to cause the disappearance of muscovite by its reaction with quartz to produce K-feldspar. (Therefore, the sillimanite zone there is the sillimanite–muscovite zone of Fig. 8.5.) However, Baker & Droop (1983) and Baker (1985) discovered a small exceptionally high-temperature area where K-feldspar occurs in metapelites.

Most metapelites in the Scottish Highlands contain a small amount of graphite or other carbonaceous matter, which should keep them in rela-tively reduced states during progressive metamorphism. Some pelitic rocks, however, are devoid of these substances, and as a result show highly variable oxidation states. Chinner (1960) described a group of gneisses of this type from a lowest temperature part of the sillimanite zone. In these gneisses, biotite becomes higher in MgO and garnet becomes higher in MnO content with increasing oxidation state (§3.6).

1.6.4 Metamorphic rocks other than pelitic in the Scottish Highlands

1.6.4.1 Metabasites in the Barrovian region. Metabasites in the Scottish Highlands may be classified into two groups: Green Beds and epidiorites. The former represent metamorphosed pyroclastic deposits of broadly basaltic composition, occasionally mixed with some clastic sedimentary materials. The pyroclastic deposits tend to suffer sorting of component grains and later chemical alteration, and so their chemical composition tends to deviate from that of the original basalt. "Epidiorite" is an old field term used by British geologists to represent metamorphosed doleritic intrusions cutting through metasediments. In low-temperature metamorphic zones, the epidiorites commonly preserve not only original igneous minerals such as pyroxene but also phenocrysts and chilled margins.

The lower half of Figure 1.5 shows the progressive mineral changes in the epidiorites in the Barrovian region of the Dalradian. The metamorphic rock types produced are greenschist in Green Beds, and greenschist and greenstone in epidiorites in low-temperature zones, and amphibolite in high-temperature zones.

1.6.4.2 Calcic metamorphic rocks. The Moine Supergroup is mainly composed of quartzo-feldspathic rocks whose mineralogical compositions are not sensitive to variation in temperature and pressure. Since metapelites are rare, zonal mapping similar to that in the Dalradian is difficult to achieve in the Moine region. Therefore, Kennedy (1948), Winchester (1974) and others undertook zonal mapping of the Moine region on the basis of mineral changes in calc-silicate rocks, which are generally scarce but widespread in the Moine, and discussed correlations with respect to P–T conditions between the Moine Supergroup and the Barrovian zones.

1.6.5 Tectonic setting and cause of the Caledonian metamorphism in the Scottish Highlands

1.6.5.1 Tectonic setting. In late Precambrian and Cambrian time, northwest Scotland was a part of the North-American–Greenland continent. On the southeast side (in the present orientation) of it, there was an ocean, called the *Iapetus*, beyond which further to the southeast there was another continental mass that included southeast Britain and the northern European continent. In Ordovician and Silurian time, the Iapetus Ocean gradually closed by subduction of the ocean floor beneath the continents on both sides. The regional metamorphism of the Scottish Highlands took place in the southeast margin of the North-

American–Greenland continent, presumably owing to the northwestward subduction of the Iapetus Ocean floor. Various interpretations have been proposed for the positions of the subduction zones and the modes of subduction. The thermal peak of the Dalradian regional metamorphism is generally believed, on the basis of radiometric data, to have occurred in Arenigian (early Ordovician) time.

The Highland metamorphism on the continental margin would have been accompanied by a belt of volcanic arc and granitoid plutons. It is conceivable that high-*P/T* ratio metamorphism took place broadly simultaneously along the subduction zone on the southeast side, resulting in the formation of paired metamorphic belts (§9.1), and that the high-*P/T* ratio metamorphic complex is hidden under the ground except the small blueschist mass in the Ballantrae Complex of the Girvan area as marked as G in Figure 1.4a (Dewey & Pankhurst 1970).

Lambert & McKerrow (1976) assumed that an active mid-oceanic ridge within the Iapetus Ocean was subducted beneath the North-American–Greenland continent in early Ordovician time, that the subducting hot lithosphere and associated basaltic magma of the ridge region heated the overlying Dalradian–Moine sedimentary pile to cause regional metamorphism, and that the basaltic magma rose later into the rock mass already undergoing metamorphism to give an additional heating effect, which resulted in the formation of the highest grade (i.e. sillimanite) zone. They claimed that the middle Ordovician gabbro masses of Aberdeenshire were formed by consolidation of the basaltic magma.

1.6.5.2 Cause of temperature increase. Barrow (1893) regarded the Dalradian regional metamorphism as having been caused by the thermal effect of intrusions, mostly still hidden, of "Older Granite". In other words, the regional metamorphism was regarded as a regional-scale contact metamorphism. On the other hand, Harker (1932) considered the regional metamorphism to have been caused by a regional-scale rise of temperature in the depths of the orogenic belt and independent of granitoid intrusions.

These two early trends of thinking have been advocated by later authors with various modifications. For example, Lambert & McKerrow (1976) considered that the regional metamorphism was caused by heating due to the subduction of hot lithosphere of a mid-oceanic ridge and to the subsequent rise of associated basaltic magma, as stated above. Harte & Hudson (1979), generally supporting Barrow's genetical view, considered that both the regional metamorphism and the formation of migmatite and older granite bodies were a consequence of extensive regional magma intrusion in the deep crust.

On the other hand, Richardson & Powell (1976) regarded the regional metamorphism as having been caused by conductive heating in the depths of the orogenic belt. They claimed that, during the regional metamorphism, the Dalradian was underlain by a thick layer of the Moine and the Lewisian, that the temperature rise in Dalradian metamorphism could well be ascribed to the spontaneous heating within the thickened crust with only a modest heat flow from the underlying Moine layer, and that the metamorphism had no genetic connection with the thermal effect due to subduction below the Scottish Highlands.

2 A new conceptual basis for metamorphic petrology: prograde and progressive metamorphism, isograds, isotherms and isobars

2.1 Temporal changes in *P–T* conditions during metamorphism

2.1.1 Prograde and retrograde stages

Although a metamorphic process generally involves temporal changes of both temperature and pressure, for simplicity we first consider the change of temperature only.

Figure 2.1 schematically illustrates a simple course of temperature change of a rock during regional metamorphism. The temperature of the rock first begins to increase at point A, and metamorphic crystallization begins at point B. Then, the temperature of the rock reaches a maximum value (C in Fig. 2.1). This is a **thermal peak** or **thermal maximum**. The period between B and C is called the **prograde stage** of metamorphism. Metamorphic crystallization in the prograde stage is simply called **prograde metamorphism**. It is followed by a gradual decrease of temperature, during which a small extent of metamorphic crystallization may occur, until point D is reached. If there is metamorphic crystallization at this stage, the period between C and D is called the **retrograde stage** of metamorphism. (Note that the prograde and retrograde stages have been defined here simply in relation to increasing and decreasing temperature, respectively, with no reference to variation in pressure.) The course and length of the temperature decrease depend mainly on the post-metamorphic history of uplift and consequent erosion of the region (e.g. England & Richardson

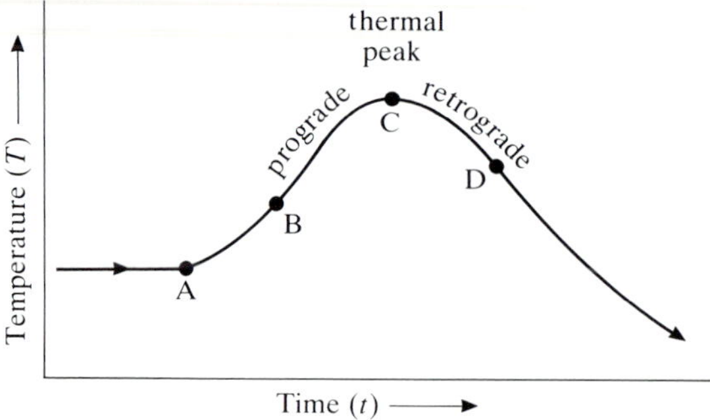

Figure 2.1 Schematic diagram showing a possible course of change of temperature that a rock undergoes during regional metamorphism.

1977). If erosion is active, the rock approaches the surface accompanied by a relatively rapid decrease of temperature, until it is exposed on the surface.

If erosion is negligible, say, in an arid region, the rock may remain virtually at the same depth in the crust, and the temperature of the rock changes toward a certain value, corresponding to the steady-state geotherm for the area at that depth. If the heating that causes the metamorphism is brought about mainly by igneous intrusion, the temperature in the vicinity of the intrusive bodies will commonly become higher than the temperature corresponding to the steady-state geotherm. So, if there is no erosion, a thermal peak occurs and then the temperature decreases toward the value corresponding to the steady-state geotherm at the depth. However, if the temperature increase that causes the metamorphism is brought about by radioactive heating of an initially cold crust, and if there is no erosion, the temperature continues to increase toward the value corresponding to the steady-state geotherm at that depth. In this case, there is no thermal peak.

Thus, all metamorphic rocks now exposed on the surface have experienced one thermal peak or more, whereas metamorphic rocks that remain at depth may or may not have experienced a thermal peak. Erosion has an essential connection with the thermal history of metamorphic rocks.

2.1.2 Mineral changes in the prograde stage and thermal peak

Metamorphic crystallization at low temperatures produces hydrous minerals such as chlorite, muscovite and epidote in metapelites as well as in metabasites. In the prograde stage, these rocks undergo a series of **dehydration reactions**, e.g. $B = D + H_2O$, where B and D represent a mineral or minerals. If carbonate-rich sedimentary rocks are present, they undergo a series of **decar-**

bonation reactions such as $B = D + CO_2$. There are reactions that involve both H_2O and CO_2, such as $B = D + H_2O + CO_2$, and $B + CO_2 = D + H_2O$ (Fig. 3.5). In some cases, a large amount of CO_2 may be produced by the oxidation of graphite or of some graphitic material in pelitic rocks, which is caused by infiltrating fluid. The H_2O and CO_2 produced by any method forms either an intergranular fluid phase or molecules adsorbed along boundaries between mineral grains. They probably move mainly upward to the surface of the Earth by mechanical flow through large or small channelways and/or by chemical diffusion through fluid and grain boundaries (Hanson 1992).

By comparing chemical analyses of pelitic rocks metamorphosed at relatively low and high temperatures, Walther & Orville (1982) estimated that the average metapelite loses a total of 5.0 wt % of volatiles (2.6% H_2O + 2.4%CO_2) during prograde metamorphism. If this amount of volatiles remained where it was produced, it would occupy 12 vol. % of the rock at 500°C and 5 kbar. Actually nearly all the volatiles escape from the rock toward the Earth's surface.

The enthalpies of devolatilization reactions at 500°C and 5 kbar are of the order of 84 kJ mol^{-1} of released $H_2O + CO_2$. The heat required for devolatilization is about 170 kJ kg^{-1} of average metapelite. If most devolatilization is assumed to occur in a temperature range of 400–600°C, the heat required for heating from 400° to 600°C is 210 kJ kg^{-1} of the rock. Thus, a total of 380 kJ of heat energy is required for the metamorphism of each kilogram of average metapelite.

In addition, there are some prograde reactions that do not involve dehydration and decarbonation, such as phase transformations between polymorphs and Mg–Fe exchange reactions between coexisting ferromagnesian minerals. The thermal effects of such reactions are small compared with dehydration and decarbonation reactions.

When mafic igneous rocks with high-temperature mineral assemblages undergo low-temperature metamorphism, the reactions are exothermic. Because such rocks are present usually in small amounts compared with pelitic and calcareous rocks in regional metamorphic complexes, regional metamorphism as a whole is normally endothermic.

Rocks undergoing regional metamorphism are not isolated in space, but are underlain and overlain by other rocks which are also undergoing metamorphism. As discussed in §2.5, the thermal peaks of rocks in a metamorphic complex are not synchronous, but are reached at different times in different parts and at different depths, and the time differences could be up to tens of millions of years. The timing of the thermal peak may be earlier or later, with increasing depth, depending on the cause of metamorphism, the structure of metamorphic belt and other factors.

For example, in a case where a swarm of plutonic intrusions at a great depth is the main source of heat, in the crust above the level of the intrusions the metamorphic rocks probably reach their thermal peak at earlier times with increasing depth (Fig. 2.9). H_2O released by prograde reactions at greater depth rises into the higher crust where the temperature of metamorphism is lower. When the higher crust reaches its thermal peak, the deeper crust is already in a retrograde stage. So, most of the H_2O that rises from deeper crust could pass through rocks above in a relatively early part of their prograde stage. It is known in some regional metamorphic complexes that a large amount of externally derived H_2O infiltrates pervasively into pelitic rocks undergoing low-grade metamorphism (Ferry 1984). Such H_2O may come from pelitic rocks undergoing metamorphism at deeper levels as well as from associated plutonic intrusions.

Because metamorphic crystallization begins at point A in Figure 2.1, the earlier history of the rock is not recorded in mineral assemblages. In cases where the thermal-peak temperature is relatively high, most, or all, of the minerals that formed in the prograde stage are obliterated or compositionally adjusted, so that the resultant rock shows a mineral assemblage approximately equilibrated at the thermal-peak condition. In such a case, it is difficult to determine the history of mineralogical changes in the prograde stage.

2.1.3 Mineral changes in the retrograde stage

In the retrograde stage, possible reactions that proceed with decreasing temperature are generally the reverse of prograde reactions, mostly producing minerals containing more H_2O and CO_2. For such reactions to take place, however, H_2O and CO_2 must be present in the rock. If all the H_2O and CO_2 evolved in the prograde stage have left the system before, and at, the thermal peak, such retrograde reactions cannot take place. In reality, usually only relatively small amounts of H_2O and CO_2 appear to remain in the rock, causing only limited retrograde reaction. New hydrous minerals produced at this stage grow, for example, along margins and cleavages of thermal-peak minerals. Such retrograde minerals can be recognized from textural and compositional relations. Therefore, the mineral assemblages equilibrated at the thermal peak can usually be recognized and discussed.

It might be thought that during the retrograde stage the peak mineral assemblages will become increasingly unstable with decreasing temperature. This is not always true. For example, if a rock undergoes a univariant dehydration reaction $A + B = C + H_2O$ in the late prograde stage, and if the released H_2O leaves the system, the mineral assemblage $A + B + C$ (without H_2O) forms. In the retrograde stage, this assemblage could continue to be stable at temperatures lower than the equilibrium temperature of the above reaction.

Some metamorphic reactions do not involve participation of volatiles such as H_2O and CO_2, e.g. polymorphic transformations, exsolution, and Mg–Fe exchange reactions. Such reactions would continue in the retrograde stage until they are halted by decreased temperature. Hence, geothermometers based on exsolution and Mg–Fe exchange reactions may record the temperature at different time points during the course of metamorphism from the mineral assemblages produced by dehydration reactions, and thus will give different values of temperature.

The amount of retrograde minerals in ordinary metamorphic rocks is usually very small, whereas very high-temperature metamorphic rocks (the granulite-facies rocks described in Ch. 11) are characteristically apt to show relatively intensive retrograde changes. This is probably partly because the early part of the retrograde stage in the latter case is still at relatively high temperatures, with a high rate of chemical reaction, and partly because H_2O is more plentifully supplied than in lower-temperature metamorphism, owing to the onset of partial melting at such high temperatures. Where partial melting occurs, a considerable part of the H_2O evolved by prograde reactions may remain in the rock, dissolved in the interstitial melt. During the course of temperature decrease, this melt is crystallized, and the H_2O dissolved in it is released, contributing to the progress of hydration reactions in the surroundings.

For example, in crust thickened by thrusting of a large sheet of crustal rocks over pre-existing crust, thermal peak is reached at later times with increasing depth (e.g. A. B. Thompson & England 1984: Fig. 5a). In such a case, the H_2O liberated by prograde reactions at great depths rises to higher crustal levels, where rocks may already be in a retrograde stage. If the rising H_2O passes *pervasively* through minute fractures whose spacing is of the order of the grain size, or through grain boundaries, it will come into contact with all the rock materials above the site of dehydration, causing intensive retrograde reactions. On the other hand, if the rising H_2O passes through widely spaced major **channelways**, it will come into contact only with the walls of the channelways, to which retrograde reactions will be confined; usually the latter appears to be the case.

Walther & Orville (1982) considered that real regional metamorphism is usually close to the latter case. The fluid produced by devolatilization is probably saturated or close to saturation in SiO_2 with respect to quartz. Because the solubility of quartz in H_2O decreases with decreasing temperature and pressure (Fig. 4.1), the aqueous fluid will precipitate quartz as it moves toward the Earth's surface. Quartz veins, which are a common feature of regional metamorphic complexes, could have been major channelways for the passage of volatiles (cf. Connolly & Thompson 1989).

In the crustal levels above the igneous intrusions that are the main cause of heating, the thermal peak of rocks is reached at a later time with decreasing depth (Fig. 2.9). In this case, the H_2O liberated by prograde reactions rises to higher crustal levels, where rocks are in prograde stages. So, the H_2O may not have a great effect on mineral equilibria, although the upward movement of such H_2O may have a marked influence on the temperature distribution and chemical migration within the crust.

2.1.4 P–T–t *paths and* T–X–t *paths of regional metamorphic rocks*

2.1.4.1 P–T–t (pressure–temperature–time*) path of metamorphic rock.* It is not only temperature but also pressure that appears generally to change during the course of regional metamorphism. Increasing pressure is caused by increasing thickness of the overlying crust, probably due either to mechanical deformation such as thrusting and folding or to magmatic intrusion into higher horizons in the crust. Decreasing pressure is caused either, by crustal deformation, or by erosion. The change over time of *P–T* conditions that a rock experiences during metamorphism is called the ***P–T–t* path** of the rock. The *P–T–t* paths of metamorphic rocks appear to be very diverse, depending on the details of the mechanical and thermal history of the region. They vary with the position and depth in which the rock in question lies within a metamorphic complex (e.g. Harte & Dempster 1987, Ghent & Stout 1988).

Figure 2.2 illustrates four of the possible types of *P–T–t* paths in relation to equilibrium curves of dehydration and solid–solid reactions shown on a *P–T* diagram. If the effects of deformation and erosion are negligibly small, and if heating is caused by intrusions into deeper levels, the heating and cooling of a rock takes place under a virtually constant pressure, as illustrated by the *P–T–t* path of case (a).

Curve (b) represents a clockwise *P–T–t* path, where the pressure of a rock located initially at a relatively shallow depth increases rapidly by thickening of the overlying crust, and this is accompanied by a slow temperature increase. While temperature is still increasing, the crust is uplifted to restore isostatic equilibrium, and so active erosion on the surface begins to result in a decrease of pressure. After the thermal peak, the rock may take one of several courses, such as B_1 and B_2, depending upon the relative magnitude of the effects of cooling and erosion. In path (b), the pressure at the thermal peak is considerably lower than the maximum pressure that the rock experiences, and a considerable decrease of pressure takes place over a small range of temperature around the thermal peak (A. B. Thompson & Ridley 1987).

Some *P–T–t* paths make the determination of thermal-peak mineral assemblages difficult. For example, if the *P–T–t* path of a rock in the vicinity of the

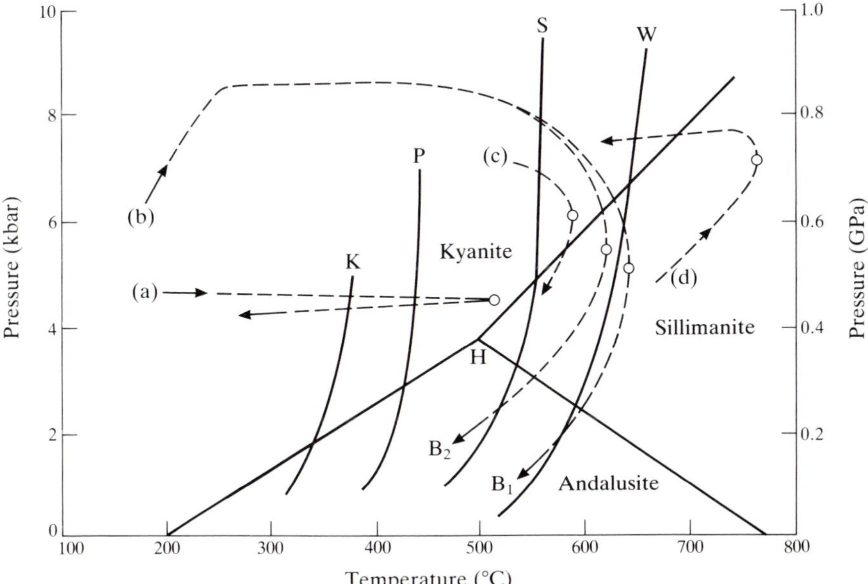

Figure 2.2 Univariant equilibrium curves for some metamorphic reactions (solid lines). Lines K, P, S and W represent equilibrium curves for four different dehydration reactions in the presence of pure H_2O fluid. The reactions are as follows: (K) kaolinite + 2 quartz = pyrophyllite + H_2O; (P) pyrophyllite = kyanite (or andalusite) + 3 quartz + H_2O; (S) staurolite + muscovite + quartz = sillimanite (or kyanite or andalusite) + garnet + biotite + H_2O; (W) hypothetical. The three straight lines meeting at point H are the boundaries between the stability fields of three polymorphs of Al_2SiO_5 (andalusite, kyanite, and sillimanite). Four of the possible *P–T–t* paths (a), (b), (c) and (d) of a metamorphic rock are shown by dashed lines. An open circle represents a thermal peak in each path.

thermal peak is like curve (c), kyanite might be transformed into sillimanite with decreasing temperature, and this sillimanite might be mistaken for a product of increasing temperature or the thermal peak. In an analogous way, some *P–T–t* paths might cause a dehydration reaction to occur with decreasing temperature.

Curve (d) shows a type of anti-clockwise *P–T–t* path commonly observed in high-temperature gneiss regions (Bohlen 1991). Such a path may be followed where basaltic magma is intruded from the mantle into the lower and middle crust, causing an increase of temperature in the crust. A rock in the lower crust experiences an increase in temperature and pressure as shown in the path. After the thermal peak, the cooling process is nearly isobaric, because the relatively high density of basaltic magmas does not cause isostatic uplift and erosion.

The minerals produced in the prograde stage usually decompose partly or completely at and before the thermal peak, whereas those produced in the retrograde stage are commonly preserved. Hence, the *P–T–t* path in the retrograde stage is easier to decipher than that in the prograde stage. In many cases, K–Ar dating of a number of minerals with different closure temperatures gives a quantitative history of post-peak cooling.

Attempts have been made to decipher the *P–T–t* paths of rocks in the prograde stage by investigating zoned crystals and inclusions in them (e.g. Atherton 1968, A. B. Thompson et al. 1977, Spear & Selverstone 1983).

In recent years it has been found in some metamorphic areas that the thermal peak is followed by a great decrease of pressure under nearly isothermal conditions. Such a stage of pressure decrease (decompression stage) should be distinguished from the retrograde stage outlined above, which is characterized by a decrease of temperature with, or without, variation in pressure.

2.1.4.2 T–X–t (temperature–composition–time) *path of intergranular fluid.* Most reactions in prograde metamorphism are dehydration and decarbonation reactions, whose equilibria are controlled not only by pressure and temperature but also by the chemical potentials of H_2O and CO_2 in the rock. At a given pressure and temperature, the chemical potentials of H_2O and CO_2 are functions of the composition of the intergranular fluid, if present, in the rock.

The composition of an intergranular fluid may change because of addition of H_2O and/or CO_2 liberated by, respectively, the dehydration and/or decarbonation reactions that occur in the rock. Moreover, fluids coming from other sources may mix with the intergranular fluid, and may drive reactions in the rock. Fluid thus formed may leave the rock and rise toward the Earth's surface, and thereafter a new intergranular fluid may form. Thus, intergranular fluid has a history of change of composition during the course of metamorphism.

The chemical reactions that take place are controlled by equilibria between fluid and rock, which vary with pressure and temperature, particularly the latter. So the history of compositional change of intergranular fluid during the course of metamorphism is usually represented in a ***T–X* diagram**, which shows temperature as the ordinate and the mole fraction of a component in fluid as the abscissa. Commonly intergranular fluids are regarded, approximately, as binary solutions composed of H_2O and CO_2. In this case, X usually denotes the mole fraction of CO_2, as for example in Figures 6.1 and 6.9.

2.1.5 Complex P–T–t *paths and plurifacial metamorphism*

In Figures 2.1 and 2.2, it is assumed that metamorphic rocks experience only one thermal peak and only one pressure maximum. In some cases, however, metamorphic rocks experience two or more thermal peaks and/or pressure maxima in the course of a single complex tectonic event. In these cases, minerals formed at an earlier time may survive a later phase of crystallization as metastable relics. Such rocks contain two or more distinct groups of minerals, which formed at different times under different P–T conditions. Metamorphic rock of this type was named **plurifacial rock** by W. P. de Roever & Nijhuis (1963). The term **plurifacial metamorphism** is used to denote the sequence of metamorphic crystallization that has produced a plurifacial rock. Some plurifacial metamorphic rocks show mineral assemblages formed in a series of genetically related crystallization stages in the course of a single orogeny, while others are the results of two or more unrelated distinct orogenies (or thermal events) that affected the same rock. The latter case is so-called **polymetamorphism**.

Retrogressive metamorphism is a special case of plurifacial metamorphism. The term retrogressive metamorphism was used in the past with the two different meanings: (a) retrograde metamorphism as defined above, that is, metamorphic crystallization with decreasing temperature immediately after the thermal peak; and (b) a younger independent metamorphic event that is lower in temperature than the older one in a polymetamorphic complex. The term provides considerable convenience for actual descriptive work, because in many cases we cannot determine with certainty to which of the above two cases the retrogressive metamorphism under investigation belongs.

2.2 Progressive metamorphism, isograds, thermal-peak isotherms and isobars, and field *P–T* curves

2.2.1 Spatially progressive metamorphism

It is common that different parts within a metamorphic region show different thermal-peak P–T conditions, and that the spatial variation of thermal-peak P–T conditions is continuous and is of a great enough magnitude to cause mineralogical changes clearly recognizable under the microscope. Within a metamorphic region, variations in temperature usually produce much greater mineralogical changes than do pressure variations. Therefore, we usually ascribe mineralogical changes across a metamorphic region mainly to a laterally (i.e. spatially) progressive increase of thermal-peak temperature. The metamorphism of such a metamorphic region is called **progressive metamorphism**. The classic progressive metamorphic region of the Scottish Highlands is described in §1.6 (Figs 1.4, 1.5).

Note that throughout this book the term progressive metamorphism is used with this meaning, that is, for mineralogical changes which occur in response to a continuous increase of thermal-peak temperature laterally across a region. In contrast, the term **prograde** metamorphism means metamorphism under increasing temperature through time, that is, mineralogical changes in the prograde stage (Fig. 2.1).

2.2.2 Zonal mapping in general and the concept of the isograd

We can divide a progressive metamorphic region into a sequence of **zones**, which show mineral assemblages that formed at successively higher ranges of thermal-peak temperature, such as the chlorite, biotite, garnet, staurolite, kyanite and sillimanite zones in the Scottish Highlands (Figs 1.4, 1.5). Minerals characterizing successive zones, such as chlorite, biotite and garnet, are called **index minerals**. Different metamorphic regions could have different sequences of index minerals. Another historically famous example of a progressive metamorphic region is illustrated in Figure 2.3. Harker (1932), in particular, emphasized the importance of such **zonal mapping**, which has since been carried out in many other regions of the world.

Although zonal mapping is usually carried out in progressive metamorphic regions, it is also sometimes done in regions across which mineralogical changes have been caused by variations in the chemical potentials of H_2O and/or CO_2, and possibly even where mineralogical changes are caused by variations in pressure, under virtually constant temperature. The boundary between two such adjacent zones is called an **isograd**. The mineralogical difference between the two zones must have been caused by a specific reaction or a set of specific reactions caused by a difference in P–T conditions and/or in activities of H_2O and CO_2, and not by a difference in bulk-rock chemical composition (with respect to components other than H_2O and CO_2). If biotite, for example, begins to form at a line in a metamorphic region, the line is usually called the biotite isograd.

Most isograds in pelitic metamorphic regions are based on dehydration reactions. The biotite and garnet isograds are examples of **dehydration-reaction isograds**. When the intergranular fluid has a nearly pure H_2O composition, the equilibrium curves of dehydration reactions have a very steep slope at pressures above 4 kbar (Fig. 2.2). So a dehydration-reaction isograd under such a condition represents a nearly constant temperature, i.e. it is nearly isothermal. The phase transformation curves between andalusite, kyanite and sillimanite (Fig.

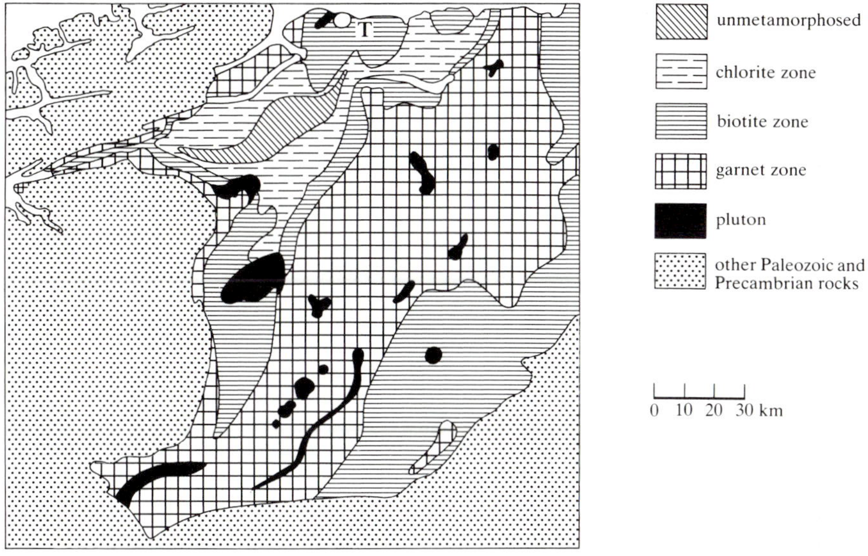

Figure 2.3 Classical Caledonian metamorphic region near Trondheim, Norway (after Goldschmidt 1915). The region is divided into the chlorite, biotite and almandine zones in the order of increasing temperature in Cambro-Silurian pelitic rocks. The metamorphic region has a high-temperature axis within the garnet zone, trending in a northeasterly direction. T indicates the city of Trondheim. The blank areas in the uppermost part are fjords leading to the Atlantic Ocean. Other Paleozoic and Precambrian rocks (dotted pattern) west and south of Trondheim include the "Basal Gneiss Complex", which represents the deepest structural level of the Scandinavian Caledonides, and is famous for having many localities of eclogite (§12.4.2).

2.2), for example, have gentle slopes. The transformation temperatures change considerably with pressure. So, on an isograd based on such a transformation, both pressure and temperature could vary considerably from point to point.

Dehydration equilibria are controlled not only by pressure and temperature but also by the chemical potential of H_2O. In a metamorphic tract having a regional-scale compositional variation of the intergranular fluid, a dehydration reaction may have taken place in one part, but not in an adjacent part of the tract, even though the *P–T* conditions are virtually constant throughout. The boundary between the two parts is an isograd (using the above definition).

If we consider a three-dimensional metamorphic complex, an index mineral begins to occur on a surface within the complex. The intersection of this surface with the Earth's surface is the isograd. The three-dimensional surface itself is called an **isogradic surface**.

Etymologically, the term isograd is derived from the term **grade of meta-**

morphism. Originally the term "grade of metamorphism" or "metamorphic grade" was used with various different meanings, but in recent years it has usually been taken to indicate the position of metamorphic rocks in a progressive sequence of metamorphism, with a higher grade usually meaning a higher temperature of formation. Low- and high-grade schists mean schists formed at low and high temperatures, respectively. If we consider that progressive metamorphism may sometimes occur owing to a regional-scale variation of the composition of intergranular fluid at an essentially constant temperature (Grambling 1986), it is sensible to define higher grade as a state showing a higher degree of dehydration or devolatilization. With increasing grade of metamorphism, pressure may increase or decrease, or may remain virtually constant.

2.2.3 Thermal-peak isotherms and thermal-peak isobars

Regional metamorphic rocks are usually exposed over a wide region in the axial part of ancient orogenic belts. The major features of the mineral assemblages of metamorphic rocks approximately represent equilibrium states at, or near, their thermal peak, which is not synchronous between different parts of a single metamorphic region (e.g. Ghent & Stout 1988). The values of temperature T and pressure P at the thermal peak should vary continuously from point to point in a region, provided that later faults do not cut it.

It is possible, in principle, to draw on a metamorphic map a series of isopleths for T as well as for P for the thermal-peak condition of every point. In this book, such isopleths are named **thermal-peak isotherms** and **thermal-peak isobars**, respectively. These terms emphasize that they are not ordinary isotherms and isobars, called **instantaneous isotherms** and **instantaneous isobars** in this book, which indicate the distribution of temperature and pressure, respectively, at a specific time.

Thermal-peak isotherms and thermal-peak isobars are the intersections of the Earth's surface with the corresponding three-dimensional thermal-peak isothermal and thermal-peak isobaric surfaces within the metamorphic complex.

Recent progress in geothermobarometric techniques based on the properties of solid-solution minerals has led to many P–T value determinations on metapelites and metabasites. This permits mapping of thermal-peak isotherms and thermal-peak isobars on the basis of P–T values determined by geothermobarometers.

Thermal-peak isotherms. The equilibrium curves for dehydration reactions usually have a very steep slope at pressures above 4 kbar in the presence of a fluid with a nearly pure H_2O composition (Fig. 2.2). Hence, isograds and isogradic surfaces based on such dehydration reactions must be approximately parallel to thermal-peak isotherms and thermal-peak isothermal surfaces, respectively, provided that the rocks contain an aqueous fluid.

However, at pressures as low as 4–2 kbar (that is, in the common range of pressure in low-pressure regional metamorphism), the temperatures of dehydration reactions usually show a considerable decrease with decreasing pressure (Figs 2.2, 3.2). There is also another factor which is apt to cause a systematic deviation of the direction of dehydration-reaction isograds in low-pressure metamorphism from that of the thermal-peak isotherms, as follows. Most commonly used isograds are based on dehydration reactions in graphite-bearing metapelites. Intergranular fluids in graphite-bearing metapelites show a decrease of the mole fraction of H_2O with decreasing pressure and increase of temperature because of the formation of CO_2, CO, etc. (Fig. 6.4). Thus, special caution is required when comparing thermal-peak isotherms and dehydration-reaction isograds in low-pressure metamorphic regions.

Helms & Labotka (1991) described a low-pressure metapelitic region, (30 km × 26 km) in size, in South Dakota, USA. They delineated a series of isograds, and further determined values of thermal-peak temperatures and pressures and activities of H_2O, based on mineral equilibria at many places over the whole region. They found that the thermal-peak isotherms were apt to be parallel to dehydration-reaction isograds. However, since the prevailing pressures were as low as 2.0–4.4 kbar, the temperatures of dehydration-reaction isograds increased considerably with increasing pressure. This resulted in a great westward increase of the distance between two adjacent isograds (staurolite and andalusite) due to a westward increase of pressure.

Thermal-peak isobars. The kyanite–andalusite equilibrium curve in Figure 2.2 has a very small positive slope, as expressed by the equation: P (kbar) = $0.0125\,T - (°C) - 2.50$. Since the dP/dT is so small, an isograd and isogradic surface based on this reaction must be in a direction close to thermal-peak isobars and thermal-peak isobaric surfaces, respectively. Use of some other solid–solid reactions with virtually no change in volume (§3.4) may in future enable us to draw isograds that are virtually parallel to thermal-peak isobars.

In the above-mentioned work in South Dakota, Helms & Labotka (1991) determined many pressure values by two geobarometric techniques. Although the values generally increased westward, they showed marked irregular variations, whose range is considerable, compared with the small range of pressure in this region. So Helms & Labotka did not draw thermal-peak isobars. They considered that many of the observed variations in pressure were within the uncertainty of the geobarometers due to poorly known activity–composition relations.

2.2.4 Thermal-peak isotherms and isobars in the Scottish Highlands

Many thermobarometric measurements have been made in the Dalradian metamorphic region of the Scottish Highlands. Fortunately, in this region the range of thermal-peak pressure is so great that thermal-peak isobars can be drawn on the basis of thermobarometric work. Thus, on the basis of a thermobarometric study, Baker (1987) drew not only thermal-peak isotherms but also thermal-peak isobars on a map (Fig. 2.4).

Comparison of Figure 2.4 with Figure 1.4 confirms the expectation that low-grade isograds based on dehydration reactions are apt to be approximately parallel to thermal-peak isotherms. However, the 750 °C thermal-peak isotherm is not parallel to any of the isograds. This may simply be the result of low accu-

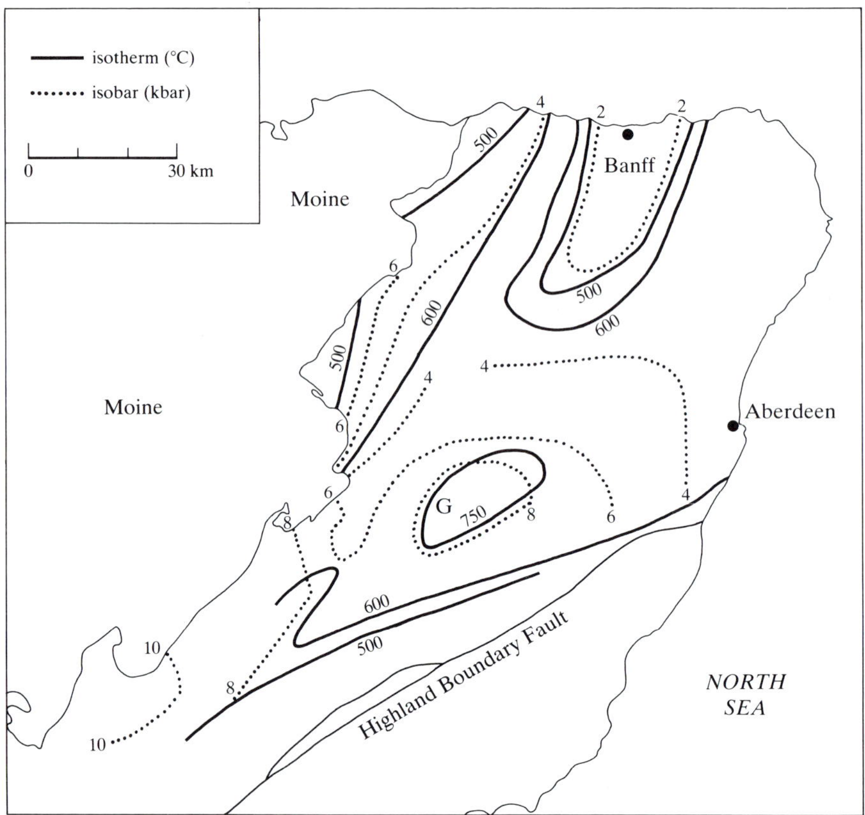

Figure 2.4 Distribution of thermal-peak temperatures and thermal-peak pressures in the Dalradian metamorphic region (including both Barrovian and Buchan regions) in the Scottish Highlands. G indicates the Glen Muick area. Based on geothermobarometric work by Baker (1987).

racy in the determination of the isotherm. (The geologic significance of this map is discussed in §9.2.2.)

2.2.5 Geometrical relations between thermal-peak isotherms and thermal-peak isobars

Figure 2.5a_1, b_1 and c_1 are schematic metamorphic maps showing possible types of relations between thermal-peak isobars and thermal-peak isotherms. Figure 2.5a_1 shows a general case where thermal-peak isobars intersect thermal-peak isotherms, making angles θ and ϕ as shown. Any direction within angle θ, as is indicated by line m, shows increasing pressure with increasing temperature, whereas any direction within angle ϕ, as is indicated by line n, shows decreasing pressure with increasing temperature, as is shown in the *P–T* diagram (a_2). Thus, whether pressure increases or decreases with temperature depends on the choice of direction on the Earth's surface.

In many cases, on the other hand, thermal-peak isobars appear to be nearly parallel to thermal-peak isotherms with θ = about 180° (Fig. 2.5b_1). If $\theta = 180°$, there is only one *P–T* curve with a positive slope (Fig. 2.5b_2). The curve is independent of the position and direction of the line on the map along which thermal-peak *P–T* conditions are measured.

Another case could exist where thermal-peak isobars are nearly parallel to thermal-peak isotherms with θ = about 0° (Fig. 2.5c_1). If $\theta = 0°$, there is only one *P–T* curve with a negative slope (Fig. 2.5c_2).

In Figure 2.5b_1 and c_1, thermal-peak isobars are parallel to thermal-peak isotherms on the present-day erosion surface, whereas thermal-peak isobaric surfaces are not generally parallel to thermal-peak isothermal surfaces within the metamorphic complex.

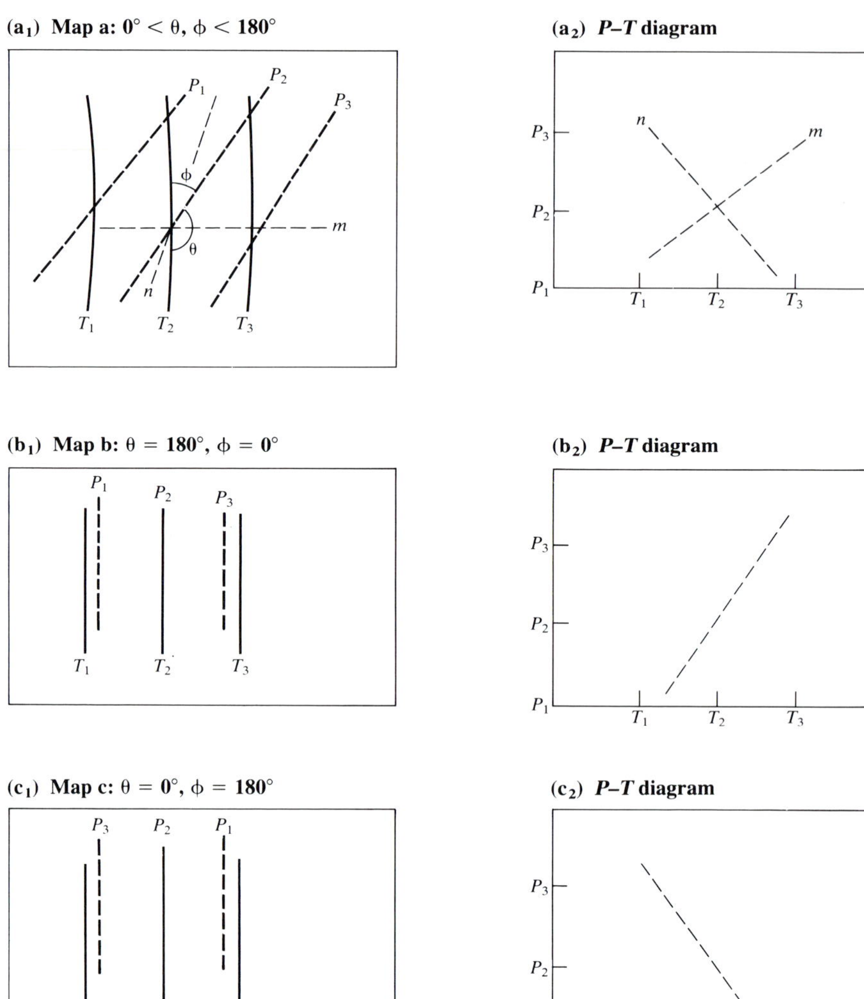

Figure 2.5 (a_1), (b_1) and (c_1) are schematic maps showing three possible geometric relations among thermal-peak isotherms (solid lines) and thermal-peak isobars (dashed lines) on the surface of a metamorphic region. Temperature increases in the order of T_1, T_2 and T_3, and pressure increases in the order of P_1, P_2 and P_3. (a_2), (b_2) and (c_2) are thermal-peak P–T diagram corresponding to (a_1), (b_1) and (c_1), respectively. In map (a_1) line m indicates an arbitrary direction within angle θ, while line n indicates an arbitrary direction within angle ϕ. In P–T diagram (a_2) m and n indicate the thermal-peak P–T conditions along lines m and n, respectively, of map (a_1).

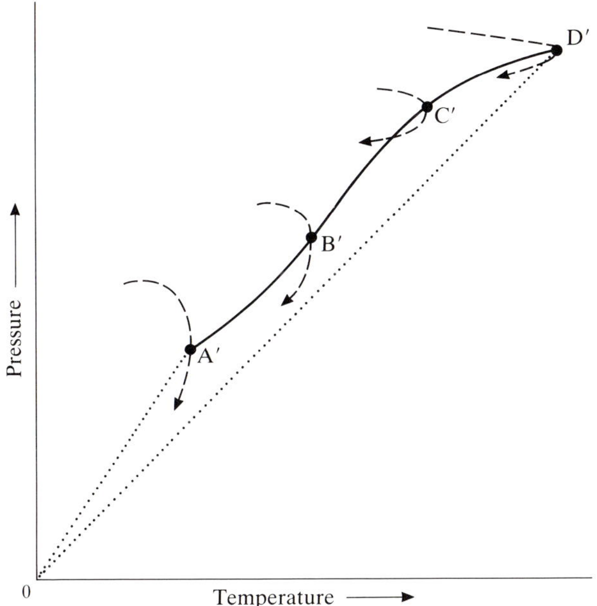

Figure 2.6 Solid line A′–B′–C′–D′ represents a field *P–T* curve, which shows the change of thermal-peak *P–T* conditions along the line passing through points A, B, C and D on the erosion surface across a progressive metamorphic region. Dashed lines with arrows show the *P–T–t* paths of rocks at each of A, B, C and D. Point O represents the *P–T* condition on the surface of the Earth (say, 1 atm and 20°C). Dotted lines correspond to the average geothermal gradients at A and D.

2.2.6 *Field* P–T *curves and their relation to* P–T–t *paths*

A field *P–T* curve is intended to express the variations of the thermal-peak *P–T* conditions across a progressive metamorphic region. On the surface of a progressive metamorphic region, we draw a line *in a direction perpendicular to thermal-peak isotherms at each point on the line.* In practice, we usually draw the line perpendicular to the dehydration-reaction isograds, because such isograds are approximately parallel to thermal-peak isotherms. The solid line in the *P–T* diagram of Figure 2.6 indicates an example of a curve showing the variation of thermal-peak *P–T* conditions along such a line. A curve of this kind on a *P–T* diagram is called a **field *P–T* curve** in this book.

We mark four places along the line on the Earth's surface as A, B, C and D. The *P–T–t* path of the rock at each of the four places A, B, C and D is shown by a dashed line with an arrow. Points A′, B′, C′ and D′ are the intersections of the paths with the field *P–T* curve, and represent thermal peaks (maximum temperatures) on the *P–T–t* paths. The average geothermal gradients of points A and D measured from the Earth's surface are shown by dotted straight lines, which connect these points with the point O that represents the *P–T* condition on the Earth's surface (say, 1 bar and 20°C). The average geothermal gradient could well differ in different parts of a progressive metamorphic region. If the thermal-peak isobars are parallel to the isotherms, as illustrated in Figure 2.5b_1, c_1, there is only one field *P–T* curve, which is independent of the position of the line along which thermal-peak conditions are measured.

Where thermal-peak isobars intersect thermal-peak isotherms as in Figure 2.5a_1, the field *P–T* curve has a positive slope if $90° < \theta$ in Figure 2.5a_1, whereas it has a negative slope if $\theta < 90°$ in Figure 2.5a_1. It is empirically known that field *P–T* curves usually show a positive slope.

2.3 Chemical reactions on isograds

2.3.1 Tentative isograds and reaction isograds

If staurolite occurs in one part of a pelitic metamorphic region, but does not occur in an adjacent part of the same region, we may tentatively consider that staurolite formed by a chemical reaction at the boundary between the two parts of the region. Then we may tentatively regard the boundary as a staurolite isograd, treating staurolite as an index mineral. Many isograds described in the literature are of this type. They are called **tentative isograds** in this book.

There are two possibilities. An index mineral of this kind may really have formed by a specific reaction caused by a continuous variation in *P–T* conditions or in the chemical potential of H_2O or CO_2 across the region. In this case the isograd is a true isograd. A true isograd, to which a specific reaction has been assigned, was termed a **reaction isograd** by Winkler (1979). In some cases, two or more reactions occur simultaneously in one mineral assemblage with increasing temperature (as discussed in §3.1.2 and §5.4). In such a case, a reaction isograd is characterized by a specific set of reactions, rather than one single reaction.

In §3.2.2, metamorphic reactions are classified into exchange and net-transfer reactions. In the former, ions are exchanged between coexisting minerals, without causing any change in the relative proportions of the minerals, whereas the latter include all reactions that consume and produce minerals. The appearance of new minerals on an isograd is the result of a net-transfer reaction or a set of net-transfer reactions. So, the term "reaction" in the above definition of isograd refers to net-transfer reactions only. Exchange reactions usually occur simultaneously in the rocks concerned, but they are not considered here.

In other cases, however, a tentative isograd does not actually represent a specific reaction or a specific set of reactions. The appearance of staurolite in only a part of the region in the above example may be the result of a subtle change in the bulk-rock chemical composition across the region under virtually the same P–T conditions. Or, what was mapped as an isograd might actually be a fault. Faults are easily overlooked and mistaken for isograds, especially when they are parallel to the schistosity (Fig. 9.6). In such cases, the tentative isograd is not a true isograd but a false isograd.

Since the change in mineral assemblage across a true isograd is caused by a reaction or reactions, it must be possible to write a balanced chemical equation or a set of balanced chemical equations to describe the differences between mineral assemblages observed on opposite sides of the isograd. When A, B, C and D denote minerals, we consider the case where the progressive metamorphic reaction $A + B = C + D + H_2O$, for example, causes the disappearance of the mineral assemblage $A + B$, and the appearance of the mineral assemblage $C + D$ at the isograd. Either C or D (but not both) may occur in rocks on the low-temperature side of the isograd, and A or B (but not both) may persist in rocks on the high-temperature side of the isograd. Which of them occurs depends on the bulk-rock chemical composition. Therefore, a reaction isograd is characterized by the appearance or disappearance of a specific mineral assemblage, rather than of a specific mineral. The mineral assemblage created at an isograd and which characterizes the zone on its high-temperature side is called a **critical mineral assemblage**.

If mineral C, for example, does not occur in rocks on the low-temperature side of the isograd, it begins to occur on the isograd, and so may be used an an index mineral (§2.2.2). However, the use of C as an index mineral may be misleading, because C could also form by other reactions, so that mineral C formed by several different reactions may occur on the high-temperature side of the isograd in the region, as shown in the next section with reference to the index mineral biotite for pelitic metamorphic rocks.

Progress of the reaction $A + B = C + D + H_2O$ decreases the amounts of both A and B simultaneously, until one of the two minerals is used up. Which of the two is used up depends on the proportion of the two minerals in the original rock. The mineral assemblage $B + C + D$ or $A + C + D$ continues to be stable after the completion of the reaction.

2.3.2 Changes of mineral assemblages caused by discontinuous and continuous reactions

Reactions that can be used to define isograds may be classified into two types: **discontinuous** and **continuous**. A discontinuous reaction takes place at a specific temperature under a given pressure in the presence of an intergranular fluid with a constant composition, and so produces discontinuous mineralogical changes with increasing temperature. For example, the reactions for curves K, P and S in Figure 2.2 are discontinuous. On the other hand, a continuous reaction takes place over a range of temperatures under a given pressure in the presence of an intergranular fluid with a constant composition, and so produces gradual and continuous mineralogical changes with increasing temperature.

If the reaction $A + B = C + D + H_2O$, for example, is discontinuous, the reaction begins and is completed on an isograd in a progressive metamorphic region. The mineral assemblage $A + B + C + D$ occurs only on the isograd. Rocks showing the assemblage $A + B$ occur on the low-temperature side of the isograd, whereas ones showing the assemblage $C + D$ occur on the high-temperature side. In the Barrovian sequence of zones in the Scottish Highlands (§1.6.3), the staurolite and sillimanite isograds are based on discontinuous reactions (so far as the metapelitic rocks are regarded as belonging to the six-component system SiO_2– Al_2O_3–FeO–MgO–K_2O–H_2O).

If the reaction $A + B = C + D + H_2O$ is continuous, the reaction begins on a line in a progressive metamorphic region, and continues to proceed across a zone (corresponding to a range of temperature) on the high-temperature side of the line. Such a line showing the beginning of the reaction is commonly used as an isograd, e.g. the biotite and garnet isograds in the Barrovian sequence. In such a case, the zone on the low-temperature side of the isograd is characterized by the mineral assemblage $A + B$, unaccompanied by $C + D$. The mineral assemblage $A + B + C + D$ makes its first appearance on the isograd, and continues to occur in a zone on the high-temperature side of it. On the isograd, the assemblage $A + B + C + D$ may occur only in relatively rare rocks whose bulk-rock chemical compositions are especially favorable for the reaction to take place, while other rocks on the isograd still show the mineral assemblage $A + B$, unaccompanied by $C + D$. With an increase of temperature, the percentage of rocks showing the mineral assemblage $A + B + C + D$ may increase. The temperature at which a continuous reaction begins in a rock varies with the bulk-rock chemical composition. Hence, the temperature of this type of isograd varies with bulk chemical composition of the rocks exposed in the region.

The distinction between the discontinuous and continuous reactions arises from a difference in the number of degrees of freedom of the reacting system in the phase rule (§3.1). Although both staurolite and garnet are solid-solution minerals, the staurolite isograd is based on a discontinuous reaction, whereas the garnet isograd is based on a continuous reaction in the Barrovian sequence. Detailed discussions on the differences between these two types of reactions in relation to the phase rule and in zonal mapping are given for metapelitic rocks in relation to the *AFM* diagram (§5.4.4, §5.4.5). The types of reactions in the Barrovian sequence are synoptically shown in Figure 10.1.

2.3.3 *A tentative isograd may be a compound of two or more reaction isograds: the case of biotite isograds in a metapelitic region*

Commonly used index minerals such as biotite and staurolite in metapelitic rocks can form by two or more different chemical reactions which take place in rocks with different antecedent mineral assemblages. Such reactions usually have different reaction temperatures, and therefore a true isograd could exist for each reaction. A tentative isograd for an index mineral could be a compound of two or more reaction isograds. In the following, a biotite isograd is taken as an example to illustrate how an index mineral characterizing a tentative isograd can form by two or more different reactions.

In many regions, pelitic rocks metamorphosed at very low temperature (i.e. just above the threshold temperature of metamorphic crystallization) are mainly muscovite–chlorite–quartz–albite schists. With increasing temperature, reaction begins to take place between muscovite and chlorite in some of the rocks with the resultant formation of biotite, as follows (with coefficients neglected):

$$\text{muscovite} + \text{chlorite} = \text{biotite} + \text{quartz} + H_2O \qquad (2.1)$$

The reactants muscovite and chlorite in reaction (2.1) are solid solutions. Under a specific temperature, pressure and H_2O activity, the compositions of these minerals vary with bulk-rock composition. The temperature of the onset of reaction (2.1) varies with the compositions of these minerals, and hence with bulk-rock composition. Therefore, the biotite isograd is the line on which biotite begins to form in metapelites that have the most favorable chemical composition for the reaction. With increasing temperature, the proportion of biotite-bearing rocks among all metapelites increases. It is usual in the biotite zone that biotite-bearing rocks still contain both reactants (muscovite + chlorite), because the reaction is continuous, proceeding gradually over a wide range of temperatures. The compositions of all these minerals change continuously as the reaction proceeds with increasing temperature. The celadonite content of muscovite decreases gradually with the progress of reaction (2.1) (e.g. Ernst 1963b, Mather 1970; see §10.1.1.)

There are many possible biotite-producing reactions in pelitic rocks other than reaction (2.1). Some pelitic rocks contain small amounts of dolomite (or ankerite), siderite, or calcite. In these rocks, biotite-producing reactions may be like the following (e.g. Bethune 1976, Graham et al. 1983):

muscovite + dolomite + quartz + H_2O = biotite + chlorite + calcite + CO_2 (2.2)

muscovite + calcite + chlorite + quartz = biotite + epidote + H_2O + CO_2 (2.3)

Symmes & Ferry (1991) consider that siderite and ankerite are wide-spread in metapelites of the chlorite zone, and suggest the possibility that biotite forms on the biotite isograd and in the biotite zone by the following types of reaction:

muscovite + siderite + quartz + H_2O = biotite + chlorite + CO_2 (2.4)

muscovite + ankerite + chlorite + quartz = biotite + plagioclase + CO_2 + H_2O (2.5)

Reaction (2.5) takes place at a temperature higher than (2.4) (Table 10.2).

Typical metapelites in such low grades contain no K-feldspar, whereas some metapelites and metapsammites do contain it, if part of the original clastic material was a decomposition product of granitoid rocks, for example. In rocks containing K-feldspar, biotite may be produced by a reaction of the following type (e.g. Ernst 1963a, J. B. Thompson & Norton 1968):

chlorite + K-feldspar = biotite + muscovite + quartz + H_2O (2.6)

This usually begins to take place at a temperature lower than reaction (2.1).

If there is a rock containing both dolomite (or ankerite) and K-feldspar, biotite may form by a reaction of the following type:

dolomite + K-feldspar + H_2O = biotite + calcite + CO_2 (2.7)

In some regional metamorphic complexes such as the Sanbagawa belt in Japan, garnet (MnO-rich almandine) begins to occur in metapelites at a lower temperature than biotite, thus creating a garnet zone on the low-temperature side of the biotite zone. In such a biotite zone, the following is also a possible biotite-forming reaction:

chlorite + muscovite + quartz = biotite + garnet + H_2O (2.8)

Thus, it is clear that metapelites in a biotite zone may contain biotites produced by two or more different reactions, which have different equilibrium temperatures. Which of these reactions occurs, depends on the antecedent mineral assemblage and also on the chemical composition of each solid-solution mineral in it. These in turn are constrained primarily by bulk-rock chemical composition. Therefore, the type of reaction that takes place is constrained by bulk-rock chemical composition.

If all the antecedent metapelites of the chlorite-zone grade are muscovite–chlorite–quartz–albite schists free of carbonate and K-feldspar, the observed biotite isograd probably represents equilibrium P–T conditions for reaction (2.1). However, if the antecedent metapelites contain some that are K-feldspar-bearing, mixed with some that are K-feldspar-free, the isograd defined by the first appearance of biotite will represent the equilibrium P–T conditions of reaction (2.6). Then, at a slightly higher temperature, reaction (2.1) would begin to produce another variety of biotite, and so the two varieties of biotite come to occur in rocks with different antecedent mineral assemblages at the same metamorphic grade.

If all the original pelitic rocks are free of carbonate and K-feldspar in a part of a region, but some contain K-feldspar in an adjacent part, the biotite isograd in the first part will be due to reaction (2.1) and that in the second part will be due to reaction (2.6). These two isograds are different reaction isograds; in routine mapping, however, the two might mistakenly be taken for one apparent biotite isograd.

The problem discussed above is that one index mineral may form by two or more different reactions which take place in rocks with different mineral assemblages in the same region. This problem should not be confused with the frequent problem where two or more reactions take place simultaneously in the same mineral assemblage (§2.3.1, §3.1.2.2).

2.4 Isograds and the composition of intergranular fluids

Most isograds are based on reactions that involve H_2O and/or CO_2, as exemplified by reactions (2.1)–(2.8). The equilibria of such reactions are controlled not only by temperature and pressure but also by the chemical potentials of H_2O and CO_2. The chemical potential is related to the composition of the intergranular fluid, if present. Hence, the nature, position and shape of isograds based on reactions that involve H_2O and/or CO_2 are closely related to the values and distribution of the chemical potentials of H_2O and CO_2. A few important cases of these relations are described below. A more detailed discussion of relevant thermodynamic relations is postponed until §3.5 and Chapters 4 and 6.

2.4.1 Isograds based on simple dehydration reactions in pelitic metamorphic regions

As a common case, we consider a middle- and high-grade pelitic metamorphic region. Here, most isograds are based on simple dehydration reactions, as expressed by a general equation: $B = D + H_2O$, where B and D represent one or more minerals (e.g. reactions (2.1), (2.6), (2.8)).

In a pelitic metamorphic region of this type, dehydration reactions produce H_2O, and so the intergranular fluid will usually be high in H_2O content. Metapelites usually contain graphite, and reaction of H_2O with graphite produces small amounts of CO_2 and other volatiles. Moreover, a fluid containing CO_2 may be produced by metamorphism of calcareous beds, and may infiltrate into

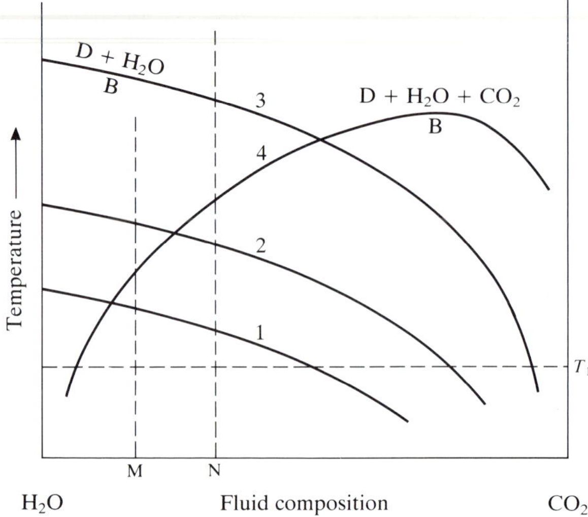

Figure 2.7 Schematic *T-X* diagram for simple dehydration reactions 1, 2 and 3 of the type: $B = D + H_2O$ at a constant pressure, where B and D represent minerals with fixed compositions. Curve 4 represents a more complicated dehydration–decarbonation reaction of the type: $B = D + H_2O + CO_2$ (with coefficients ignored). M and N indicate fluid compositions in two different areas within a metamorphic region. See text and Figure 3.5.

the nearby pelitic rocks. So it is likely that the intergranular fluid is generally rich in H_2O but contains a small amount of CO_2, the concentration of which may vary from rock to rock, even over a small area.

At a constant pressure, the equilibrium temperature of a simple dehydration reaction decreases with decreasing concentration of H_2O. Schematic equilibrium curves for three imaginary reactions of this type are illustrated in Figure 2.7 by curves 1, 2 and 3.

Fluids may have a nearly pure H_2O composition in some parts of the region, and may have slightly deviated compositions like M and N in Figure 2.7 in other parts. Such a small difference in fluid composition causes only slight differences in reaction temperature, and hardly causes a change in the order of reactions in prograde metamorphism.

2.4.2 Other types of dehydration–decarbonation reactions and intersecting isograds

There are other types of reactions that involve H_2O and/or CO_2 in metamorphic rocks (§3.5). Here we consider only the type of reactions that involve both H_2O

and CO_2: $B = D + H_2O + CO_2$ (with coefficients ignored). Reactions (2.3) and (2.5) provide examples. This type of reaction occurs in carbonate-bearing rocks. The equilibrium temperature of an imaginary reaction of this type shows a maximum temperature at an intermediate H_2O/CO_2 ratio, as shown by curve 4 in Figure 2.7.

Let us imagine a metamorphic region where all the reactions shown by curves 1, 2, 3 and 4 occur, and where intergranular fluids vary slightly in composition from area to area, as shown by M and N in Figure 2.7. In such a case, the order of reaction in progressive metamorphism is 1, 2, 3 in an area of pure H_2O, 1, 4, 2, 3 in an area of composition M, and 1, 2, 4, 3 in an area of composition N. Thus, the order of isograds may change sensitively with composition of the intergranular fluid. In a metamorphic region like this, the isograd for reaction 4 intersects the other isograds. Generally speaking, isograds based on different types of reaction involving H_2O and/or CO_2 commonly intersect one another. Carmichael (1970) has described an instructive example of intersecting isograds in the Whetstone Lake area of the Grenville tectonic province of the Canadian Shield.

2.4.3 Isograds formed under constant P–T *conditions*

Let us imagine a metamorphic region where the exposed rocks represent virtually constant pressure and temperature. If there is a regional-scale variation of fluid composition such as the one shown by a horizontal line at T_1 in Figure 2.7, dehydration reactions 1, 2 and 3 should occur in this order, with decreasing concentration of H_2O. In such a metamorphic region, the sequence of observed index-mineral zones due to the decreasing concentration of H_2O would be very similar to progressive metamorphic zones produced by increasing temperature at a constant fluid composition such as N.

Grambling (1986) discovered an example of a series of dehydration-reaction isograds caused by a decrease of the concentration of H_2O in the intergranular fluid at virtually constant pressure and temperature in the Pecos Baldy area, New Mexico.

2.4.4 Internally buffered isograds

In all the cases discussed above, relatively smooth regional-scale isograds occur because the composition of the intergranular fluid is either uniform through the region, or varies only gradually from area to area on a regional scale. Such a condition would be realized by mixing of, and diffusion in, intergranular fluid and infiltration of an externally derived fluid. However, Rice (1977) showed the existence of an entirely different type of isograd. In his contact-metamorphic aureole, the compositions of fluids were controlled by reactions with the minerals of individual host rocks, and so were a function of temperature (§6.6.2). In this type of case, the fluid composition is said to be internally buffered (Ch. 6). In such a region, a set of isograds could form, corresponding to the course of compositional change of internally buffered fluid with increasing temperature.

2.5 Isograds and instantaneous isotherms, and the polychronous origin of isograds

2.5.1 Relations of isograds to instantaneous isotherms

As a preliminary to a general discussion of the polychronous origin of isograds, we discuss here the semi-quantitative relations between an isograd and the relevant instantaneous isotherms in small metamorphic regions where the lateral shift of the area of deformation, uplift and plutonic intrusion with time need not be taken into consideration.

Most common isograds are based on dehydration reactions, and so to a first approximation represent thermal-peak isotherms in the presence of an aqueous fluid in ordinary regional metamorphism, i.e. above 4 kbar pressure (§2.2.3). Therefore, for simplicity we regard a dehydration-reaction isograd as a thermal-peak isotherm. In pelitic rocks of the Barrovian sequence, staurolite forms by a discontinuous dehydration reaction. The temperature for staurolite formation is here denoted as T, and thus T represents the temperature of the staurolite isograd in the following discussions.

2.5.1.1 Isograd formed by continuously changing temperature distribution. In the general case, the instantaneous isotherm for temperature T changes its shape and position with time during metamorphism. Figure 2.8a is a schematic map showing the relationship of instantaneous isotherms for T at three successive times t_1, t_2, and t_3, and the resulting isograd. We assume that at each time, temperature increases toward the lower right direction in the figure, and so staurolite forms in the area on the lower right side of each isotherm. Eventually staurolite comes to be contained in metapelites in the whole area on the lower right side of the envelope B–B, which is tangential to every member of the set of consecutive instantaneous isotherms for temperature T. Hence, the envelope B–B represents the staurolite isograd.

On this isograd, staurolite forms only at one specific point at any specific time, and that point moves continuously to the left along line B–B with time. In other words, the isograd is not synchronous. At point M_2 on the isograd, for example, temperature is lower than T at time t_1, equal to T at t_2, and again lower than T at t_3. Thus, point M_2 reaches a thermal peak at t_2.

2.5.1.2 Effect of syn-metamorphic folding. The shapes of isograds – as well as the *P–T–t* paths of metamorphic rocks – are influenced by syn-metamorphic folding and faulting. A simplest case of syn-metamorphic folding is schematically illustrated in Figure 2.8b.

The figure represents a vertical cross-section of a metamorphic complex, for which the instantaneous isotherms of temperature T at four successive times t_1, t_2, t_3 and t_4 are shown by dashed lines. At the beginning (time t_1), the instantaneous isothermal surface is horizontal. Then, relatively fast folding occurs, producing a folded isotherm. The figure shows the state of maximum folding that is reached at time t_2. Thermal relaxation follows, and so the shape of the instantaneous isotherm of T changes and finally returns to a horizontal position at time t_4. The temperature at t_4 is assumed to have become higher than that at t_1 owing to some change in the environment.

The staurolite isograd in this cross section follow the thermal-peak isotherm for T, and is an envelope passing through M, O and Q, as shown by a solid line. The time of formation of the isograd is t_2 at point M, and t_4 at point O. After the metamorphism, the region may be tilted. Later differential erosion may expose the folded structure and associated isograd of Figure 2.8b on the surface.

Thermal perturbations caused by small folds and faults would be relaxed too fast to cause metamorphic crystallization in the surroundings. If the wavelength of folding is about 10 km, the relaxation will occur in a period of about 0.25 million years, long enough to cause metamorphic crystallization (Sleep 1979, Chamberlain & Karabinos 1987).

Rocks in the thermal antiform M and thermal synform S show contrasting *P–T–t* paths. Immediately after the folding, a rock at M shows a decrease of temperature, whereas a rock at S shows an increase. Thus, *P–T–t* paths of rocks must differ from place to place within a folded metamorphic region (Chamberlain & Karabinos 1987).

2.5.1.3 Isograds formed by a horizontal intrusion at middle depth. Hamilton & Myers (1967) have suggested that large batholiths such as the Mesozoic ones in the western United States may actually be flat plate-like intrusions, laterally wide but vertically thin, formed at relatively shallow depths. De Yoreo et al. (1989) have given geologic and geophysical evidence that granitic bodies in the

Acadian regional metamorphic complex of the northern Appalachians were horizontal plate-like masses that were originally intruded into depths of about 7–15 km (Fig. 9.10). For simplicity, we consider a case where an infinitely large, horizontal plate-like igneous mass is instantaneously intruded into the middle depth of the crust, and discuss the temperature distribution and isograds within the crust above and below the intrusion. All isothermal surfaces at any time during metamorphism are horizontal.

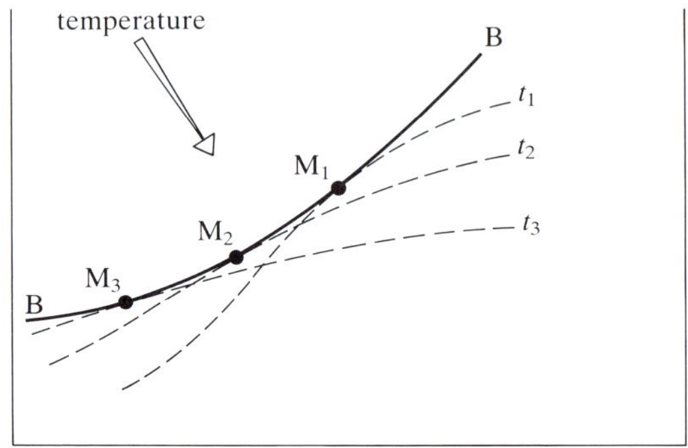

(b)

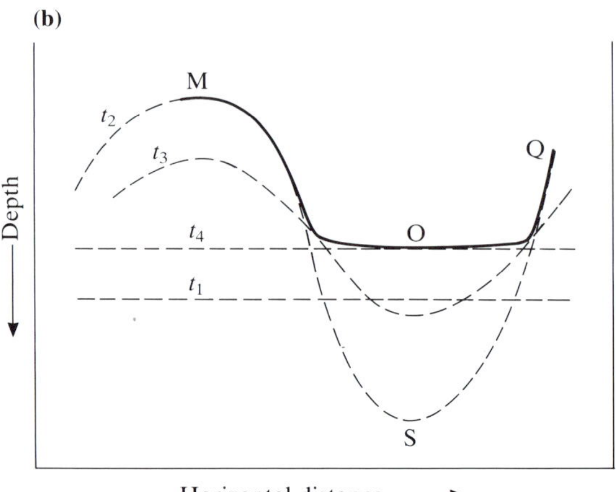

Figure 2.8 Schematic relations of instantaneous isotherms to dehydration-reaction isograds. (a) Map showing the instantaneous isotherms that indicate the temperature of the formation of staurolite (T) at three times t_1, t_2, and t_3 in order of passing time, and the resultant staurolite isograd B–B. The resultant staurolite isograd (B–B) is an envelope tangential to the successive isotherms. (b) Vertical cross section of a folded metamorphic complex showing instantaneous isotherms for temperature T at four times t_1, t_2, t_3 and t_4 in order of passing time. The resultant isograd between M and Q is shown by a solid line.

Figure 2.9 schematically shows a possible temporal change of temperature distribution in the crust. We assume that at time t_1 prior to the intrusion the crust has a steady-state geotherm. Then, at t_2 a horizontal intrusion forms at depth M, creating a high-temperature layer. The thermal relaxation that follows decreases the temperature of this layer and increases that of the overlying and underlying layers (successively t_3 and t_4). After a long time, the geotherm returns to its initial shape, if the increase of the crustal thickness by the intrusion is negligible.

The maximum temperature each depth has experienced is represented by an envelope tangential to every consecutive temperature–depth curve, as shown by a solid line. This envelope controls the mineral assemblages at various depths and the depths of all isogradic surfaces.

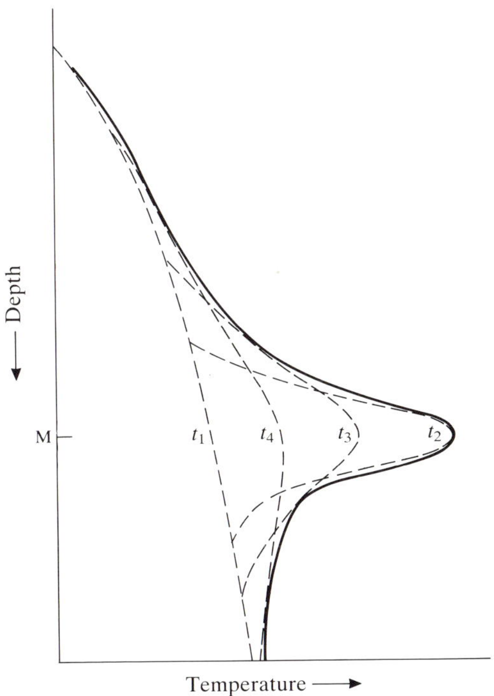

Figure 2.9 Depth–temperature relation in a continental crust, the middle level of which is intruded and heated by a large flat plutonic body. Instantaneous temperature distributions at four successive times t_1, t_2, t_3 and t_4, are shown by dashed lines and the corresponding envelope by a solid line. Qualitatively adapted from Jaeger (1968). (Depths near M are actually inside the intrusion, and so the temperature distribution there is more complicated than shown here.) For a more quantitative relation, see De Yoreo et al. (1989).

After metamorphism, the region may be tilted so that all isogradic surfaces are exposed as isograds. In the case of isograds formed at crustal levels shallower than M, the lower the temperature of an isograd, the younger the age of formation of the isograd. The two staurolite isograds shallower and deeper than M form at different times.

In this case, the thermal-peak temperature increases downwards from the Earth's surface to depth M and then decreases for some distance below it.

2.5.2 Polychronous origins of isograds

In the above discussions, we considered relatively small metamorphic areas. Over much larger regions, on the other hand, the areas where deformation, uplift and plutonic intrusion take place shift laterally with time within the region during metamorphism. This could be the major cause for a large variation in the age of formation of isograds within a metamorphic region.

In response to complicated variations of thermal history within the crust in general, metamorphic rocks within a large complex should reach their thermal peaks at different times in different places. Since isograds represent the equilibrium states at the thermal peaks of individual rocks, different isograds in a metamorphic region will usually form at different times, and different parts of an isograd will also form at different times.

The age of a biotite isograd is the age of the formation of biotite on the isograd and at the same time the age of the thermal peak of rocks on the isograd (see Fig. 2.8a). It differs from the K–Ar age of the biotite on the isograd. The K–Ar age of biotite shows the time that has elapsed since the mineral was cooled through the Ar diffusion closure temperature (about 310°C), and so is younger than the age of formation of the biotite. The Ar diffusion closure temperature for muscovite is about 350°C, and the K–Ar ages for muscovite of rocks on a biotite isograd are also younger than the age of the isograd.

Although the K–Ar ages for biotite and muscovite on a biotite isograd do not give the age of the isograd, the distribution of K–Ar ages in a metamorphic region gives a clue to the variation of thermal history within the region. Figure 2.10 shows the distribution of K–Ar ages for biotite in the southeastern Scottish Highlands. Dewey & Pankhurst (1970) found that the K–Ar biotite ages show a regular, regional-scale pattern of variation, increasing gradually to the east and southeast. They drew a set of isopleths, called **thermochrons**, on the map. These thermochrons intersect all the isograds. The biotite isograd intersects the thermochrons for 460, 450, 440, 430, and 420 Ma. In other words, the K–Ar ages of biotite range from 420 to 460 Ma along the biotite isograd. Biotites along the biotite isograd were cooled through 310°C at greatly different times ranging from 460 to 420 Ma. A similar relation has been found for K–Ar muscovite ages.

Although such a cooling process would be strongly influenced by the time and distribution of uplift within the region, Figure 2.10 suggests a general tendency for the central and deep parts of an orogenic pile to be more slowly cooled than the marginal and shallow parts. The intersection of the thermochrons and isograds in the Scottish Highlands suggests that, generally, the cooling process bears little relationship to thermal peak.

As for the biotite isograd, however, the temperature of formation of the isograd would be near 400 °C, and this is relatively close to the Ar closure temperatures of biotite and muscovite. So, a considerable part of the observed range of the K–Ar age of rocks on the biotite isograd might result from variation in the age of formation of the biotite isograd.

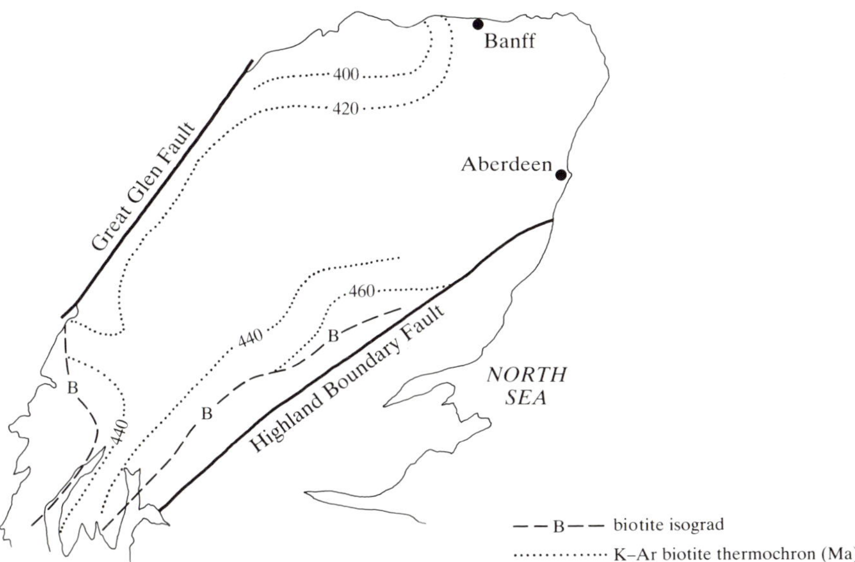

Figure 2.10 Distribution of K–Ar ages for biotite in a metamorphic region of the Scottish Highlands, the same region as in Figure 1.4b. After Dewey & Pankhurst (1970).

2.6 General models for isothermal and isogradic surfaces in the crust

2.6.1 Thermal models with horizontal isothermal surfaces

2.6.1.1 Grubenmann model. Grubenmann (1904–06) proposed a scheme of classification of regional metamorphic rocks, based on the assumption that the mineralogical compositions of metamorphic rocks depend on the depth of their formation as well as on their bulk-rock chemical composition. In other words, it assumed that the *P–T* conditions vary with depth within the crust. Thus, both temperature and pressure are functions of, and increase with, depth, and so isothermal surfaces within the crust are horizontal and stationary during regional metamorphism (P. H. Thompson 1977; Fig. 2.11a). In such a case, temperature is a function of pressure. Then, the *P–T* conditions of all cases of regional metamorphism are represented by a single curve in the *P–T* plane (Fig. 2.11b).

In this model, the present surface of the Earth must have been formed by post-metamorphic uplift and erosion, the amounts of which are usually different in different parts of a region. So, in the crustal cross section of Figure 2.11a, the present erosion surface corresponds to a fortuitous line such as N—E or N—F. All such lines correspond to a single curve, with a positive slope, on the *P–T* diagram (Fig. 2.11b).

This model was widely used in its original form or in a slightly modified form, which might be called a geosynclinal model, until about 1960. However, it is no longer tenable, because tectonic processes that cause regional metamorphism have clearly been shown to involve crustal movement and magmatic activity to produce uneven and temporally changing isothermal surfaces.

2.6.1.2 Recent one-dimensional models. For the past 20 years, one-dimensional thermal models have been proposed for discussions of the temporal changes of *P–T* conditions in regional metamorphic belts with special reference to the continental collision zone in the eastern Alps (Oxburgh & Turcotte 1974, England & Richardson 1977, England & Thompson 1984). These models made a great contribution in drawing attention to temporal changes of *P–T* conditions. However, they also assume that all rock properties and temperature and pressure are functions of depth alone. In other words, isothermal surfaces are assumed to be always horizontal, although they move upward or downward with time. This is evidently not valid in real cases (e.g. Harte & Dempster 1987).

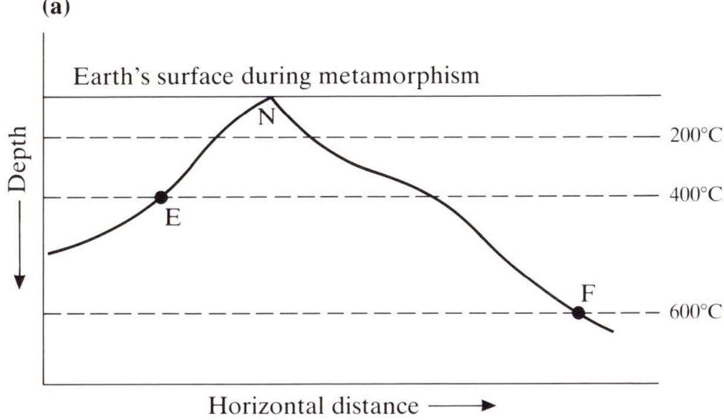

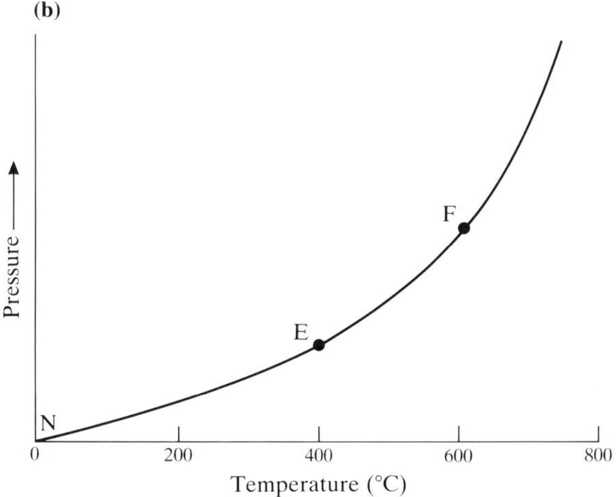

Figure 2.11 Grubenmann crustal model with stationary horizontal isothermal surfaces during metamorphism. (a) A vertical cross section of the crust during metamorphism. Dashed lines indicate isothermal surfaces. Solid lines N–E and N–F indicate two of the possible positions during metamorphism of the present-day surface of the Earth. (b) A corresponding *P–T* diagram.

2.6.2 Uneven shapes of instantaneous isothermal surfaces and isogradic surfaces in the orogenic crust

2.6.2.1 Instantaneous isothermal surfaces. If there are no crustal movements, no intrusions, and no erosion for a long time (say, more than 100 million years) the temperature distribution in the crust approaches a steady state. In such a state, crustal heat flow is made up of two parts: radioactive heat production in the crust, and heat flow coming from the upper mantle. The geothermal gradient is related not only to the heat flow but also to the thermal conductivity of rocks.

In orogenic regions, however, temperature distribution will usually be far from any steady state. Because crustal movements are controlled by plate-tectonic mechanisms, many large-scale movements will take place at a rate comparable with plate movements (usually a few to several centimeters per year). This rate is much faster than that of heat conduction. So, tectonic movements usually cause a perturbation of geotherms.

Thus, during orogeny, the temperature distribution within the crust is perturbed by various mechanical and magmatic movements. Thermal relaxation follows. The course of relaxation varies with the values of heat supply and thermal conductivity, as well as with the subsequent deformation, uplift and erosion. Moreover, a new tectonic event may begin during thermal relaxation to create a new perturbation of temperature. Thus, during orogeny, the temperature distribution within the crust is not stationary but transient, and the isothermal surfaces at any specific time during metamorphism are not horizontal but uneven.

Figure 2.12a illustrates a possible simple case of uneven isothermal surfaces in a crustal cross section at a specific time during regional metamorphism. Temperature is assumed to increase not only downward but also horizontally to the axial zone of the metamorphic belt.

In a case like this, the geothermal gradient must be distinguished from the temperature gradient (e.g. P. H. Thompson 1977). The **geothermal gradient** is the rate of increase of temperature with depth, that is, measured vertically downward within the crust, at a specific time, while the **temperature gradient** at a point within the crust is the rate of increase of temperature measured in the direction perpendicular to the isothermal surface passing through the point at a specific time. The temperature gradient as defined here should not be confused with the petrographically determined metamorphic thermal gradient to be defined below, for the latter is related to thermal-peak P–T conditions.

The **pressure gradient** may also be defined analogously as the rate of increase of pressure in the direction perpendicular to the isobaric surface passing through the point. If the Earth's surface is horizontal and there is no lateral variation in the density of crustal rocks, this direction is vertically downward, so far as pressure is regarded as a function of depth. If there is topographical relief, the direction must show some deviation from the vertical.

Within the crust undergoing orogeny, temperature may even increase upward for a limited time, because, for example, a hot thrust sheet may be pushed over a cold rock mass, or a hot magma may be intruded into a high horizon in the crust. Any thermal relaxation following could produce complicated transient temperature distributions. Two-dimensional thermal models for some such cases have been calculated by Fowler & Nisbet (1988).

If a region is folded at a rate faster than the rate of heat conduction, isothermal surfaces are folded. After the folding ceases, thermal relaxation occurs. This causes cooling in the antiformal parts of folds and heating in synformal parts. Hence, *P–T–t* paths of rocks differ between adjacent antiform and synform.

Orogenic belts have a lateral variation in structure and a limited width. So isothermal surfaces must show a shape corresponding to the belt. At the margin of an orogenic belt, isothermal surfaces dip outward. The western United States is a young orogenic belt. There, observed values of the geothermal gradient in the shallow continental crust show great lateral variation. Excluding some strongly deviated values, the geothermal gradients in the uppermost two kilometers of crust range from 15° to about 70°C km^{-1} (Nathenson & Guffanti 1988).

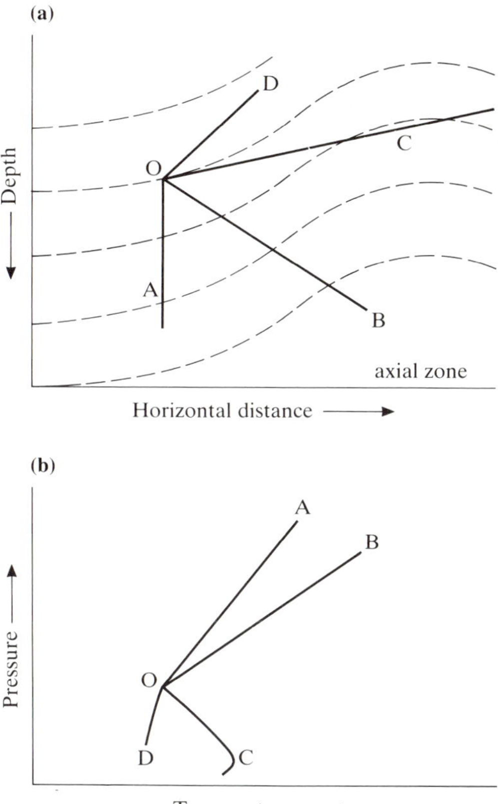

Figure 2.12 (a) A vertical cross section of the crust showing possible uneven shapes of instantaneous or thermal peak isothermal surfaces (dashed lines). Temperature is assumed to increase downward. The present-day surface of the Earth was in position O–A, O–B, O–C, or O–D during metamorphism. (b) The corresponding *P–T* relation. The temperature and pressure in this diagram refer to those either at any specific time or at the thermal peak.

2.6.2.2 Thermal-peak isothermal surfaces. Some (not all) such transient irregular temperature distributions produce irregular thermal-peak isothermal surfaces, which in turn lead to the formation of correspondingly irregular isogradic surfaces. The thermal peak is reached at different times in different parts of a metamorphic region (e.g. Ghent & Stout 1988). Isogradic surfaces based on mineral assemblages as well as thermal gradients calculated from observed mineral equilibria pertain to *P-T* conditions at the thermal peak at each point in the crust. An isogradic surface based on a dehydration reaction is usually an approximate thermal-peak isotherm. The rate of increase of thermal-peak temperature in a direction perpendicular to the isogradic surfaces based on a dehydration reaction has been termed the **metamorphic thermal gradient** (De Yoreo et al. 1991). It is an observable quantity characterizing a metamorphic complex.

Figure 2.12a represents the irregular shape of thermal-peak isothermal surfaces. The present surface of a metamorphic region may have lain in any direction within the crust during metamorphism, as schematically indicated by lines O—A, O—B, O—C and O—D in Figure 2.12a. Correspondingly, the field *P-T* curve of progressive metamorphism observed on the erosion surface may have taken various directions, with a positive or negative slope, as shown in Figure 2.12b.

Thermal-peak isobaric and isothermal surfaces are commonly deformed by post-metamorphic folding. Shapes actually observed in the Scottish Highlands are described in §9.2.

2.6.3 Thermal effects of fluid flows

2.6.3.1 Introduction. Thermal models of regional metamorphism, based on heat transfer by conduction alone, have been successfully postulated by many authors. However, flows of fluid appear to occur in most, or all, cases of regional metamorphism. It is supposed that such a fluid flow may considerably modify the results of thermal-model studies based on conduction alone in some cases. The possible sources of fluids in metamorphic rocks include: (a) H_2O that originally filled interspaces between mineral grains of sedimentary protoliths; (b) H_2O and CO_2 produced by dehydration–decarbonation reactions in the prograde stage; (c) H_2O released from crystallizing magmas in syn-metamorphic intrusions; (d) H_2O that formed by dehydration within subducting slabs and rising through the overlying crust; and (e) H_2O and/or CO_2 produced by degassing of the mantle.

The shallow crust, particularly in the vicinity of igneous intrusions, may have a high porosity and permeability with a network of interconnected large channelways filled with an aqueous fluid. This fluid may show a hydrostatic pressure gradient distinct from the lithostatic pressure gradient of the surrounding solid rocks. In such a case, if part of the network of fluid channels is heated, for example, by an igneous intrusion in the vicinity, **thermal convection** may occur in the channels, possibly leading to the transfer of a large amount of heat, so that thermal models based on conduction alone become inapplicable. Although some authors have advocated the application of this type of convective circulation model to regional metamorphism in the deeper crust, the validity of the idea is highly questionable. In the following section, the possibility of thermal convection of fluids in regional metamorphic complexes is ignored.

It is likely that **forced flows** of fluids occur in most, or all, cases of regional metamorphism. Such flows may be either pervasive or channelized. The direction of flow will generally be upward, because the fluids are less dense than the surrounding rocks, although in some cases fluids will flow laterally or obliquely through preferred structural paths and by local pressure differences.

2.6.3.2 Brady's (1988) analysis. Brady has made a quantitative estimate of the thermal effects of forced flows during regional metamorphism. If the crust experiences a steady flow, pervasive or channelized, of fluid rising from depth toward the surface, temperatures are elevated compared with the geotherms produced by thermal conduction alone. As temperatures rise, geothermal gradients near the surface and conductive heat loss at the surface must increase. If the flow of fluids continues long enough, temperatures will rise until a steady state is achieved, in which the heat added by conduction and fluid flow from below is exactly balanced by heat loss at the surface by conduction and fluid flow. A steady-state solution to the heat conduction equation modified for fluid flow provides a value for the maximum expected increase of temperature due to fluid flow.

The increase of temperature by fluid flows increases with the fluid flux, which means the amount of fluid passing through a unit area per unit time. The flux of fluid caused by dehydration–decarbonation reactions can be roughly estimated from the content of volatile components in relevant rocks and the duration of devolatilization reactions. The maximum value of the fluid flux by this process was estimated at about 300 or 400 $m^3 m^{-2} Ma^{-1}$, if the fluid flow is distributed rather uniformly over a metamorphic region. Brady has found that such an amount of flux cannot cause a significant increase in temperature in the region. This is true even when the crustal thickness is greatly increased by overthrusting. The same conclusion also holds when the H_2O released from a subducting slab rises through the crust. Thus, in these cases thermal models based on conduction alone are completely justified.

If fluid flows are focused into narrow zones in the crust, on the other hand, the value of fluid flux and, therefore, its thermal effects should increase. In such cases, fluid flows may cause a significant increase of metamorphic grade, even where the fluids are created by dehydration and decarbonation reactions. Additional H_2O may be supplied from crystallizing magmas in syn-metamorphic intrusions. Crystallization of magma may occur rapidly in a short time compared with regional metamorphism, and so may result in a relatively high rate of production of H_2O, and hence a relatively high thermal pulse of short duration. This may cause an increase in metamorphic grades if the flow is focused in a narrow area.

Some examples apparently showing a marked increase of metamorphic grade caused by focused flow of fluid have been described in recent years. Chamberlain & Rumble (1988) have found the occurrence of a number of separate small areas, each a few kilometers across, showing anomalously high-temperature mineral assemblages within a relatively high-grade gneiss region in New Hampshire. They ascribed the formation of the anomalous areas to the transport of a large amount of heat by flows of fluid through a network of fractures (§6.8).

Schuiling & Kreulen (1979) and Kreulen (1988) considered that the metamorphic thermal dome (about 7 km × 25 km) in the island of Naxos in Greece may have been produced by a large flux of CO_2-rich fluid rising from the mantle.

2.6.4 So-called inverted metamorphic zones

The term **inverted metamorphic zones** indicates a state where a higher-grade index-mineral zone overlies one of lower grade *in the present-day crust*. Since Tilley (1925) described the occurrence of the flat-lying garnet zone over the biotite zone in part of the Scottish Highlands, such structures have been reported to occur in many places in the world. Many of the inverted metamorphic zones were definitely formed by a post-metamorphic overturning of an originally normal superposition of index-mineral zones, in which lower-grade zones overlie higher-grade zones (e.g. Rosenfeld 1968).

However, some of the inverted metamorphic zones have been claimed to have formed as an originally inverted superposition of index-mineral zones. This is possible: Figure 2.9 shows the creation of an upward increase of thermal-peak temperature on the lower side of an intrusion. It should be noted that an originally inverted superposition of index-mineral zones is an expression of an upward increase of thermal-peak temperature and not of an upward increase of temperature at any specific time.

The formation of originally inverted superposition of metamorphic zones has been attributed in the literature to many different mechanisms. For example, a structure of this type was ascribed to igneous intrusion by St-Onge (1981), to mechanical transport of high-temperature rock masses into relatively high levels by Le Fort (1975), to movement along an inclined shear zone by Mason (1984), to shear heating along a thrust fault by Graham & England (1976), and to a preferential flow of fluid by Watkins (1985) and Peacock (1987a,b).

A most magnificent example of inverted metamorphic zones occurs along the whole length of the Himalayas. The origin of the structure was ascribed by many geologists to syn-metamorphic thrusting or folding, by others to post-metamorphic thrusting or folding, and to the effects of intrusions or other factors by still others. Some even doubt the existence of inverted metamorphic zones there. For a brief review of recent diverse opinions, see Barnicoat & Treloar (1989). The diversity of opinions illustrates the difficulty in interpretation of the origin of such structures.

Many of the existing descriptions of the inverted metamorphic zones may well be doubted. Regional metamorphic complexes are usually made up of lithostratigraphic units with clearly different pre-metamorphic bulk-rock chemical compositions. The chemical differences lead to the formation of layers with different mineralogical compositions in metamorphic regions. Hence, isogradic surfaces are apt to be assumed to agree with boundaries between such layers. However, this assumption could be misleading in general. In many regions, the presence of inverted metamorphic zones has been claimed on the basis of this assumption.

3 Thermodynamic properties of metamorphic reactions

This chapter is intended to provide a concise summary of the main results of thermodynamic and experimental investigations of metamorphic reactions for students of metamorphic geology. Derivation of thermodynamic equations is omitted in many cases, and instead the significance of the results of these investigations to field petrology is explained.

3.1 Phase rule, components and the number of reactions

3.1.1 Gibbs' phase rule

3.1.1.1 Extensive and intensive properties. The thermodynamic properties of a system may be divided into two categories, extensive and intensive. An **extensive** property is one whose total value is equal to the sum of its values in each of the parts of the system. For example, if we divide a system into parts, the mass and volume of the system as a whole are the sums of the masses and volumes of the respective parts, and thus mass and volume are extensive properties. On the other hand, properties that do not depend on the amount of matter in the system are called **intensive**. For example, temperature, pressure, density, and concentrations of components in each phase are intensive properties. The condition of chemical equilibrium depends only on the intensive properties.

3.1.1.2 The phase rule. Let us consider a small volume of rock that is undergoing metamorphic crystallization within the crust. In the rock, the intensive properties temperature and pressure are uniform, and we may assume that chemical equilibrium holds (§1.1.3). For such a system, the phase rule of J. Willard Gibbs is written as follows:

$$F = c + 2 - p \geq 0 \qquad (3.1)$$

where F is the number of degrees of freedom, or variance (thermodynamic variance), c is the number of independent components necessary to define the compositions of all phases in the system, and p is the number of phases.

The variance of the system is the maximum number of intensive variables to which we can assign arbitrary values, or which can be changed independently, without causing a change in the number of phases in the system in equilibrium. If we assign specific values to F intensive variables, the state of the system is defined, and so all the other intensive variables of the system also have specific values.

The word "independent" in the case of independent components means that the composition of any one component cannot be expressed in terms of other components, as is illustrated in the following section. Although the number of independent components in a system is definite, we can commonly choose from two or more alternative sets of independent components. If andalusite and sillimanite have the definite composition Al_2SiO_5, the number of independent components (c) of a system composed of andalusite alone, sillimanite alone, or andalusite and sillimanite is one from the viewpoint of the phase rule, although the minerals may be described as composed of the elements Al, Si and O, or the oxides Al_2O_3 and SiO_2. If quartz is associated with Al_2SiO_5, the system has two components: these may be either Al_2SiO_5 and SiO_2, or Al_2O_3 and SiO_2. The phase transformation reaction from andalusite to sillimanite in the presence of quartz occurs in a two-component, three-phase system. However, since quartz does not participate in the reaction, we may ignore this mineral together with the component SiO_2, and thus treat a partial system made up of one component and two phases. The variance is 1 in either case. If the Al in andalusite and sillimanite is partly replaced by a small and variable amount of Fe^{3+}, the minerals may then be described as being composed of two components, Al_2SiO_5 and Fe_2SiO_5 (or Al_2SiO_5 and $AlFeSiO_5$). The number of phases (p) means the number of kinds of minerals, plus the number of fluid phases, if present.

The variance in the Gibbs' phase rule is relevant to the variability of only the intensive parameters of the system. The state of the system as represented by the intensive variables (including the compositions of solid-solution minerals) becomes specific only when specific values are assigned to F intensive variables, where F is the phase-rule variance. In this case, the compositions of solid-solution minerals in a metamorphic rock could be calculated for such a specific state of the system by solving a set of equations representing equilibrium conditions.

3.1.1.3 Derivation of the phase rule. The phase rule is derived simply by counting the number of unknown variables and the number of relations to determine them, as follows. A set of variables needed to describe the state of the system are pressure, temperature and the mole fraction of each component in every phase. So, a system of c components with p phases has $(2 + cp)$ variables. All the mole fractions are not mutually independent, because for each phase the sum of the mole fractions of all components must be unity. This gives p relations between mole fractions. Equilibrium requires that the chemical potential of each component should be the same in all phases, giving rise to $c(p - 1)$ equations. Therefore, the number of variables we can assign arbitrary values to is: $(2 + cp) - p - c(p - 1) = c + 2 - p$. This is the variance of the system.

3.1.1.4 Phase rules in special cases. Although the phase rule (3.1) holds true in most common equilibrium systems, modifications become necessary in some special cases. For example, let us consider a system composed of solid rock and a fluid-filled fissure in equilibrium. If the walls of the fissure are mechanically strong , the solid and the fluid may be at different pressures so that the whole system has two different pressures. Since this increases the number of intensive variables by one, the variance is given by: $F = c + 3 - p$.

3.1.2 System components, phase components and the number of reactions

3.1.2.1 System components and phase components. As an example, let us take the assemblage of olivine $(Mg,Fe)_2SiO_4$ and orthopyroxene $(Mg,Fe)SiO_3$. This system has three independent components, e.g. MgO, FeO and SiO_2. Although the number of independent components is definite, the choice of components is not unique. For example, we could regard the system as being made up of the three components: Mg_2SiO_4, Fe_2SiO_4 and SiO_2, because the composition of orthopyroxene can be expressed by those of olivine and SiO_2 since there is a stoichiometric relation: $2(Mg,Fe)SiO_3 = (Mg,Fe)_2SiO_4 + SiO_2$. Alternatively, we may use $MgSiO_3$, $FeSiO_3$ and SiO_2 as three components, provided we can assign negative values to the amounts of components, because $(Mg,Fe)_2SiO_4 = 2(Mg,Fe)SiO_3 - SiO_2$. All these sets of three components are possible examples of the independent components of the system that appear in the expression of the phase rule. Such components are sometimes called **system components**.

On the other hand, the composition of olivine solid solution can be most properly expressed in terms of Mg_2SiO_4 and Fe_2SiO_4, while that of orthopyroxene can be expressed in terms of $MgSiO_3$ and $FeSiO_3$. In this respect, it is practical to use these four components to describe all phases of the system. Such components are called **phase components**. The above four phase components are not mutually "independent", because there is a stoichiometric relation among them: $Mg_2SiO_4 + 2\ FeSiO_3 = Fe_2SiO_4 + 2\ MgSiO_3$. Thus, only three of them are independent. As a set of system components, we may use any three out of the four phase components.

The *phase components of a system* include all phase components of all phases in the system. In some cases, components in two phases in the system have the same formulas. For example, $MgSiO_3$ may be used as a phase component in orthopyroxene as well as in coexisting clinopyroxene. Each of them is counted as a different phase component of the system.

3.1.2.2 Number of reactions. By analogy to the above example, the following relation holds for any system (J. B. Thompson 1982b, Spear et al. 1982):

number of independent stoichiometric relations among phase components
= (number of phase components) − (number of system components) (3.2)

Each of the stoichiometric relations represents a possible reaction in the system. In the equilibrium state of the system, an equilibrium condition holds for each such equation.

3.1.3 Duhem's theorem

We now consider a system in which the mass of each of the system components is specified. Thus, the system has a definite total mass. In this case, a set of intensive and extensive properties necessary for the description of the system includes temperature, pressure, the concentrations of all phase components in each phase, and the mass of each phase. Hence, the total number of intensive and extensive variables of the system is: 2 + (number of all phase components of all phases) + (number of phases).

A set of independent equations among these variables includes the equilibrium conditions for all independent reactions between phase components, the condition that the sum of the weight fractions of all phase components in each phase is unity, and the relations that specify the mass of each system component. The total number of such equations is:

(number of independent reactions) + (number of phases) + (number of system components)
= (number of all phase components of all phases) − (number of system components) + (number of phases) + (number of system components)
= (number of all phase components of all phases) + (number of phases)

Hence, the variance of the system in this case = (total number of intensive and extensive variables) − (total number of equations) = 2.

Thus, in any system in which the mass of each of the system components is specified, the equilibrium state of the system is completely determined by two independent variables. This is called **Duhem's theorem** (Prigogine & Defay 1954).

In a system with a phase-rule variance equal to 2 or greater, the two independent variables mentioned above could be chosen from intensive parameters. If temperature and pressure are specified in a system of this kind, the compositions and *masses* of all phases are determined. In a system with a phase-rule variance of 1 or 0, on the other hand, one or both of the two independent variables mentioned above must be extensive.

3.1.4 Additive components and exchange components

So far we have expressed the composition of solid-solution minerals by the traditional method in terms of end-members. This method becomes very complicated when it is applied to minerals in which many different types of atomic substitution occur.

J. B. Thompson (1982a,b), on the other hand, has proposed a new method of expressing the substitution of an Fe atom for an Mg atom in crystals by the symbol $FeMg_{-1}$. The composition of orthopyroxene, for example, can be expressed by the formula of the Mg end-member $MgSiO_3$ and the extent of substitution $FeMg_{-1}$ per mole of $MgSiO_3$. For this meaning, $FeMg_{-1}$ may be regarded as a component, called an **exchange component**, whereas conventional components such as $MgSiO_3$ and $FeSiO_3$ are called **additive components**. In an analogous way, K–Na, NaSi–CaAl, and Tschermak substitutions may be expressed by NaK_{-1}, $CaAlNa_{-1}Si_{-1}$, and $Al_2Mg_{-1}Si_{-1}$, respectively.

Thus, if orthopyroxene is regarded as a binary Mg–Fe solid solution, we may use $MgSiO_3$ as the additive component and express the composition of a member with 50 mol % $FeSiO_3$ as: $MgSiO_3 + 0.5\ FeMg_{-1}$ ($= Mg_{0.5}Fe_{0.5}SiO_3$). When both Mg–Fe and Tschermak substitution occur, the composition of an orthopyroxene may be expressed as a sum of $MgSiO_3$ and some amounts of $FeMg_{-1}$ and $Al_2Mg_{-1}Si_{-1}$.

When we discuss heterogeneous equilibria, it is convenient to use the same types of exchange components for as many coexisting phases as possible. The same type of exchange component, such as $FeMg_{-1}$, in different minerals is counted as a different component.

The method of using exchange components has clear advantages compared with the traditional method of using end-members alone. It simplifies the method of counting the number of independent components in minerals and moreover simplifies the method of derivation of reaction equations. For example, when both Mg–Fe and Tschermak substitution are considered to occur in a mineral, the composition of the solid solution has traditionally been expressed in a square diagram, in which the horizontal and vertical axes represent the extents of the two types of substitution. This might give the impression that the mineral is composed of four components located at the four corners, although actually only three of the four components are independent. If we use Thompson's method, the mineral is evidently composed of three independent components: one additive component and $FeMg_{-1}$ and $Al_2Mg_{-1}Si_{-1}$. If K–Na and Mg–Mn substitutions occur in addition, many more end-members are needed in the traditional method, and it is troublesome to consider how many of them are independent. Thompson's method gives the independent components directly: one additive component and $FeMg_{-1}$, $Al_2Mg_{-1}Si_{-1}$, NaK_{-1} and $MnMg_{-1}$.

3.2 Classifications of metamorphic reactions

3.2.1 General classification

Some metamorphic reactions take place among solid phases alone with no participation of volatiles. An example is: $A + B = C + D$, where A, B, C and D represent solid minerals. Such reactions are called **solid–solid reactions**.

There is another type of metamorphic reaction, which involves H_2O and/or CO_2. Examples are: $A + B = C + H_2O$; $A + B = C + CO_2$; $A + B + H_2O = C + D + CO_2$. The first of the examples involves liberation of H_2O alone, and is called a **simple dehydration reaction**. The second involves liberation of CO_2 alone, and is called a **simple decarbonation reaction**. (Sometimes such reactions are collectively termed **devolatilization reactions**.) The third of the examples above is a more complicated reaction.

Many rock-forming minerals contain iron, which may be transferred by **oxidation–reduction reactions** between uncombined, divalent and trivalent states in minerals of the Earth's crust. An example is: 6 hematite = 4 magnetite + O_2. This type of reaction is important in defining the stabilities of iron-bearing minerals.

In many metamorphic rocks, an Fe–Mg exchange reaction can take place between two coexisting ferromagnesian minerals. In the reaction, Fe atoms migrate from a first mineral to a second, and this is compensated by the migration of the same number of Mg atoms in the opposite direction, thus causing no change in the mole number of each mineral. More generally, when two coexisting minerals show the same type of atomic substitution, an **exchange reaction** can occur with respect to the substitution. This type of reaction is a special kind of solid–solid reaction, because there is no participation of volatiles such as H_2O, CO_2 and O_2. Since the volume change caused by this type of reaction is negligibly small, the effect of pressure on the equilibrium is also very small.

3.2.2 Exchange reactions and net-transfer reactions

Exchange reactions do not cause an increase or decrease of the numbers of the unit cells of participating minerals. In contrast, all other types of reactions described above cause an increase in the number of unit cells of a mineral or minerals and a decrease in the number of unit cells of another mineral or other minerals. In other words, they cause a net transfer of matter from one mineral or minerals to another mineral or minerals. Such reactions were called **net-transfer reactions** by J. B. Thompson (1982b). There are reactions that cause a net transfer of matter from phase to phase, but in addition involve exchange of atoms between coexisting phases. Such reactions are also classed as net-transfer reactions.

We could write a number of stoichiometric relations among components of phases coexisting in a metamorphic rock. All such stoichiometric relations represent possible reactions in the system. Any linear combination of two of these reactions is another reaction. In such a case, the number of reactions is not limited. However, the total number of independent reactions is definite, being given by equation (3.2). Depending on the change in *P–T* conditions, all of these reactions will progress at different rates.

Thus, there is commonly a wide range of choice in selecting a set of independent reactions to describe the changes in a metamorphic rock. When such independent reactions are classified into exchange and net-transfer reactions, the numbers of independent exchange reactions and of independent net-transfer reactions are not necessarily definite, as will be illustrated later by the set of equations (5.8), (5.9) and (5.10). The overall reaction in a metamorphic rock is a linear combination of all the reactions in a set, which includes both exchange and net-transfer reactions.

If the same type of atomic substitution occurs in two coexisting minerals, there is one exchange reaction with regard to the substitution between the two minerals. Hence the maximum number of independent exchange reactions can easily be counted. J. B. Thompson (1982a,b) has proposed a strategy, taking as many reactions as possible as exchange reactions. In this case, the minimum number of net-transfer reactions is obtained by subtracting the maximum number of exchange reactions from the total number of stoichiometric relations given by equation (3.2).

Since natural metamorphic rocks commonly have many components and many phases, actual reactions may be very complex. Investigations of these problems have been made on isograds in metamorphosed carbonate rocks in the Augusta–Waterville area (Maine, USA) by Ferry (1983b), and then on the biotite isograd in metapelites of the same area by Ferry (1984). In his discussion of the biotite isograd in metapelitic rocks, Ferry (1984) assumed that the system components of the pelitic rocks + intergranular fluid are 13 in number: SiO_2, Al_2O_3, FeO, MgO, MnO, CaO, Na_2O, K_2O, TiO_2, C, O, H and S. Thirteen constituent minerals are considered: quartz, muscovite, biotite, chlorite, ankerite, siderite, calcite, plagioclase, ilmenite, rutile, pyrite, pyrrhotite and graphite. In this case, the number of net-transfer reactions is at least 11, and there is a wide latitude in the choice of a particular set of net-transfer reactions.

3.3 Thermodynamic symbols, concepts and units

a_i **Activity** of component i in a solution. If we choose a pure phase of component i at the pressure and temperature under consideration as standard state, $\mu_i = G(P,T) + RT \ln a_i$, where G is the molar Gibbs' energy of the pure phase. Then, $a_i = X_i$ in an ideal solution, where X_i is the mole fraction of component i. For other solutions, $a_i = X_i\gamma_i$. The activity is a dimensionless intensive property.

C_p **Molar heat capacity** (molar heat) at constant pressure (usually measured in J mol^{-1} K^{-1}).

ΔC_p (Sum of the heat capacities of the products) – (sum of the heat capacities of the reactants).

c Number of independent **components** (or system components) in the phase rule. (See §3.1.)

F **Variance**, or **number of degrees of freedom**, in the phase rule. Usually $F = c + 2 - p$. (See §3.1.)

f *and* f_i° **Fugacity of a pure gas**, denoted by i (measured in pressure units such as bar or Pa).The fugacity of a pure gas is a quantity directly related to the Gibbs' energy and is a function of the P and T of the gas. In an ideal gas, the fugacity is equal to the pressure. In a non-ideal gas, it can be calculated from the measured P-T-V relationships. The ratio f/P is called the **fugacity coefficient**. In the ordinary P-T range of regional metamorphism, the fugacity coefficient of H_2O is in the range 0.1–1.5, and that of CO_2 is in the range 2–50.

f_i **Fugacity of component** i in a gas mixture (measured in pressure units such as bar or Pa). The fugacity of component i is a quantity directly related to the chemical potential of i in the gas mixture, and is a function of P, T and the composition of the mixture. In an ideal gas mixture, it is equal to the partial pressure of component i. In a non-ideal gas mixture, it can be calculated from the measured relationship between P, T, and the partial molar volume of i in the mixture. The ratio f_i/P_i is called the **fugacity coefficient** of component i in the mixture. (See §4.1.5.) In

H_2O–CO_2 mixtures at 3–5 kbar and 400–600 °C, the fugacity coefficient of H_2O is in the range 0.2–0.6, and that of CO_2 is in the range 2.6–50 (Bowers et al. 1984).

G **Gibbs' energy** or Gibbs' free energy (measured in energy units such as J or kJ). 1 J = 10 cm^3 bar = 0.2390 cal. $G = H - TS = U + PV - TS$.

ΔG Gibbs' energy change. $\Delta G = \Delta H - T\Delta S$.

$\Delta G°$ **Standard Gibbs' energy of reaction** (measured in energy units such as J or kJ). It means the Gibbs' energy change for the process of transforming stoichiometric numbers of moles of the *pure* separated reactants, each in its standard state, to stoichiometric numbers of moles of the *pure* separated products, each in its standard state. This quantity is related to the equilibrium constant K by the relation:

$$\Delta G° = -RT \ln K. \text{ (See below.)}$$

H **Enthalpy** (measured in energy units such as J or kJ). 1 J = 10 cm^3 bar = 0.2390 cal. $H = U + PV$.

ΔH Enthalpy change.

$\Delta H°$ **Standard enthalpy of reaction** or **standard heat of reaction** (measured in energy units such as J or kJ). It means that the enthalpy change for the process of transforming stoichiometric numbers of moles of the *pure* separated reactants, each in its standard state, to stoichiometric numbers of moles of the *pure* separated products, each in its standard state.

K **Equilibrium constant** for a chemical reaction. If the reaction is expressed by the equation: $bB + cC = dD + eE$, where B, C, D and E represent phase components, and b, c, d and e denote coefficients, then the equilibrium constant is given by $K = a_D^d\, a_E^e / a_B^b\, a_C^c$. If pure components at the temperature and pressure under consideration (TK, P bar) are chosen as standard state, then $\Delta G° = -RT \ln K$. The equilibrium constant is dimensionless.

K_D **Distribution coefficient** for an exchange reaction. (See §3.7.)

P **Pressure**. The SI unit of pressure is Nm^{-2} or Pa. However, in this book bar and kilobar are used in accordance with an old-fashioned custom in the geologic community. 1 bar = 10^5 Pa = 0.1 MPa = 0.98692 atm = 14.50 psi. 1 kbar = 0.1 GPa.

P_i **Partial pressure** of component i in an ideal or non-ideal gas mixture. It is defined by the relation: $P_i = X_i P$. In an ideal gas mixture, it agrees with the pressure that gas i exerts if it alone is present in the container. In a non-ideal gas mixture, it does not agree.

p Number of **phases** in the phase rule. (See §3.1.)

R **Gas constant** per mole. It is:

$$8.3144\ \text{J mol}^{-1}\,\text{K}^{-1} = 83.144\ \text{cm}^3\ \text{bar mol}^{-1}\,\text{K}^{-1}.$$

S **Entropy** (usually measured in J K^{-1}).

ΔS Entropy change.

T **Temperature** Kelvin (K) in thermodynamic equations unless otherwise stated. When temperature in degrees Celsius (°C) is used, it is explicitly stated. Temperature in degrees Celsius is equal to temperature K -273.15. Room temperature 25° corresponds to 298.15 K, usually indicated as 298. In other contexts such as in P–T diagrams and P–T–t paths, T means temperature in any scale.

U **Internal energy** (measured in energy units such as J or kJ). $1\,J = 10\,cm^3\,bar = 0.2390\,cal$.

V **Volume** (measured in cm^3, m^3 or $J\,bar^{-1}$). $1\,J\,bar^{-1} = 10\,cm^3$.

ΔV Volume change.

ΔV_s Volume change of solid phases alone by a reaction.

X_i **Mole fraction** of component i. In a gaseous mixture, the mole fraction means the ratio (number of molecules of species i)/(total number of all molecules). The mole fraction of a phase component in a crystalline solution, which is used in the calculation of the activity, depends on the mode of mixing of replaceable atoms in the structure. (See, for example, Ch. 4 in Powell (1978).)

γ_i **Activity coefficient** of component i in a solution, as defined by $a_i = X_i \gamma_i$. In an ideal solution $\gamma_i = 1$.

μ_i **Chemical potential** of component i in a solution (usually measured in $J\,mol^{-1}$ or $kJ\,mol^{-1}$). It is the rate of increase of Gibbs' energy of the solution with a constant composition for addition of 1 mole of component i at a constant pressure and temperature.

$$G \text{ (of a solution)} = \sum_i n_i \mu_i$$

where n_i is the number of moles of component i in the solution. In a pure phase composed of one component only, the chemical potential of the component is equal to the molar Gibbs' energy of the substance.

3.4 Solid–solid reactions

3.4.1 Equilibria among pure solid phases

The following are examples of solid–solid reactions in metamorphic rocks:

$$\text{kyanite} = \text{sillimanite} \tag{3.3}$$

$$\text{andalusite} = \text{sillimanite} \tag{3.4}$$

$$\text{jadeite} + \text{quartz} = \text{albite} \tag{3.5}$$

$$2\,\text{jadeite} = \text{albite} + \text{nepheline} \tag{3.6}$$

Reactions (3.3) and (3.4) are phase transformations between polymorphs of Al_2SiO_5; special cases of solid–solid reactions. Reactions (3.5) and (3.6) represent reactions that change jadeite into albite in the presence and absence of quartz, respectively.

The volume changes of minerals with temperature and pressure are usually negligibly small. Moreover, thermodynamic equations for solid–solid reactions may be simplified by the following relations. First, ΔC_p (heat capacity difference between the products and reactants) as well as ΔS (entropy difference) are usually negligibly small at temperatures above 25°C. Secondly, the molar heat as well as the molar entropy of a silicate are approximately equal to the sums of those of the constituent oxides.

In a solid–solid reaction, if all the solid phases are pure, that is to say, composed of only one component, the system is univariant. In this case, the following approximate relation holds:

$$\Delta G^\circ(P,T) = \Delta H^\circ(1\text{bar}, 298) - T\Delta S^\circ(1\text{bar}, 298) + (P - 1)\,\Delta V_s = 0 \quad (3.7)$$

where ΔG° denotes the standard Gibbs' energy of reaction. This equation represents a straight line in the *P–T* diagram. Thus, equilibrium curves of solid–solid reactions among pure solid phases are usually straight.

The slope of such a univariant curve is given by the **Clausius–Clapeyron equation** as follows:

$$\mathrm{d}P/\mathrm{d}T = \Delta H/T\Delta V = \Delta S/\Delta V \quad (3.8)$$

where ΔH, ΔS and ΔV denote the enthalpy, entropy and volume changes, respectively, by the reaction at pressure and temperature on the curve.

Most of the curves have a positive slope ($\mathrm{d}P/\mathrm{d}T$), because the high-temperature side of a reaction usually has a larger entropy and volume than the low-temperature side. It should be noted that the positive slopes of such equilibrium curves mean that an increase of pressure has an opposite effect to that produced by an increase of temperature (J. B. Thompson 1955). A marked exception is the andalusite–sillimanite transformation (reaction 3.4), which has a negative slope, because sillimanite has a smaller molar volume than andalusite (Fig. 3.1).

Most equilibrium curves of solid–solid reactions have a gentle slope, showing a value of $\mathrm{d}P/\mathrm{d}T$ in the range of 1–3 kbar per 100°C. In other words, the equilibrium temperatures of many solid–solid reactions are relatively sensitive to pressure variation. For example, the $\mathrm{d}P/\mathrm{d}T$ of the kyanite–sillimanite transformation reaction is 2.02 kbar per 100°C, and that of the kyanite–andalusite transformation reaction is 1.25 kbar per 100°C (Fig. 3.1).

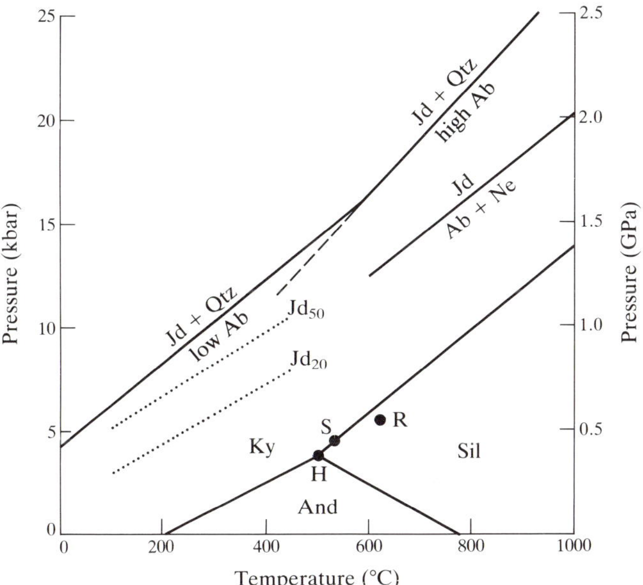

Figure 3.1 Phase relations of Al_2SiO_5 polymorphs after Holdaway (1971), and the equilibrium curves for the reactions: jadeite + quartz = high or low albite, and 2 jadeite = albite + nepheline (see text). Points H and R indicate the location of the triple point of Al_2SiO_5 polymorphs as determined by Holdaway (1971) and Richardson et al. (1969), respectively. Point S indicates the triple point for coarse-grained sillimanite determined by Salje (1986).

3.4.2 Solid–solid reactions involving solid-solution minerals

Many solid–solid reactions involve solid-solution minerals. The equilibria of such reactions can also be calculated for the relevant stoichiometric relations among phase components.

For example, natural jadeite and albite are usually solid solutions. Natural jadeite may contain diopside and acmite components, which should decrease the pressure necessary for the stabilization of jadeite (Gasparik 1985). In clinopyroxene containing a jadeite component (omphacite), cation ordering may occur, which has a considerable effect on the relevant equilibrium relations. Albite also shows complicated structural changes.

The following stoichiometric relation represents a breakdown reaction of anorthite, which occurs at very high pressures (20–30 kbar at 1000–1400°C), if all the terms in the equation represent pure phases.

$$3\,\text{anorthite} = \text{grossular} + 2\,Al_2SiO_5 + \text{quartz} \qquad (3.9)$$

In metapelites, the coexistence of plagioclase, garnet, sillimanite (or kyanite or andalusite) and quartz is common. In this case, the equilibrium of reaction (3.9) holds for the pertinent phase components in the relevant solid-solution minerals. This equilibrium holds in the ordinary pressure range of regional metamorphism, and is sensitive to pressure. Ghent (1976) proposed its use as a geothermobarometer. If the compositions of all the relevant solid-solution minerals are known, the equilibrium condition holds on a curve in the P–T diagram. If the equilibrium temperature is known from some other method such as the garnet–biotite Fe–Mg exchange geothermometer (§3.7), the equilibrium pressure can be read from the curve. This method has proved to be very successful. For calibration of this geothermobarometer (see Koziol & Newton 1988).

There are many solid–solid reactions that involve hydrous minerals. An example is almandine + muscovite = annite + 2 kyanite + quartz. In this reaction, the left- and right-hand sides have the same number of hydroxyl ions. The muscovite and biotite formulas have the same K/(OH) ratio. So, if such an equation is balanced with respect to K, it is automatically balanced with respect to OH, too.

The following solid–solid reaction (involving hydrous minerals) hold in quartz-bearing garnet amphibolite:

$$\underset{\text{anorthite}}{6\,CaAl_2Si_2O_8} + \underset{\text{tremolite}}{3\,Ca_2Mg_5Si_8O_{22}(OH)_2} = \underset{\text{grossular}}{2\,Ca_3Al_2Si_3O_{12}}$$

$$+ \underset{\text{pyrope}}{Mg_3Al_2Si_3O_{12}} + \underset{\text{tschermakite}}{3\,Ca_2Mg_4Al_2Si_7O_{22}(OH)_2} + \underset{\text{quartz}}{6\,SiO_2} \qquad (3.10)$$

Kohn & Spear (1990) have found that the equilibrium constant for this reaction varies considerably with pressure but is nearly independent of temperature, and so the equilibrium constant could be a good geobarometer. Hoisch (1990) has calibrated equilibria of six solid–solid reactions that occur in the muscovite + biotite + garnet + plagioclase + quartz assemblage, e.g. almandine + grossular + muscovite = 3 anorthite + annite.

3.4.3 Stability relations of the Al_2SiO_5 polymorphs and the jadeite + quartz assemblage

The stability relations of the Al_2SiO_5 polymorphs (andalusite, kyanite and sillimanite) and of the jadeite + quartz assemblage form a basic reference framework for the determination of the P–T conditions of metamorphism (Miyashiro 1961), as discussed in Chapter 8. Because of their importance, the experimental data on these relationships will be briefly reviewed below.

3.4.3.1 Stability relations of the Al_2SiO_5 polymorphs. The three polymorphs of Al_2SiO_5 are connected by solid–solid reactions, like reactions (3.3) and (3.4). Miyashiro (1949) proposed the existence of a triple point at which the three polymorphs coexist in equilibrium. Experimental determination of the phase relations was made first by Clark et al. (1957) for the reaction kyanite = sillimanite, and then by Newton (1966) on the triple point. Richardson et al. (1969) located the triple point at about 620°C and 5.5 kbar, whereas Holdaway (1971) placed it at about 500°C and 3.8 kbar (Fig. 3.1). A more recent determination by Hemingway et al. (1991) gave 511°C and 3.87 kbar, values similar to Holdaway's.

Holdaway ascribed the differences in the location of the triple point found by different experimenters to the effects of intense grinding and the extremely fine grain sizes of samples as well as the disorder between Al and Si in the tetrahedral sites of sillimanite. Moreover, the three polymorphs are actually solid solutions, although their composition ranges are small. Andalusite usually contains a greater amount of Fe_2O_3 than coexisting sillimanite. The presence of the Fe_2O_3 component could cause the stable coexistence of andalusite and sillimanite over a considerable range of temperature at a specific pressure (Grambling & Williams 1985, Kerrick & Spear 1988).

Salje (1986) has shown that very fine-grained (or fibrolitic) sillimanite has a higher heat capacity than coarse-grained sillimanite at the same temperature, that this difference causes a great difference in the temperature of the stability boundary between andalusite and sillimanite, and that it is the main cause of the discrepancy in the location of experimentally determined triple point. Salje's triple points determined with two samples of coarse-grained sillimanite are located near 530°C and 4.4 kbar (point S in Fig. 3.1), that is, close to that of Holdaway, whereas his triple point determined with fibrolite is located near that of Richardson et al. For a detailed review, see Kerrick (1990).

3.4.3.2 Low-pressure stability limit of the jadeite + quartz assemblage. The jadeite + quartz assemblage changes to albite with increasing temperature and/or decreasing pressure, as shown by reaction (3.5). This reaction represents the low-pressure stability limit for the jadeite + quartz assemblage. The equilibrium curve was experimentally determined first by Birch & LeComte (1960), and then by Newton & Smith (1967), Holland (1980), and others.

Albite undergoes a continuous phase transformation over a temperature range around 600°C. This causes a change in the slope of the equilibrium curve. In Figure 3.1, the equilibrium curve for temperatures above 600°C is based on Holland, and passes 600°C at 16.25 kbar and 1000°C at 26.85 kbar (with a slope of 26.5 bar per °C). For simplicity, the phase transformation of albite is assumed to take place at 600°C, and the equilibrium curve at lower temperatures is assumed to have a slope of 20 bar per °C. The resulting curve for temperatures below 600°C is close to those of Newton & Smith (1967) and of Hlabse & Kleppa (1968).

Usually, natural jadeites are not pure, but contain aegirine and diopside components. Such pyroxene solid solutions coexist with quartz and albite at lower pressures than pure jadeite. Approximate *P–T* conditions for the coexistence of these pyroxenes with 50% and 20% jadeite component are shown by dotted lines in Figure 3.1 on the basis of ideal solution models. The jadeite–aegirine series is close to ideal (Fig. 12.2), whereas the jadeite–diopside series is not so at low temperatures (Fig. 12.3). In the absence of quartz, reaction (3.6) can occur. This equilibrium holds at much lower pressures than reaction (3.5), as shown in Figure 3.1.

3.5 Reactions that involve H_2O and/or CO_2

The reactions that involve H_2O and/or CO_2 show different characteristics from solid–solid reactions for the following reason: first, at relatively low pressures H_2O and CO_2 have large molar volumes and high compressibilities as compared with solids. So they show a large change of volume with temperature and pressure. Secondly, these reactions are accompanied by large heats and entropies of reaction compared with solid–solid reactions. Thirdly, H_2O and CO_2 may be highly or considerably mobile during metamorphism, as discussed in Chapters 4 and 6.

3.5.1 Simple dehydration and decarbonation reactions involving pure solid phases in the presence of pure H_2O or pure CO_2, respectively

In this book, the terms *simple dehydration reactions* and *simple decarbonation reactions* mean reactions in which H_2O or CO_2, respectively, participates as the only volatile substance. The following are examples, in which solid phases have fixed compositions.

$$\text{pyrophyllite} = \text{kyanite} + 3\ \text{quartz} + H_2O \tag{3.11}$$

$$\text{muscovite} + \text{quartz} = \text{sillimanite} + \text{K-feldspar} + H_2O \tag{3.12}$$

$$\text{calcite} + \text{quartz} = \text{wollastonite} + CO_2 \tag{3.13}$$

$$\text{dolomite} = \text{calcite} + \text{periclase} + CO_2 \tag{3.14}$$

In reactions such as these, the side having liberated H_2O or CO_2 is stable at higher temperatures in almost all cases, because the entropy of the liberated H_2O and CO_2 is usually very great.

If all the relevant solid phases have fixed compositions, and the fluid phase has pure H_2O or pure CO_2 composition at the same pressure as the solids, the number of components is equal to the number of solid phases, and hence the system is univariant.

There are many experimental data on simple dehydration and decarbonation reactions in the presence of a fluid having practically pure H_2O or CO_2 compositions, respectively. In such a case, the following relation holds:

$$\Delta G^\circ(1\text{bar}, T) + (P - 1)\,\Delta V_s = -RT \ln K = -RT \ln f \qquad (3.15)$$

where f denotes the fugacity of H_2O or CO_2 at the pressure and temperature of interest. Using this equation, we can calculate the equilibrium curve from thermochemical data, or obtain the Gibbs' energy of reaction from an experimentally determined equilibrium curve. In some natural conditions, the intergranular fluid will be nearly pure H_2O, and so the above relations may hold approximately. Nearly pure CO_2 is rare in nature, even in metamorphism of carbonate rocks, with the possible exception of granulite-facies metamorphism.

The slope of such a univariant curve is given by the Clausius–Clapeyron equation (eq. 3.8). In simple dehydration and decarbonation reactions, ΔH, ΔS and ΔV are usually positive in the normal range of pressure of regional metamorphism. So, univariant curves usually have a positive slope. Note that in such cases the volume change for solids + fluid is usually positive, while the volume change of the solids alone is usually negative (J. B. Thompson 1955).

Four examples of univariant curves for simple dehydration reactions in the presence of pure H_2O fluid are shown in Figure 3.2. At pressures as low as a few hundred bar, the molar volumes of H_2O and CO_2 are very large, and ΔS is positive. Hence, the slope of the curve (dP/dT) is very small and positive. With increasing pressure, ΔV becomes smaller, and so the slope becomes larger, and the univariant curve is usually strongly bent in the pressure range up to 2 kbar. With a further increase of pressure, the compressibilities of H_2O and CO_2 decrease and so the equilibrium curves become straighter. Above 3 kbar, many dehydration curves are nearly straight. In this pressure range, equilibrium curves for different reactions show appreciably different slopes. Dehydration reaction curves K, P and S in Figure 2.2 are very steep. Some dehydration reaction curves are nearly vertical at several kilobar. All dehydration reaction curves at pressures above a few kilobar are steeper than the equilibrium curves of most solid–solid reactions (Fig. 3.2). This is because dehydration and decarbonation reactions have a much greater value of ΔS than solid–solid reactions.

At still higher pressures, H_2O is so strongly compressed that some simple dehydration reactions that result in a decrease in the volume of solid phases, also cause a decrease in the total volume of solid + fluid. In such a case, the univariant curve has a negative slope, as illustrated by the high-pressure part of curve (4) in Figure 3.2.

At pressures above 10 kbar, the univariant curves of many, or most, simple dehydration reactions may have a negative slope owing to the strong compression of H_2O. Figure 3.3a shows this type of transition of an equilibrium curve from a positive slope (segment A) to a negative slope (segment B) with increasing pressure.

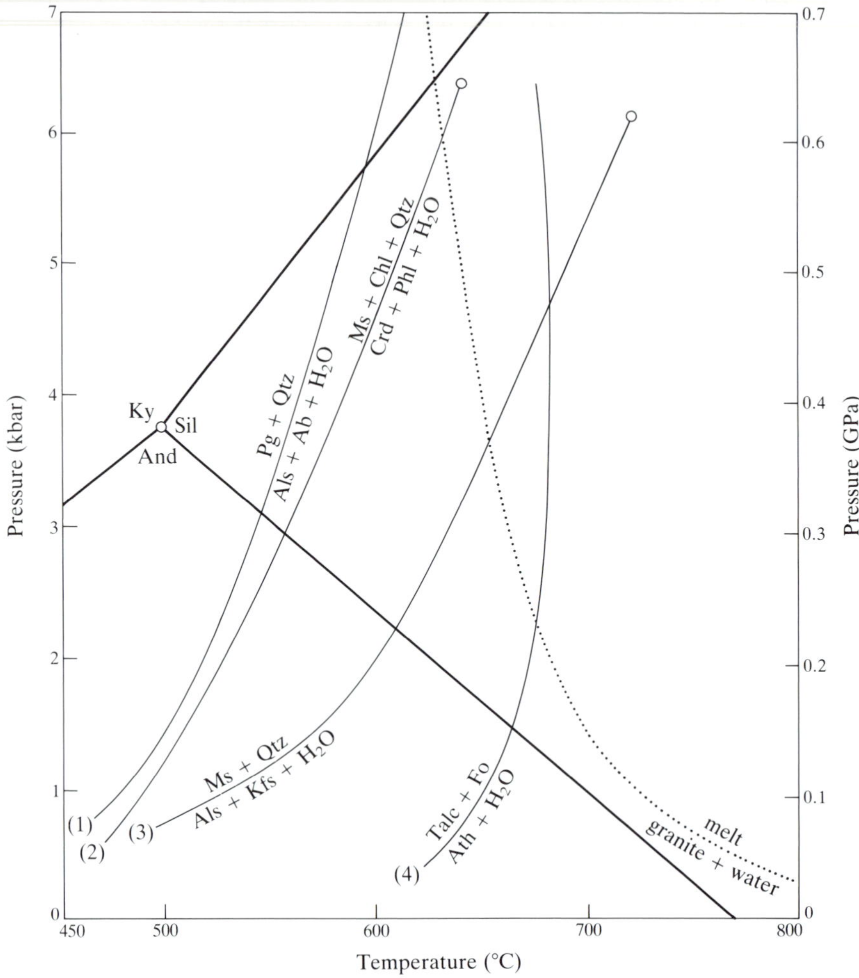

Figure 3.2 Univariant curves for simple dehydration reactions (thin solid lines):

(1) paragonite + quartz = Al_2SiO_5 + albite + H_2O (Chatterjee 1970, 1972).

(2) muscovite + Mg-chlorite + 2 quartz = Mg-cordierite + phlogopite + 4 H_2O (Seifert 1970).

(3) muscovite + quartz = Al_2SiO_5 + K-feldspar + H_2O (Kerrick 1972).

(4) 9 talc + 4 forsterite = 5 anthophyllite + 4 H_2O (Greenwood 1976).

Two of the dehydration reaction curves end at an invariant point as indicated by an open circle. The thick solid lines show boundaries between the stability fields of andalusite, kyanite and sillimanite after Holdaway (1971). The dotted line is the melting curve of granite (Luth et al. 1964).

The equilibrium curves of some simple dehydration reactions show a negative slope for the whole length of their stable part. In other words, increase in H_2O pressure causes dehydration reactions. Figure 3.3b shows an example of a curve of this type for the reaction: paragonite = jadeite + kyanite + H_2O under very high H_2O pressures.

Since zeolites have exceptionally low densities, decomposition reactions of some zeolites show a decrease of the total volume of solid + fluid at very low pressure, resulting in a negative slope of the equilibrium curves. An example is: analcime + quartz = albite + H_2O (Fig. 11.1).

Some dehydration reactions show not only a decrease of volume but also of entropy. Equilibrium curves for these reactions have a positive slope with the dehydrated assemblage on the low-temperature side. Examples have been observed in reactions involving zeolites, e.g. analcime = jadeite + H_2O (Fig. 11.1), and laumontite = lawsonite + 2 quartz + 2 H_2O (Fig. 11.2). Figure 3.3a shows such a curve with a positive slope (segment C) possibly continuous with the curve of the same reaction with a negative slope (segment B).

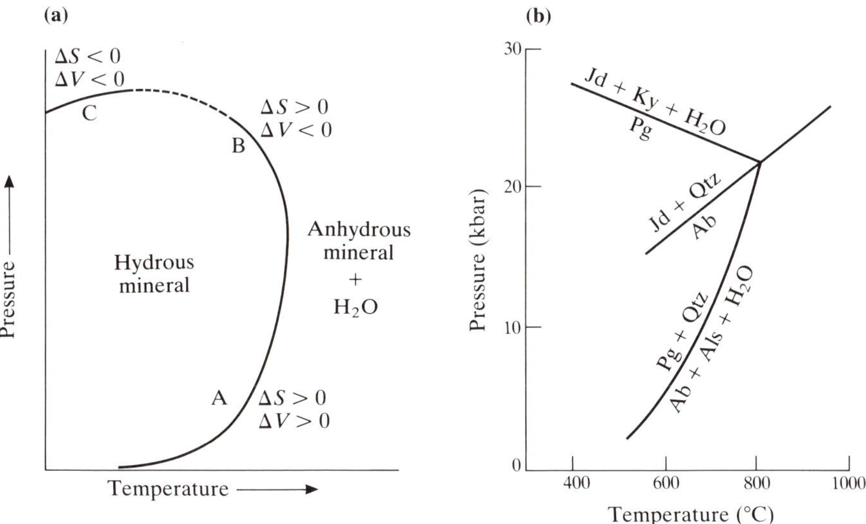

Figure 3.3 (a) Schematic equilibrium curve for a simple dehydration reaction showing continuous change from a maximum temperature to a maximum pressure in the presence of an aqueous fluid. See text for segments A, B and C. (b) Simple dehydration equilibrium curves with a positive and a negative slope. The equilibrium curve for the reaction: paragonite + quartz = albite + Al_2SiO_5 + H_2O, has a positive slope (Chatterjee 1972, Holland 1979a), whereas that for the reaction: paragonite = jadeite + kyanite + H_2O, has a negative slope (Holland 1979a).

Univariant curves for simple decarbonation reactions in the presence of pure CO_2 show essentially the same type of shape as dehydration curves of Figure 3.2.

3.5.2 Simple dehydration and decarbonation reactions in the presence of impure fluid

3.5.2.1 When all solid phases have fixed compositions. In real metamorphism, the fluid is usually a mixture of H_2O, CO_2 and other components. Let us consider dehydration and decarbonation reactions in the presence of this type of impure fluid. We begin with the case where all solid phases have fixed compositions.

To a first approximation, we may assume that the impure fluid obeys the rule of ideal mixing of non-ideal gases (the so-called Lewis–Randall fugacity rule). Under this assumption, the following relation holds:

$$\Delta G^\circ(P,T) = -RT \ln a_i = -RT \ln X_i \qquad (3.16)$$

Then:

$$\Delta G^\circ(1\text{bar}, T) + (P-1)\,\Delta V_s = -RT \ln f_i = -RT \ln X_i f_i^\circ \qquad (3.17)$$

where f_i° denotes the fugacity of pure H_2O or pure CO_2 at the pressure and temperature of interest.

The Gibbs' energy of a dehydration or decarbonation reaction may be calculated from thermochemical data, or from an equilibrium curve determined synthetically in the presence of a pure fluid. The above equations give the relationship between the decrease of the mole fraction of H_2O or CO_2 in the fluid and the decrease of the equilibrium temperature.

Figure 3.4 illustrates these relations in a simple dehydration reaction. At a given pressure (total pressure), the equilibrium temperature decreases with decreasing mole fraction of H_2O. For a fixed value of mole fraction of H_2O, the equilibrium temperature increases with increasing total pressure (e.g. Wyllie 1962). Analogous relations hold for simple decarbonation reactions, if the mole fraction of CO_2 is used in place of that of H_2O. Jacobs & Kerrick (1981) used activities of H_2O and CO_2 obtained from a modified Redlich–Kwong equation in their calculation of decarbonation equilibrium curves.

3.5.2.2 When solid-solution minerals are involved. Commonly some, or all, minerals involved in dehydration and decarbonation reactions are solid solutions. Equilibria in these reactions can also be calculated from the relevant stoichiometric relations among phase components. The equations obtained show the relationships of pressure, temperature, activities of components in solids, and activities or fugacities of H_2O and CO_2. If the prevailing pressure and tempera-

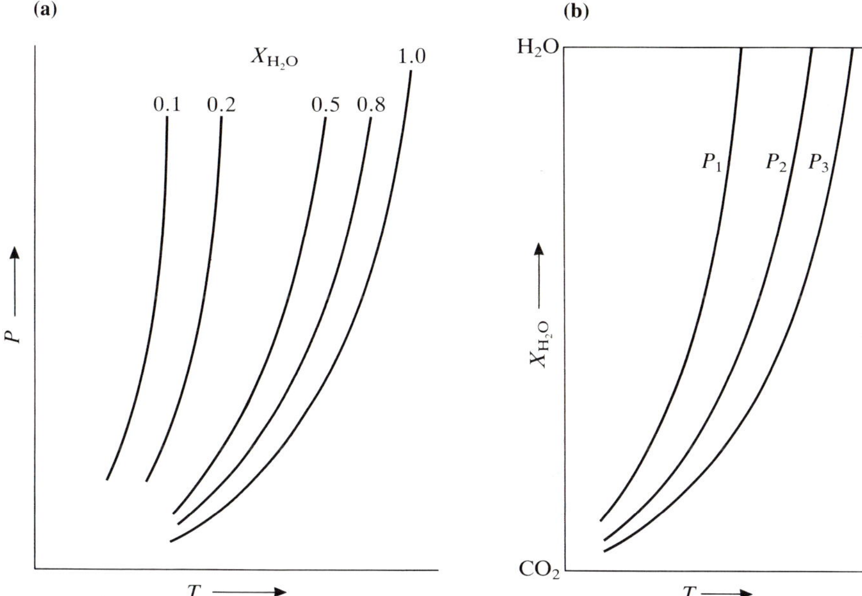

Figure 3.4 Schematic equilibrium curves for a simple dehydration reaction in the presence of an $H_2O + CO_2$ fluid. The equilibrium temperature decreases with decreasing pressure as well as with decreasing mole fraction of H_2O in the fluid. (a) *P–T* diagram; (b) *T–X* diagram. Here, *X* represents the mole fraction of H_2O. Pressure increases in the order of P_1, P_2 and P_3.

ture and the compositions of participating minerals are known, the equations give the activity or fugacity of H_2O or CO_2.

Many examples of dehydration and decarbonation reactions in the presence of H_2O–CO_2 fluid have been discussed by Kerrick (1974), Skippen (1974) and Ferry (1976b, 1987).

3.5.3 Greenwood's (1967a) classification of the reactions that involve H_2O and/or CO_2

We consider any reaction that involves H_2O and/or CO_2:

$$bB + \ldots = dD + \ldots + m\,H_2O + n\,CO_2 \qquad (3.18)$$

where $B, \ldots, D, \ldots$ represent solid minerals with a fixed composition, and $b, \ldots, d, \ldots, m$ and n are stoichiometric coefficients. The equation is written so that the right-hand side is stable at higher temperatures. The coefficients m and n may be positive, zero or negative. If they are negative, H_2O and CO_2 are on the left-hand side (reactant side). The fluid is assumed to be a binary mixture of H_2O and CO_2.

Greenwood (1967a) classified the reactions that involve H_2O and/or CO_2 into types according to the values of m and n, and showed the relationship between the types and the shapes of equilibrium curves on the T–X diagram under a constant pressure, as shown below. For the calculation, see, for example, Powell (1978: 62–4). In the following discussion, the mole fractions of H_2O and CO_2 are denoted as X_1 and X_2, respectively.

(a) If $m = n = 0$, equation (3.18) represents a solid–solid reaction. The equilibrium temperature is independent of the fluid composition. The equilibrium curve is a straight horizontal line on a T–X diagram, like curve (1) in Figure 3.5.

(b) If $m > 0$ and $n = 0$, equation (3.18) represents a simple dehydration reaction. The equilibrium temperature is highest for $X_1 = 1$, and decreases infinitely with decreasing mole fraction of H_2O, as is exemplified by curve (2) in Figure 3.5. The equilibrium T–X curve is convex upward.

(c) If $m = 0$ and $n > 0$, equation (3.18) represents a simple decarbonation reaction. The equilibrium temperature is highest for $X_2 = 1$, and decreases infinitely with decreasing mole fraction of CO_2, as is exemplified by curve (3) in Figure 3.5. The equilibrium T–X curve is convex upward.

(d) If $m > 0$ and $n > 0$, the reaction involves both dehydration and decarbonation. The equilibrium curve is convex upward, and shows a maximum temperature at the fluid composition of $X_2 = n/(m + n)$, as shown by curve (4) in Figure 3.5.

(e) If $m > 0$ and $n < 0$, the reaction consumes CO_2 and evolves H_2O with increasing temperature. The equilibrium T–X curve has an S-shape, as shown by curve (5) in Figure 3.5. The left-hand part of the curve is convex downward, and the equilibrium temperature increases infinitely as X_2 decreases. The right-hand part of the curve is convex upward, and the equilibrium temperature decreases infinitely as X_2 increases.

(f) If $m < 0$ and $n > 0$, the reaction consumes H_2O and evolves CO_2 with increasing temperature. The equilibrium T–X curve is a mirror image of that of the preceding case with respect to H_2O and CO_2, as shown by curve (6) in Figure 3.5.

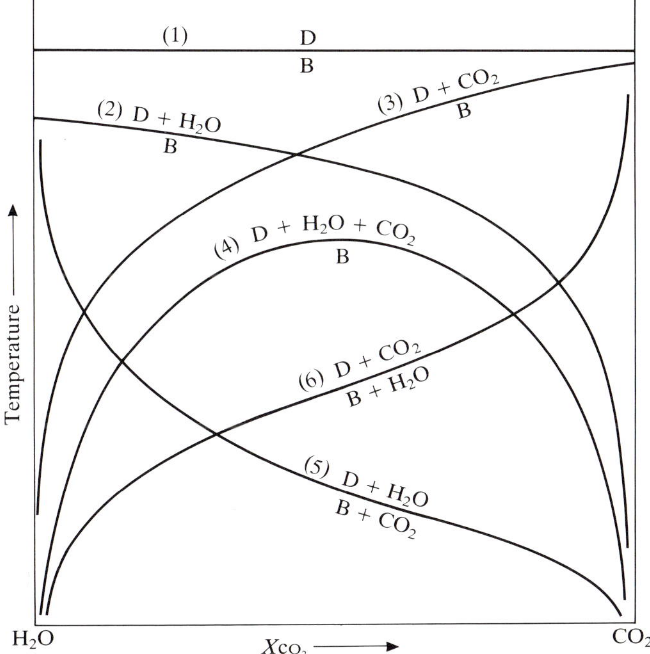

Figure 3.5 Schematic *T–X* diagram showing the general shapes of equilibrium curves for the following types of reactions at a constant pressure (Greenwood 1967a). The presence of a fluid phase composed of H_2O and CO_2 is assumed. The following reactions are written so that the right-hand side represents higher temperature (with coefficients ignored): (1) solid–solid reaction: $B = D$; (2) simple dehydration reaction: $B = D + H_2O$; (3) simple decarbonation reaction: $B = D + CO_2$; (4) $B = D + H_2O + CO_2$; (5) $B + CO_2 = D + H_2O$; (6) $B + H_2O = D + CO_2$ (B and D represent solid phases with fixed compositions).

3.6 Oxidation–reduction reactions

3.6.1 Oxide minerals and O_2 fugacity

3.6.1.1 O_2 buffers for synthetic experiments. In synthetic experiments involving iron-bearing minerals, the O_2 fugacity of the environment is usually controlled by O_2 buffers, which are based on reactions such as:

$$\underset{\text{hematite}}{6\,Fe_2O_3} = \underset{\text{magnetite}}{4\,Fe_3O_4} + O_2 \qquad \text{(HM)}$$

$$2\,NiO = 2\,Ni + O_2 \qquad \text{(NNO)}$$

$$\underset{\text{magnetite}}{2\,Fe_3O_4} + \underset{\text{quartz}}{3\,SiO_2} = \underset{\text{fayalite}}{3\,Fe_2SiO_4} + O_2 \qquad \text{(QFM)}$$

$$\underset{\text{magnetite}}{2\,Fe_3O_4} = \underset{\text{wustite}}{6\,FeO} + O_2 \qquad \text{(MW)}$$

$$\underset{\text{wustite}}{2\,FeO} = \underset{\text{iron}}{2\,Fe} + O_2 \qquad \text{(WI)}$$

$$\underset{\text{magnetite}}{1/2\,Fe_3O_4} = \underset{\text{iron}}{3/2\,Fe} + O_2 \qquad \text{(MI)}$$

$$\underset{\text{fayalite}}{Fe_2SiO_4} = \underset{\text{iron}}{2\,Fe} + \underset{\text{quartz}}{SiO_2} + O_2 \qquad \text{(QFI)}$$

These reactions are listed in the order of decreasing equilibrium fugacity of O_2, as shown in Figure 3.6a. Each reaction is written so that it has 1 mole of O_2 on the right-hand side. The assemblages on the left-hand and right-hand sides of each reaction are stable above and below the equilibrium curve, respectively. In parentheses at the right margin are commonly used symbols for individual O_2 buffers.

When the solid phases have fixed compositions, the following relation holds for each of the above reactions:

$$\Delta G°(1\text{bar}, T) + (P-1)\,\Delta V_s = -RT\ln f \qquad (3.19)$$

where f denotes the fugacity of O_2. An increase of pressure causes a slight increase of O_2 fugacity in all the above buffers.

Wüstite is shown above as FeO for simplicity. Actually it is a solid-solution mineral Fe_xO with $x = 0.95$–0.83. Some positions of Fe in the wüstite structure are vacant, and the charge balance is maintained by the presence of Fe^{3+}.

H_2O tends to show equilibrium for the following dissociation reaction:

$$H_2O = H_2 + 1/2\ O_2 \qquad (3.20)$$

$$\Delta G°(1\text{bar}, T) = -RT\ln K$$

$$= -RT\ln\frac{f_{H_2}f_{O_2}^{1/2}}{f_{H_2O}} \qquad (3.21)$$

where for the ordinary range of temperature of regional metamorphism $\Delta G°$ is given by the experimental formula:

$$\Delta G° = 246.12 - 0.05356\,T\,\text{kJ mol}^{-1}.$$

In a fluid initially having pure H_2O composition, the mole fraction of O_2 formed by dissociation is one half of that of H_2, and the total pressure of the fluid is the sum of the partial pressures of H_2O, H_2 and O_2. The calculated O_2 fugacity of this pure water is also shown in Figure 3.6a for comparison. It is much higher than that of any of the above-discussed buffers at the same temperature.

If H_2O is added to one of the buffer assemblages listed above, equilibrium is reached when the O_2 fugacity of the aqueous fluid becomes equal to that of the pertinent buffer assemblage. As the O_2 in the aqueous fluid is consumed by reaction with a solid or solids, the mole fraction of O_2 remaining in the fluid is

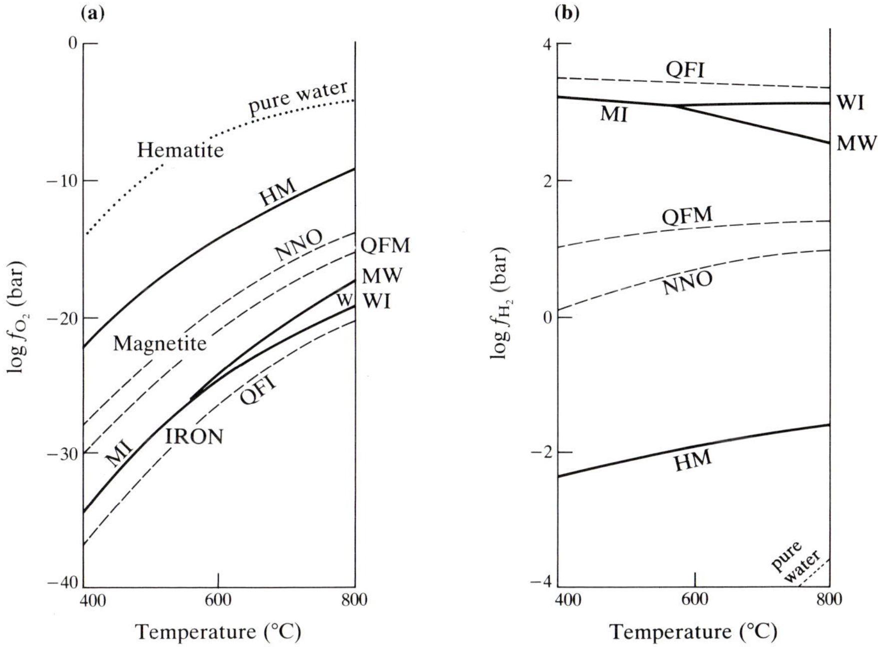

Figure 3.6 (a) Solid lines show the stability fields of hematite, magnetite, wüstite (indicated by W), and native iron in the Fe–O systen with respect to temperature and O_2 fugacity at a pressure of 2 kbar. The O_2 fugacity curves of the NNO and QFM buffers (dashed lines) lie within the magnetite field, whereas that of the QFI buffer (dashed line) lies within the native iron field (Eugster & Wones 1962, Table 2). The O_2 fugacity of fluid initially with pure H_2O composition at 2 kbar is shown by a dotted line for comparison.

(b) Fugacity of H_2 in the O_2 buffers mixed with H_2O (Eugster & Skippen 1967, Table 1).

smaller than one half of that of H_2. Because of equation (3.21), such buffers mixed with H_2O show a specific H_2 fugacity at a given pressure and temperature, as shown in Figure 3.6b. Buffers with higher O_2 fugacities show lower H_2 fugacities. A fluid initially with pure H_2O composition shows a H_2 fugacity lower than those of any of the buffers shown. It should be noted that the fugacities, and hence the mole fractions, of O_2 are many orders of magnitude smaller than those of H_2 in buffered fluids of this kind. In fluids showing O_2 fugacities higher than the QFM buffer, the mole fractions of both H_2 and O_2 are very small, and the fluids are almost entirely composed of H_2O. Hence, the fugacities of H_2O in these fluids are very close to that of H_2O in pure H_2O.

In synthetic experiments with O_2 buffers, minerals are synthesized in the presence of H_2O in a closed tube, made of Pt or AgPd alloys, which is immersed in an O_2 buffer assemblage mixed with H_2O. H_2 can penetrate through the tube wall at temperatures above about 400°C, maintaining the same H_2

fugacity between the interior of the tube and the surrounding buffer. Since H_2O is present on both sides of the wall, the O_2 fugacity inside the tube is also equal to that of the buffer.

3.6.1.2 Oxides of the FeO–Fe_2O_3–TiO_2 system in natural rocks. Natural metamorphic rocks commonly contain magnetite and/or hematite, but not wüstite nor native iron. This means that the lower limit of the O_2 fugacity in metamorphic rocks is represented by the curves of the MI and MW buffers in Figure 3.6a.

So far we have treated iron oxides as belonging to the binary system Fe–O. In natural rocks, magnetite and hematite are commonly solid solutions containing a considerable amount of TiO_2, and are commonly accompanied by ilmenite. All these belong approximately to the ternary system FeO–Fe_2O_3–TiO_2.

In regional metamorphic rocks, magnetite has a practically constant composition, whereas ilmenite and hematite make a solid-solution series, which is separated, by a miscibility gap, into two segments, ilmenite solid solution and hematite solid solution, as shown in Figure 3.7. In this ternary system, an assemblage of two solid phases is trivariant, and so under a given pressure and temperature the solid-solution minerals in it show variable compositions, and correspondingly the O_2 fugacity is also variable. On the other hand, an assemblage of three solid phases is divariant, and so under a given pressure and temperature solid-solution minerals in it show a specific composition, and the O_2 fugacity is also specific.

Figure 3.7a shows schematic paragenetic relations at relatively low temperatures (e.g. greenschist and blueschist facies). Ilmenite can coexist with rutile or magnetite or both at relatively low O_2 fugacities in rocks such as graphite-bearing schists. In rocks with higher O_2 fugacities, hematite can coexist with rutile or magnetite or both (Mielke & Schreyer 1972). Figure 3.7b shows schematic paragenetic relations at high temperatures (e.g. amphibolite and granulite facies). Here, the ilmenite–hematite tieline is stable in place of the rutile–magnetite tieline (Chinner 1960, Rumble 1976b). With increasing O_2 fugacity, the Fe_2O_3 content of ilmenite–hematite solid solution increases.

Actually ilmenite in metapelites commonly contains a considerable amount of MnO, up to about 3 wt % (e.g. Zen 1981, Table 3). So, the four-mineral assemblage rutile + magnetite + ilmenite + hematite is divariant, being stable over a range of temperatures at a given pressure.

In many metamorphic areas, rutile is absent, and in its place sphene is the main Ti-bearing mineral. Rutile and sphene are connected by the following stoichiometric relation: rutile + calcite + quartz = sphene + CO_2. This indicates that under a very low activity of CO_2, sphene forms in place of rutile + calcite + quartz. The formation of sphene takes place at a mole ratio $CO_2/(H_2O + CO_2)$ smaller than about 0.05 at 500°C (Ernst 1972a).

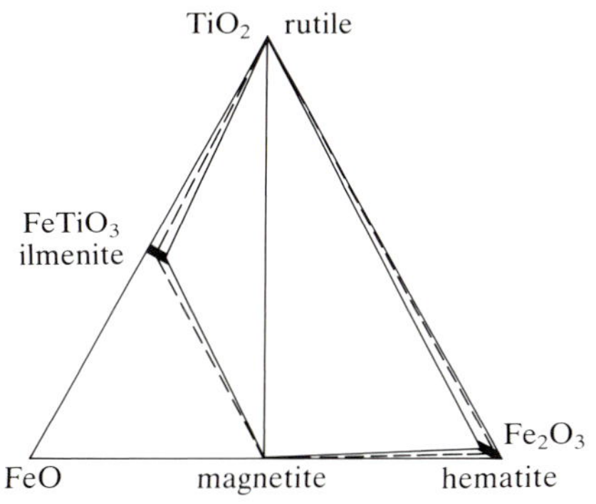

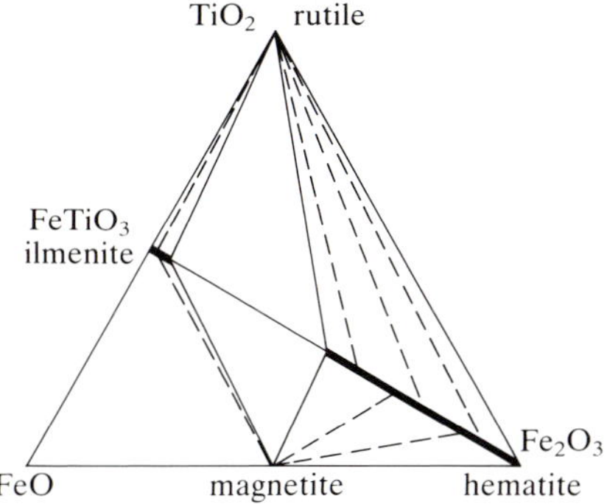

Figure 3.7 Schematic diagrams showing paragenetic relations in the FeO–Fe_2O_3–TiO_2 system (a) in low-temperature metamorphism, and (b) in middle- and high-temperature metamorphism.

3.6.2 Stability of ferromagnesian silicates with respect to O_2 fugacity

Although the bulk-rock Mg/Fe ratio is a main factor controlling the Mg/Fe ratio of ferromagnesian minerals in a rock, the O_2 fugacity limits the Mg/Fe ratio of the solid-solution minerals. As an example, we consider olivine.

In Figure 3.6a, fayalite can be stable in the field between the curves of the QFM and QFI buffers. If the O_2 fugacity is higher than that of the QFM buffer, fayalite reacts with O_2 and decomposes into magnetite + quartz, whereas if it is lower than that of the QFI buffer, fayalite releases O_2 and decomposes into native iron + quartz + O_2.

Olivine solid solutions have a wider stability field than pure fayalite with respect to O_2 fugacity, because decreasing mole fraction of the fayalite component in olivine causes a decrease of the chemical potential of the component. If an olivine solid solution occurs in place of fayalite in the QFM buffer assemblage, the following equilibrium holds:

$$\underset{\text{magnetite}}{2\,Fe_3O_4} + \underset{\text{quartz}}{3\,SiO_2} = \underset{\text{Fa in olivine}}{3\,Fe_2SiO_4} + O_2 \tag{3.22}$$

$$\Delta G^\circ(1\text{bar},\ T) + (P-1)\Delta V_s = -RT\ln a_{Fa}{}^3 f = -RT\ln X_{Fa}{}^6 f \tag{3.23}$$

where X_{Fa} denotes the mole fraction of the fayalite component in olivine, and f denotes the fugacity of O_2. At a given temperature and pressure (pressure on solid phases), the left-hand side of equation (3.23) is constant, and so a decrease of the mole fraction of the fayalite component in olivine causes an increase of the equilibrium fugacity of O_2. In other words, an olivine solid solution can be stable not only in the field between the curves of the QFM and QFI buffers, but also up to a certain higher value of O_2 fugacity. This upper limit of the O_2 fugacity for the stability of olivine becomes higher with decrease of the mole fraction of fayalite in olivine. (Pure fayalite and olivines with compositions close to fayalite are stable in association with quartz in the ordinary *P–T* range of metamorphism, whereas more Mg-rich olivine may react with quartz to produce orthopyroxene so that the quartz + olivine + magnetite assemblage becomes unstable. This complication is ignored for the present argument in which we are considering olivine only as an example of ferromagnesian minerals in general.)

If an olivine solid solution occurs in place of fayalite of the QFI buffer, the following equilibrium holds:

$$\underset{\text{Fa in olivine}}{Fe_2SiO_4} = \underset{\text{iron}}{2\,Fe} + \underset{\text{quartz}}{SiO_2} + O_2 \tag{3.24}$$

$$\Delta G°(1\text{bar}, T) + (P-1)\,\Delta V_s = -RT \ln f/X_{Fa}{}^2 \tag{3.25}$$

Thus, at a fixed pressure and temperature, a decrease of the mole fraction of the fayalite component in olivine causes a decrease of the equilibrium fugacity of O_2.

In a similar way, the composition ranges of other Fe–Mg solid-solution minerals are also limited by O_2 fugacity. Under a very high O_2 fugacity, they decompose and produce a ferric iron-bearing mineral assemblage, while under an exceptionally low O_2 fugacity, they may decompose and produce an assemblage with native iron. In natural metamorphic rocks, conditions are not so reducing as to form native iron, and so the decomposition of Fe–Mg solid-solution minerals under too low an O_2 fugacity to produce native iron need not be considered.

Figure 3.8 shows experimental data indicating the decrease of the high Fe/Mg limit of Fe–Mg solid-solution series of olivine, orthoamphibole and biotite, with increase of O_2 fugacity beyond that of the QFM buffer.

For example, the decomposition reaction of biotite for the QFM and NNO buffers is:

$$\underset{\text{annite in biotite}}{KFe_3AlSi_3O_{10}(OH)_2} + \tfrac{1}{2}\,O_2 = \underset{\text{anidine}}{KAlSi_3O_8} + \underset{\text{magnetite}}{Fe_3O_4} + H_2O \tag{3.26}$$

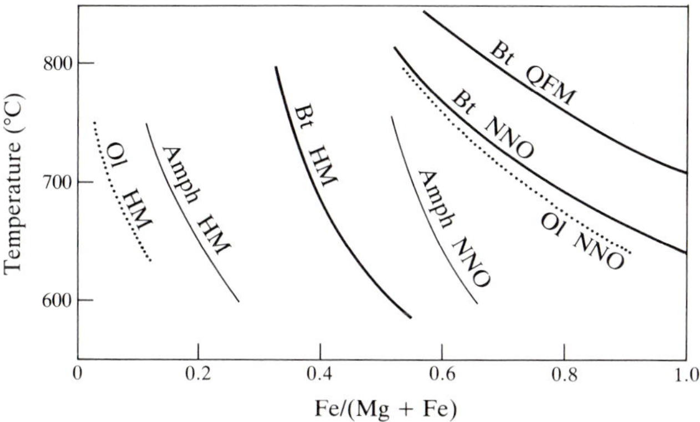

Figure 3.8 Maximum Fe/(Fe + Mg) ratios of olivine, orthorhombic Mg–Fe-amphibole, and biotite at O_2 fugacities defined by the QFM, NNO and HM buffers at 2 kbar H_2O pressure (Wones & Eugster 1965, Popp et al. 1977).

3.6.3 Graphite-bearing versus graphite-free metapelitic rocks

Pelitic metamorphic rocks usually contain a small amount of graphite or some related substance. The fugacity of O_2 in such a rock is controlled by equilibrium with graphite (or related substance) according to a reaction like: $C + O_2 = CO_2$. Because of the reaction, the fugacity of O_2 in graphite-bearing rocks is kept in the magnetite field of Figure 3.6a, and the O_2 fugacity is limited by that of CO_2 fugacity in the same rock, as discussed in §6.2. For this reason, metapelitic rocks within an area usually show a relatively uniform state of oxidation, and this contributes to the development of relatively regular progressive mineralogical changes in them (Miyashiro 1964).

However, some metapelites contain no graphite. Graphite-free rocks can show a great variation in O_2 fugacity. Some rocks of this kind may contain magnetite, but commonly contain hematite. O_2 molecules are not mobile during metamorphism (§6.4), and so the O_2 fugacity of a metamorphic rock is largely controlled by the initial oxygen content of the rock. Even at the same pressure and temperature, the variation in the initial oxygen content of rocks causes a variation in the O_2 fugacity at the thermal peak, which in turn causes variations in the compositions of silicate solid solutions. A beautiful, instructive example of such mineralogical variations was described by Chinner (1960) from the lowest temperature part of the sillimanite zone in Glen Clova in the Scottish Highlands.

3.7 Exchange reactions

3.7.1 Thermodynamics of exchange reactions

The thermodynamic study of exchange reactions was initiated by Ramberg & DeVore (1951) in their formulation for the Fe–Mg distribution between coexisting olivine and orthopyroxene, and was extended to other ferromagnesian mineral pairs particularly by Mueller and by Kretz (e.g. Mueller 1962). The thermodynamic treatment of solid-solution minerals was developed gradually in the 1960s and 1970s.

Exchange reactions occur not only for the Fe–Mg substitution but also for any other type of substitution, only if the same type of substitution occurs in two or more coexisting minerals. For example, a K–Na exchange reaction occurs between coexisting muscovite and K-feldspar, and a NaSi–CaAl exchange reaction occurs between coexisting plagioclase and hornblende.

We can derive conventional equations for exchange reactions using J. B. Thompson's (1982a,b) exchange components (§3.1.4). For example, since $FeMg_{-1}$ occurs in oxides, pyroxenes and olivines, we have relations like:

$$\begin{aligned} FeMg_{-1} &= (FeO\text{–}MgO) \text{ (oxide)} \\ &= (FeSiO_3\text{–}MgSiO_3) \text{ (pyroxene)} \\ &= 1/2\,(Fe_2SiO_4\text{–}Mg_2SiO_4) \text{ (olivine)} \end{aligned} \tag{3.27}$$

By rewriting, we get conventional equations for exchange reactions:

$$\overset{\text{in opx}}{MgSiO_3} + \overset{\text{in oxide}}{FeO} = \overset{\text{in opx}}{FeSiO_3} + \overset{\text{in oxide}}{MgO} \tag{3.28}$$

$$\overset{\text{in ol}}{Mg_2SiO_4} + 2\overset{\text{in opx}}{FeSiO_3} = \overset{\text{in ol}}{Fe_2SiO_4} + 2\overset{\text{in opx}}{MgSiO_3} \tag{3.29}$$

If the same type of exchange component occurs in n coexisting phases, there are $(n-1)$ independent exchange reactions between the phases.

For reaction (3.29), for example, we have the following relation:

$$\Delta G^\circ(P,T) = -RT\ln K \tag{3.30}$$

where

$$K = \frac{a_{Fa}a_{En}^2}{a_{Fo}a_{Fs}^2} \tag{3.31}$$

If we assume random ionic mixing of Mg and Fe in olivine and in orthopyroxene, we can rewrite the activities as follows:

$$a_{Fo} = X_{Fo}^2\,\gamma_{Fo}^2, \quad a_{Fa} = X_{Fa}^2\,\gamma_{Fa}^2,$$
$$a_{En} = X_{En}\,\gamma_{En}, \quad a_{Fs} = X_{Fs}\,\gamma_{Fs}$$

where X denotes mole fraction, and the **activity coefficient** γ has been introduced to express a deviation from ideality. Then,

$$K = \left(\frac{X_{Fa}X_{En}}{X_{Fo}X_{Fs}}\right)^2\left(\frac{\gamma_{Fa}\gamma_{En}}{\gamma_{Fo}\gamma_{Fs}}\right)^2 = K_D^2\left(\frac{\gamma_{Fa}\gamma_{En}}{\gamma_{Fo}\gamma_{Fs}}\right)^2 \tag{3.32}$$

K_D is usually called the **distribution coefficient**. It is a function of pressure, temperature and composition. It may be rewritten in more easily memorable forms as follows:

$$K_D = \left(\frac{X_{Fa}}{X_{Fo}}\right)/\left(\frac{X_{Fs}}{X_{En}}\right) = \left(\frac{Fe}{Mg}\right)_{ol}/\left(\frac{Fe}{Mg}\right)_{opx} \tag{3.33}$$

If both minerals are ideal solutions, all the activity coefficients are unity, and K_D is independent of composition.

In place of reaction (3.29), we may start from one half of the equation as follows:

$$MgSi_{0.5}O_2 + FeSiO_3 = FeSi_{0.5}O_2 + MgSiO_3 \tag{3.34}$$

In this case, we get the following:

$$\begin{aligned}\Delta G^\circ(P,T) &= -RT\ln K = -RT\ln\left(\frac{a_{Fa}a_{En}}{a_{Fo}a_{Fs}}\right) \\ &= -RT\ln\left(\frac{X_{Fa}X_{En}}{X_{Fo}X_{Fs}}\right)\left(\frac{\gamma_{Fa}\gamma_{En}}{\gamma_{Fo}\gamma_{Fs}}\right) \\ &= -RT\ln K_D\left(\frac{\gamma_{Fa}\gamma_{En}}{\gamma_{Fo}\gamma_{Fs}}\right)\end{aligned} \tag{3.35}$$

Thus, the distribution coefficient here is the same as that of reaction (3.29), whereas the standard Gibbs' energy of reaction here is one half of that case.

In exchange reactions between non-ideal crystalline solutions, the distribution coefficient K_D varies with composition at constant pressure and temperature. Figure 3.9 illustrates how the distribution coefficient varies with composition in the olivine–orthopyroxene pair under three different sets of pressure and temperature.

Although each mineral shows a small change of volume by atomic substitution, the two minerals involved in the same exchange reaction usually show volume changes of similar magnitude with opposite signs. So the total volume changes in exchange reactions are usually very small as compared with those in net-transfer reactions. Hence, the effect of pressure on the equilibrium constants of exchange reactions is usually very small. Exchange equilibria are essentially temperature dependent, and so can commonly be used as good geothermometers. However, some exchange equilibria are so insensitive, not only to pressure but also to temperature, that they are inappropriate as geothermometers.

Many silicate solid solutions based on Mg–Fe substitution have been found to be close to ideal, although there are some showing a fairly considerable deviation from ideality. In recent years, precise investigations have been made on the deviation from ideality in many solid-solution series. Solid solutions based on other types of substitution show various different degrees of deviation.

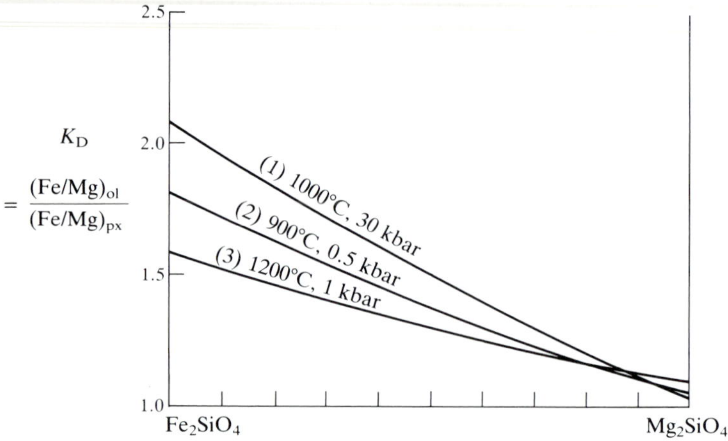

Figure 3.9 Change of the Fe/Mg distribution coefficient between olivine and Mg–Fe-pyroxene with the Fe/Mg ratio of olivine under three different sets of temperature and pressure: (1) 1000 °C and 30 kbar; (2) 900 °C and 0.5 kbar; (3) 1200 °C and 1 kbar (Matsui & Nishizawa 1974).

3.7.2 Fe–Mg exchange reaction between garnet and biotite as an example

As a well investigated example of exchange equilibria that is insensitive to pressure but sensitive to temperature, we examine the Fe–Mg exchange reaction between garnet and biotite.

3.7.2.1 Experimental calibration. Ferry & Spear (1978) studied, by synthetic experiment, the relationship of the following Fe–Mg exchange equilibrium between garnet and biotite to temperature:

$$\overset{\text{pyrope}}{Mg_3Al_2Si_3O_{12}} + \overset{\text{annite}}{KFe_3AlSi_3O_{10}(OH)_2}$$

$$= \overset{\text{almandine}}{Fe_3Al_2Si_3O_{12}} + \overset{\text{phlogopite}}{KMg_3AlSi_3O_{10}(OH)_2} \qquad (3.36)$$

As the two minerals are close to ideal solid solutions, the distribution coefficient is approximately a function of temperature. At 2 kbar and T (K), it is given by:

$$\ln K_D = \ln (Fe/Mg)_{grt}/(Fe/Mg)_{bt} = 2109/T - 0.782 \qquad (3.37)$$

The value of K_D is about 10.5 at 400 °C, and about 3.3 at 800 °C.

The effect of pressure was calculated from the volume change. Thus, K_D is given as a function of pressure (P bar) and temperature (TK) as follows:

$$\ln K_D = (2089 + 0.0096\,P)/T - 0.782 \qquad (3.38)$$

At 400 °C, a pressure increase of 1 kbar causes an increase of 0.14 in K_D, which is equal to the change in K_D caused by a temperature decrease of 4°C. Thus, this distribution coefficient is a good geothermometer insensitive to pressure.

In the Fe–Mg exchange reaction (3.36), the biotite belongs to the phlogopite–annite series. In middle-grade pelitic schists, biotites usually show a little higher Al and lower Fe + Mg content than this because of the Tschermak substitution. When biotite is assumed to have a composition of this type, the standard Gibbs' energy of reaction, and hence, the distribution coefficient are influenced by it, although these changes may be small.

For an accurate geothermometer, other components, e.g. Ca and Mn in garnet, and Ti in biotite, as well as the deviations of the relevant minerals from ideality must be taken into consideration (Hodges & Spear 1982, Ganguly & Saxena 1984, Indares & Martignole 1985).

3.7.2.2 Use as a geothermometer. The garnet–biotite pair may occur in a number of metapelite specimens collected in a small area. In such a case, the FeO and MgO contents of garnet and biotite in the pairs vary with the bulk composition of the rocks. If we assume that these rocks were equilibrated virtually at the same pressure and temperature, and that the two minerals are ideal solutions, the distribution coefficient must be constant among the rocks:

$$K_D = (\mathrm{Fe/Mg})_{grt}/(\mathrm{Fe/Mg})_{bt} = \text{constant} \qquad (3.39)$$

If the Fe/Mg ratios of coexisting garnet and biotite are taken as co-ordinates in a rectangular diagram as in Figure 3.10, the condition of equation (3.39) represents a straight line passing through the origin. The slope of the line varies with the value of the distribution coefficient. We may plot the analytical data of coexisting garnets and biotites in a diagram of this type, and determine the slope of the line and hence the temperature.

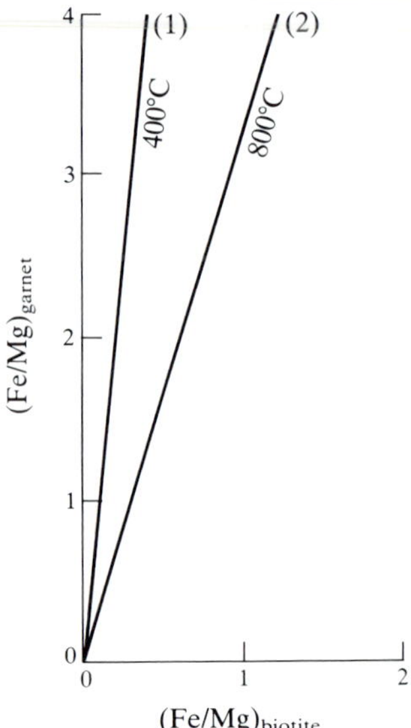

Figure 3.10 A graphical method for empirical determination of Fe/Mg distribution coefficient between garnet and biotite, where $K_D = (Fe/Mg)_{grt}/(Fe/Mg)_{bt}$: (1) 400°C, (2) 800°C.

4 Volatile substances in regional metamorphism

4.1 States of H_2O, CO_2 and other volatiles in rocks undergoing metamorphism

4.1.1 Introduction

Unmetamorphosed pelitic sediments commonly contain an aqueous fluid between solid grains. Dehydration and decarbonation reactions during prograde metamorphism evolve a large amount of H_2O and CO_2 from mineral grains. In some cases, a considerable amount of CO_2 may be produced by oxidation of graphite (or graphite-like carbonaceous materials) in pelitic rocks. Hence, rocks undergoing prograde metamorphism must contain H_2O, CO_2 and other volatile substances in some states. These substances tend to move through the crust generally upwards toward the Earth's surface. However, our knowledge of their states and methods of movement is meager and mostly speculative.

In the following section, the two most important states of existence of H_2O and CO_2 in rocks are discussed: an intergranular fluid as a true thermodynamic phase, and adsorbed molecules and ions along the grain boundaries between minerals (although there may be intermediate states between them). In addition, very small amounts of H_2O and CO_2 are contained in channels, open spaces, mosaic boundaries and so on in the atomic structures of minerals.

Before about 1940, virtually all petrologists believed that all rocks undergoing metamorphism contained a fluid phase (usually an aqueous fluid), and that all chemical reactions and migration took place in it. In the period 1940–55 when the granite controversy moved the geologic community, some radical advocates of granitization hypotheses came to deny the existence of an aqueous fluid in deep crust and to ascribe regional metamorphism and granitization to direct reactions among solid phases and long-range diffusion of a large amount of rock-forming substances mainly along grain boundaries (e.g. Perrin & Roubault 1949, Ramberg 1952).

This view is no longer tenable, and virtually all recent authors have once again emphasized the presence and importance of a fluid phase during metamorphism. Nevertheless, it is still likely that although metamorphism occurs in many, or most, cases in the presence of a fluid phase (**fluid-present metamorphism**), there are cases where it occurs in the absence of a fluid phase (**fluid-absent metamorphism**). In fluid-absent metamorphism, molecules of H_2O, CO_2 and other volatile substances are probably still present in an adsorbed state along grain boundaries and play an important rôle in some cases. Therefore, two possible states of volatile substances in rocks undergoing metamorphism are discussed below: intergranular fluid phase and intergranular adsorption films.

4.1.2 Intergranular fluid phase

4.1.2.1 Critical point, supercritical aqueous fluids, and solubility of quartz in H_2O. At relatively low pressures and low temperatures, fluids show a distinction between liquid and gas. The thick solid line in Figure 4.1 represents the equilibrium curve for the coexistence of liquid and gas phases, both with pure H_2O composition. The curve ends at a critical point, which lies at 0.22 kbar and 374 °C. Natural aqueous fluids of geologic importance commonly contain a considerable amount of NaCl and other substances in solution. The critical point for H_2O–NaCl solutions becomes higher with increasing content of NaCl, and lies near 1 kbar and 600 °C for 20 wt % NaCl (Sourirajan & Kennedy 1962). The critical point for pure CO_2 lies at 0.07 kbar and 31 °C.

At pressures and temperatures above the critical point, there is no distinction between liquid and gas, and instead there is only one type of fluid, called a **supercritical fluid**. Regional metamorphism occurs in the P–T range of supercritical aqueous fluids.

The solubility of quartz in H_2O generally tends to increase with pressure and temperature (Fig. 4.1). Hence, an aqueous fluid rising through continental crust tends to deposit quartz. However, the solubility shows an anomalous behavior under the P–T conditions near the critical point. Thus, the solubility of quartz in H_2O increases with decreasing temperature at pressures near, and below, 1 kbar and in the temperature range of about 350–700 °C (Fig. 4.1).

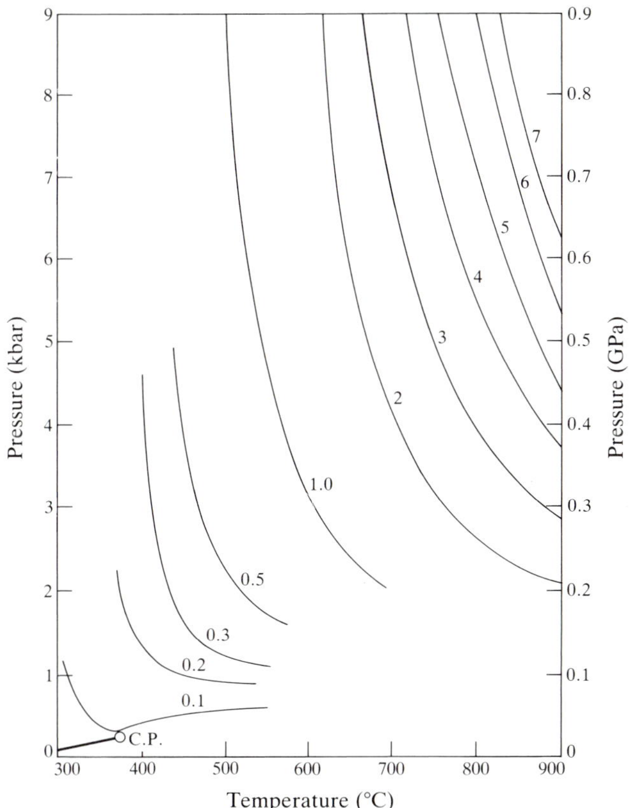

Figure 4.1 The thick solid line shows the equilibrium curve for coexisting liquid and gas (vapor) phases of pure H_2O composition, and this equilibrium curve ends at a critical point (C.P.). Thin lines show the solubility (in wt %) of quartz in H_2O. After Weill & Fyfe (1964) and Anderson & Burnham (1965).

4.1.2.2 Intergranular fluid phase in ordinary regional metamorphism.
Since rocks undergoing prograde metamorphism generally evolve a large amount of H_2O and CO_2, these substances usually form an intergranular fluid phase, so far as they are not very efficiently removed.

When a dehydration reaction occurs in the middle or low pressure range of regional metamorphism (i.e. at several kilobar or lower pressure), the total volume of the products (i.e. dehydrated solids + H_2O) is usually larger than that of the reactants. This promotes the formation of microcracks, through which the resultant fluid can flow away. Metamorphism is commonly accompanied by deformations, which produce long interconnected networks of tensile microcracks along grain boundaries as well as through grains to facilitate fluid flow.

When the total volume of the products is larger than that of the reactants, the dehydration equilibrium curve has a positive slope. With increasing pressure, the H_2O is more readily compressed than solids, and so at some high pressure the total volume of the products may become smaller than that of the reactants. In this case the equilibrium curves of these reactions begin to have a negative slope (Fig. 3.3a). Under this condition, dehydration reactions do not cause fracturing.

In prograde regional metamorphism of pelitic rocks, the amount of volatile evolved is usually very great at low temperatures, because of dehydration of highly hydrous minerals, particularly chlorite, possibly accompanied by decarbonation of admixed carbonate minerals. With increasing temperature, the volatiles that are evolved decrease in amount. In the retrograde stage, no more volatiles are produced, and so if deformation does not continue, new microcracks will not form any more. Under these conditions, metamorphic crystallization cuts long fluid-filled channels and fractures into many isolated fluid pores or inclusions, each of which is completely surrounded by solid. The pressure in the fluid will be equal to the pressure of the surrounding solid phases. In the gravity field, these fluid bodies tend to rise by means of solution of wall-rock minerals at the top accompanied by precipitation of the same minerals at the bottom of each body. Moreover, the effective crushing strength of minerals becomes small at high temperatures, and so the buoyancy of the fluid bodies tends to move them upwards mechanically. Thus, a fluid phase may completely disappear from the rock. Even under this kind of condition, however, a fluid phase may continue to be present as discrete bubbles along grain boundaries and at three- and four-grain junctions (cf. White & White 1981, Hay & Evans 1988).

Although the presence of a fluid phase in metamorphism has been very widely supported, evidence for it is very meager. A relatively direct line of evidence is that minerals of regional metamorphic rocks commonly contain fluid inclusions, some of which appear to have formed at, or near, the thermal peak of the metamorphism, although others were probably formed after the peak, or even after the metamorphism. So, study of fluid inclusions gives a clue to the properties of metamorphic fluids. Available data on fluid inclusions in metamorphic rocks are briefly reviewed in the next section (§4.2).

Some authors calculated the partial pressures of the abundant volatile species that would have been in equilibrium with the mineral assemblages in regional metamorphic rocks. The sum of these partial pressures in the same rock was found to be essentially equal to the estimated lithostatic pressure of the region in some cases. This tends to support the presence of a fluid phase. However, the sums of partial pressures were much smaller than the lithostatic pressure in other cases.

4.1.2.3 Intergranular fluid phase at shallow depths. Fluid-filled fractures may be connected with one another for a long vertical length at shallow depths. Such channels of fluid may or may not reach the Earth's surface. If a channel reaches the surface, the pressure in the fluid phase at depth h km is hydrostatic pressure $d'gh/10$ bar, where d' is the average density of the overlying fluid in g cm^{-3}. In other words, the fluid pressure could be less than a half of the lithostatic pressure acting on the surrounding solid rocks. The difference in pressure between the fluid and the surrounding rock must be maintained by the effective crushing strength of the mineral grains. A mineral surrounded by fluid inside such a fracture is at a much lower pressure than the same mineral in the adjacent rock. The former has a lower molar Gibbs energy than the latter, because the molar Gibbs energy increases with pressure as shown by the relation: $(\partial G/\partial P)_T = V > 0$. This tends to cause solution of a mineral in a rock, accompanied by precipitation of the same mineral in a nearby fracture, and so will result in the filling up of the fracture. So, such interconnected channels with a large vertical length can be present for a long time only at low temperatures near the surface, owing to the low reaction rate. The depth at which such a condition can hold varies with conditions. In drilling operations of oil wells in the Middle East and other places, this condition has been found approximately to hold at depths of less than 1.5 km. The observed fluid pressure is substantially smaller than the pressure on adjacent solid rocks, but may still be greater than the hydrostatic pressure (e.g. Fyfe et al. 1978: Ch. 9).

4.1.3 Intergranular adsorption film

Along grain boundaries (grain interfaces) between minerals in metamorphic rocks, two crystals of different minerals or of the same minerals in different orientations are in contact with each other. So the structural regularity of each crystal is disturbed by the influence of the crystal in contact with it. Thus, grain boundaries are thin layers of molecular dimensions with relatively high energy, where the rate of diffusion is greater than in the normal part of the same minerals. They can accommodate H_2O, CO_2 and other simple molecules and OH groups and other ions in partly structurally combined states. We term these thin layers **intergranular adsorption films**. Recent discussions of the physical chemistry of adsorption films are given, for example, in J. B. Thompson (1987) and Hochella & White's (1990) book.

Adsorbed substances can migrate along grain boundaries by chemical diffusion down the gradient of chemical potential of each substance in the rock. Although diffusion can occur in addition within some crystals (for example, through the structural channels along the c axes of quartz and cordierite), diffusion in intergranular films will usually transport much more material, and both these two types of diffusion are usually much less efficient transport mechanisms than chemical diffusion in an intergranular aqueous fluid, if it is present.

The total amount of H_2O and CO_2 in intergranular adsorption films per unit volume of rock is believed to be extremely small. Unless their diffusivity is high, the substances will not be able to cause an appreciable amount of mineralogical changes. If they can diffuse actively through rocks, on the other hand, the substances may continue to come into the pertinent rock from a much larger volume surrounding it, so far as the chemical potential gradient necessary for the diffusion is maintained by the mineral assemblage of the rock. This continued diffusion and reaction may result in a large amount of mineralogical change. Thus, metamorphic processes are partly controlled by the magnitude of diffusivity of substances in intergranular adsorption films.

If we take a specific intergranular adsorption film, the chemical potential of H_2O and CO_2 therein must increase with the concentrations of these components. The chemical potential of each volatile is connected to its fugacity and partial pressure. As the concentrations of the volatiles in the film increase, the sum of their partial pressures may become as large as the pressure of the rock (solid). Once the sum of partial pressures reaches the rock pressure, a fluid phase appears and is in mechanical and chemical equilibrium with the minerals of the surroundings.

A detailed discussion of phase relations in fluid-absent metamorphism has been given by A. B. Thompson (1983).

4.1.4 Flow of fluids during regional metamorphism

For a long time, the existence of flows of fluids during regional metamorphism has been imagined or assumed by many authors. These flows were considered to transport heat and chemical substances to the extent that they influenced the temperature of metamorphism, the shape of isograds, metasomatism, and other features of a metamorphic complex. Long-range upward flows of aqueous fluids in orogenic belts were assumed, by Wegmann (1935) and others, to cause regional migmatitization of orogenic crust. The Na, K, Si and other substances transported by flowing fluids were assumed to be the main agent of granitization and migmatitization (e.g. Sederholm 1923, 1926, Wegmann 1935, Read 1957).

However, reliable observation and analysis of evidence for the existence, and properties of these flows began to appear only relatively recently. In the past 20 years it has become clear that such flows of fluids can produce regional-scale heterogeneity of intergranular fluid which will cause intersecting isograds, transport chemical substances and heat, and drive chemical reactions even under a constant pressure and temperature. In other words, flow of fluid has been found to be an important agent of metamorphism in some cases.

As early as 1970, Carmichael gave evidence for the existence of flows of fluids and a consequent regional-scale variation of the CO_2/H_2O ratio in intergranular fluids during metamorphism in the Whetstone Lake area in the Grenville province of Canada (§2.4.2). This metamorphic region is made up of dominantly calcareous metasedimentary rocks, which produced CO_2-rich fluids during prograde metamorphism. In his interpretation, H_2O-rich fluids that were released from a syn-metamorphic granitic pluton infiltrated into the surrounding rocks undergoing metamorphism and mixed with the internally produced CO_2-rich fluids. Fluids of the two different origins were well mixed to form locally virtually homogeneous intergranular fluids, which pervaded interbedded pelitic and calcareous rocks. On a regional scale, however, the mixed fluids showed a marked decrease of CO_2/H_2O ratio towards the granitic pluton over a distance of several kilometers. Owing to the regional-scale compositional variation of fluids, isograds based on dehydration reactions intersect an isograd based on a reaction that involves decarbonation (Carmichael 1970).

Ferry (e.g. 1976a, 1984, 1987) has clarified the reactions driven by, and the amount of, infiltrating fluids in argillaceous carbonate rocks, and related chemical migration in the Augusta–Waterville area, Maine (§6.7).

Grambling (1986) discovered a strange case of progressive metamorphism that took place under a virtually constant temperature in the Pecos Baldy area, New Mexico. The progressive mineralogical changes appear to have been caused by fluid flow accompanied by fluid–rock reactions (§2.4.3).

4.1.5 *Chemical potentials, activities and fugacities of H_2O and CO_2 in intergranular fluid*

The chemical potentials, activities and fugacities of H_2O and CO_2 are quantities controlling the equilibria between these substances and the minerals of the rock as well as the migration of these substances by diffusion.

The **chemical potential** (μ_i) of any component i (H_2O or CO_2 in this case) in a solution phase is a quantity that is defined in direct relation to the Gibbs energy of the solution, and is a function of pressure P, temperature T, and composition of the solution.

On the other hand, the quantity called **activity** (a_i) of component i is defined in relation to the chemical potential of the same component, but in addition, on the basis of an arbitrarily chosen standard state for it. A commonly used standard state is a pure phase of the component (such as pure H_2O or pure CO_2) at the pressure and temperature under consideration (P bar and T K). In this case, the following relation holds:

$$\mu_i = G_i^\circ(P,T) + RT \ln a_i \tag{4.1}$$

where μ_i and a_i denote the chemical potential and activity of the component i (such as H_2O or CO_2) in the solution, respectively, and $G_i^\circ(P,T)$ denotes the molar Gibbs energy of the pure phase of i at P bar and T K. Thus, a_i increases and decreases with μ_i. Because of this relation, we may use activity in place of chemical potential to specify the thermodynamic state of the component in an intergranular fluid. If the solution is an ideal gas mixture, the activity of a component is equal to its mole fraction. If the solution is a non-ideal mixture, it is what may be called a thermodynamically corrected mole fraction, varying between 1 and 0. Thus, $a_i = 1$ for the pure phase (standard state).

For a gaseous or supercritical solution, it is common to use another standard state, which is a pure phase (such as pure H_2O or pure CO_2) at 1 bar and the temperature under consideration (T K). In this case, the following relation holds:

$$\mu_i = G_i^\circ(1 \text{ bar}, T) + RT \ln f_i \tag{4.2}$$

where f_i is the **fugacity** of H_2O or CO_2 in the solution at P bar and T K. In an ideal gas mixture, the fugacity of a component is equal to its partial pressure.

If there is an intergranular fluid with pure H_2O composition in a rock, the chemical potential of H_2O in this case takes the maximum possible value, equal to the molar Gibbs energy of pure H_2O at the pressure and temperature under consideration. The activity of H_2O in it is unity. A real intergranular fluid is impure, because it contains dissolved rock-forming elements in addition to CO_2 and other volatiles. So the chemical potential of H_2O in it is smaller than the molar Gibbs energy of pure H_2O, and the activity of H_2O is smaller than 1. In any stable solution, as the mole fraction of a component decreases, the chemical potential and activity of the component also decrease. As the mole fraction of H_2O approaches zero, the chemical potential of H_2O decreases infinitely (i.e. $\mu \rightarrow -\infty$).

In the absence of an intergranular fluid phase, the chemical potentials of H_2O and of CO_2 are smaller than the molar Gibbs energies of pure H_2O and pure CO_2, respectively, at the pressure and temperature under consideration.

4.2 Fluid inclusions in metamorphic rocks

Fluid inclusions occur in minerals of metamorphic rocks. They are small droplets of fluid, usually microscopic in size and usually rich in H_2O in composition. They are commonly made up of two phases, containing liquid and a bubble of gas or vapor. H_2O-rich fluids trapped at high temperatures decrease the volume on cooling to surface temperatures so that a bubble of aqueous vapor may form. Or, such bubbles may contain large amounts of other gases, such as CO_2, that were originally in solution in the fluids. Fluids trapped at high temperatures commonly contain NaCl and other salts dissolved in them, and upon cooling to surface temperatures precipitate crystals of these salts. Fluid inclusions give evidence of the compositions of fluids during metamorphism as well as the temperatures of mineral formation.

This section is mainly based on reviews by Hollister & Crawford (1981), Crawford & Hollister (1986) and Roedder (1984).

4.2.1 Modes of occurrence

In metamorphic rocks, fluid inclusions occur most commonly in quartz grains, but also in garnet, feldspar and other minerals. Quartz may form at various stages over a long period of time in metamorphism, and quartz grains formed at different stages may trap different fluids. Quartz grains, once formed, may suffer later deformation and metamorphic crystallization, which may cause alteration of pre-existing fluid inclusions and the trapping of new ones. Some fluid inclusions are incorporated during the original growth of a mineral on its surface. They are called **primary inclusions**. Other fluid inclusions are trapped in pre-existing crystals, for example, by healing of fluid-filled fractures. They are called **secondary inclusions**. In metamorphic rocks, most fluid inclusions appear to be secondary. In a mineral, cracks may form, along which trails of secondary fluid inclusions can form. Such trails of inclusions tend to migrate later from their original positions to a more irregular or random distribution.

Since the volume of a fluid increases with increasing temperature and decreasing pressure, a curve representing a constant specific volume of fluid has a positive slope in the P–T diagram. This type of curve is called an **isochore**. In the prograde metamorphic stage or at the thermal peak, if the rate of increase of rock pressure with temperature is smaller than that of dP/dT for the isochore of fluid inclusions, the fluid can expand and rupture the inclusion walls if the pressure difference between the fluid and the solid exceeds a certain value (of the order of 1 kbar in quartz). In P–T–t paths like (a) or (b) in Figure 2.2, therefore, fluid inclusions formed in the prograde stage will mostly be destroyed before, or near, the thermal peak. Otherwise, such early inclusions may be destroyed simply by intensive metamorphic crystallization at, and near, the thermal peak. It is generally considered that the fluid inclusions observed in metamorphic minerals are those trapped during, and after, the thermal peak. Some rocks contain secondary fluid inclusions that formed at different stages, among which the earliest may be close to the thermal peak.

H_2O and CO_2 are completely miscible above 374 °C. At lower temperatures, an H_2O–CO_2 mixture may separate into coexisting H_2O-rich and CO_2-rich fluids. The low-temperature limit for complete mixing varies with pressure, showing a minimum value of 265 °C at about 2.15 kbar. Such a separation into two fluids may occur either before or after trapping of fluids in inclusions.

4.2.2 Fluid inclusions in low- and middle-grade metamorphic rocks

The major volatile components in fluid inclusions in low- and middle-grade metamorphic rocks are H_2O, CO_2, CH_4 and N_2, usually in this order of abundance. The $CO_2/(H_2O + CO_2)$ ratio of inclusions is usually very small in metapelites, but may be highly variable in calc-silicate rocks.

Fluid in sedimentary rocks may contain CH_4, C_2H_6 and higher hydrocarbons produced by decomposition of solid organic substances. Such hydrocarbons, however, usually decompose during metamorphism when the temperature exceeds 300 °C, except for CH_4, whereas solid organic substances are gradually changed into graphite through intermediate states. A most beautiful correlation of the composition of fluid inclusions with metamorphic grade was found in a north–south traverse across the regional metamorphic complex of the Lepontine Alps in Switzerland (Frey et al. 1980).

Fluid inclusions usually contain some NaCl and less commonly $CaCl_2$. The observed NaCl content is usually of the order of 2–6 wt % in metapelites and 20–25 wt % in calcareous rocks.

4.2.3 Fluid inclusions in high-grade metamorphic rocks

Touret (1971, 1981) discovered that in middle-grade metamorphic rocks, most fluid inclusions are usually composed mainly of H_2O, whereas in granulite-facies regions a high proportion of fluid inclusions are mainly composed of dense CO_2. The increase of the proportion of CO_2-dominant fluid inclusions is gradual and begins in the higher amphibolite facies. Even in granulite-facies regions, however, there still remain many H_2O-dominant fluid inclusions (e.g. Lamb et al. 1991).

Since H_2O is much more soluble than CO_2 in silicate melt, if an H_2O–CO_2 mixture comes into contact with a silicate melt, a CO_2-rich fluid forms. Hence, partial melting in high-grade metamorphism is a possible factor in the formation of CO_2-dominant fluid in granulite-facies regions. It has also been claimed that a major part of the CO_2 may have risen from the mantle (Newton et al. 1980). Peridotite inclusions in basalt are widely believed to be fragments of upper mantle rocks brought up by the rising basalt magmas. Olivine and pyroxene in such inclusions contain fluid inclusions, many of which are nearly pure liquid CO_2 under pressure (Roedder 1965). This, together with the formation of carbonatite magma, probably in the mantle, suggests that CO_2 is abundant in some parts of the upper mantle.

5 Composition–paragenesis diagrams

5.1 Mineralogical phase rules

As a preliminary to the description of composition–paragenesis diagrams to follow, the so-called mineralogical phase rules are introduced in this section.

5.1.1 Goldschmidt's mineralogical phase rule for the closed system

Gibbs' phase rule is given by the equation: $F = c + 2 - p$, where F denotes the variance (number of degrees of freedom), c the number of system components, and p the number of phases, of the system (§3.1).

If each rock undergoing metamorphism is a closed system with no exchange of substance with the environment, the mass of each element in our system is constant. The state of the system changes with pressure P and temperature T, both being controlled by conditions external to the system. Goldschmidt (1912b) applied the phase rule to this model of rocks undergoing metamorphism.

The thermal-peak temperature and pressure of metamorphic rocks vary continuously from place to place over a metamorphic region, provided that the region was not cut by post-metamorphic faults. If rocks showing a specific mineral assemblage occur in various places through an area in the region, the mineral assemblage must have a variance equal to, or greater than, 2. Thus,

$$F = c + 2 - p \geq 2$$

Therefore,

$$c \geq p \tag{5.1}$$

If the rock has no intergranular fluid phase, p represents the number of minerals. If the rock has a fluid phase, the number of minerals is equal to $p-1$. Thus, the number of minerals that could coexist in stable equilibrium is equal to, or less than, the number of independent components. This is called **Goldschmidt's mineralogical phase rule** (Goldschmidt 1912b).

The mineralogical phase rule holds not only for the whole rock system, but also for the partial systems of the rocks. For the purpose of illustration, we consider a one-component system with Al_2SiO_5 composition as a partial system of pelitic metamorphic rocks. According to Goldschmidt's mineralogical phase rule, only one mineral (kyanite, sillimanite or andalusite) is stable through a metamorphic area. This corresponds to the fact that kyanite is stable in the kyanite zone, whereas sillimanite is stable in the sillimanite zone, of the Barrovian sequence. However, kyanite and sillimanite could coexist stably at the transformation temperature, which corresponds to the sillimanite isograd. Thus, the mineral assemblage of rocks at an isograd may have a greater number of phases by 1 than the number the mineralogical phase rule permits.

If we choose a rock fortuitously from a continuous metamorphic region that includes both kyanite and sillimanite zones, the probability of taking the rock exactly from the sillimanite isograd is negligibly small. Hence, fortuitously chosen metamorphic rocks should usually show either stable kyanite or stable sillimanite, but not both. Thus, fortuitously chosen metamorphic rocks usually obey the mineralogical phase rule.

5.1.2 Korzhinskii's mineralogical phase rule for rocks under the externally controlled chemical potentials of certain volatile components

When rocks undergoing metamorphism are open to certain volatile components such as H_2O and CO_2, the components can enter or leave each rock system until chemical equilibrium is reached. Let us consider a group of metamorphic rocks in which not only temperature and pressure but also the chemical potentials of these components are externally controlled at constant values. Components of this type are called **externally buffered components** (Ch. 6). The mechanism that causes the uniformity of chemical potentials may be active flow and infiltration of a fluid with a constant composition, or active diffusion through the intergranular fluid. The components whose masses in each rock are constant, and whose chemical potentials are not externally buffered, are called **fixed components**.

The number of externally buffered components is denoted by c_e, and that of the fixed components by c_x. For equilibrium to hold at fortuitous values of externally controlled temperature, pressure and chemical potentials of the externally buffered components, the variance of the system must normally be equal to, or greater than, $2 + c_e$. Hence,

$$c = c_e + c_x$$

$$F = (c_e + c_x) + 2 - p \geq 2 + c_e$$

So,

$$c_x \geq p \qquad (5.2)$$

Thus, the number of minerals that can coexist in stable equilibrium in an open system is generally equal to, or less than, the number of fixed components. This is called **Korzhinskii's mineralogical phase rule** (Korzhinskii 1950, 1959, 1965).

5.2 Composition–paragenesis diagrams in general

5.2.1 Two idealized cases of metamorphic conditions in a small area

An important problem in metamorphic petrology is the explanation of how the observed diversity of mineralogical compositions of metamorphic rocks have formed in terms of temperature, pressure and other physicochemical factors. To analyze this problem, we take an area in a metamorphic region small enough for temperature and pressure of metamorphism to be regarded as being virtually uniform throughout the area. The temperature and pressure are externally controlled, that is, determined by conditions outside the area. Some other variables (e.g. the chemical potential of H_2O) may also be externally controlled and uniform throughout.

Even this small area usually shows a diversity of metamorphic rocks with various different mineralogical compositions. This diversity cannot be ascribed to the externally controlled conditions such as temperature and pressure, and must be ascribed to another factor that is intrinsic to individual rocks, i.e. bulk-rock chemical composition. In accordance with these conditions, we may try to construct diagrams that show the relationships between the chemical composition and the mineralogical composition of rocks under constant values of temperature, pressure and possibly other externally controlled conditions of the area. Such diagrams are called **composition–paragenesis diagrams**.

For the construction of composition–paragenesis diagrams, we may define two simple, idealized cases as follows. Composition–paragenesis diagrams that are based on rigorous thermodynamic logic can be easily constructed in these cases.

Case I: closed systems. In this case, each metamorphic rock in the small area under consideration is a closed system. In other words, the bulk chemical composition of each rock remains unchanged during metamorphism. This would apply to rocks in which all possible reactions are solid–solid reactions without involving volatile components. Even when metamorphic rocks undergo dehydration reactions, the closed-system condition may apply to a partial system observed in the rocks.

In a closed system of this kind, Duhem's theorem (§3.1.3) holds, and so the state of the system, including the kinds and proportions of minerals and their compositions, is determined by two independent variables, which may be regarded as pressure and temperature, if the phase-rule variance is 2, or greater. Goldschmidt's mineralogical phase rule also holds. Triangular composition–paragenesis diagrams in this case are discussed in §5.2.2 below.

Case II: externally controlled chemical potentials of certain volatile components. In this case, each rock is an open system with respect to the volatile components whose chemical potentials are externally controlled. Korzhinskii's mineralogical phase rule holds. Triangular composition–paragenesis diagrams in this case are discussed in §5.2.3 below.

Equilibria of dehydration and decarbonation reactions are controlled not only by temperature and pressure but also by the chemical potential of H_2O and CO_2. Then, under a specific temperature and pressure and specific chemical potentials of H_2O and CO_2, the mineralogical compositions of rocks that undergo not only solid–solid but also dehydration and decarbonation reactions depend on bulk-rock composition, excluding H_2O and CO_2 contents.

Case II applies approximately to many, but not all, small areas of metapelitic and metabasic rocks, as discussed in Chapter 6.

5.2.2 Triangular composition–paragenesis diagrams for closed-system rocks

5.2.2.1 Diagrams with minerals with fixed compositions. According to Goldschmidt's mineralogical phase rule, a rock composed of three components can contain three, two, or one mineral(s) over a range of externally controlled temperatures and pressures. When the mineralogical compositions of such rocks are represented in a triangular composition–paragenesis diagram, the paragenetic relations can be shown by dividing the diagram into a family of subtriangles. Figure 5.1 illustrates this type of diagram for the SiO_2–Al_2O_3–MgO system under two different sets of P–T conditions.

In this case, all minerals are assumed to have fixed compositions, and each diagram is divided into subtriangles indicating parageneses to be observed. A point representing the bulk composition of a rock may plot within, or on a side of, the subtriangles. In these cases, the rock should show coexistence of three or two minerals, respectively. A point plotting at a corner of any subtriangles represents the presence of only one mineral.

The mineral paragenesis of diagram (b) in Figure 5.1 is stable at lower temperatures and higher pressures than that of diagram (a). The two diagrams are connected by the reaction:

$$\text{cordierite} + 3\,\text{spinel} = 5\,\text{enstatite} + 5\,\text{corundum}$$

Merely for the purpose of discussion, we assume that the paragenesis of (a) holds

(a) High-temperature contact metamorphism (pyroxene-hornfels facies)

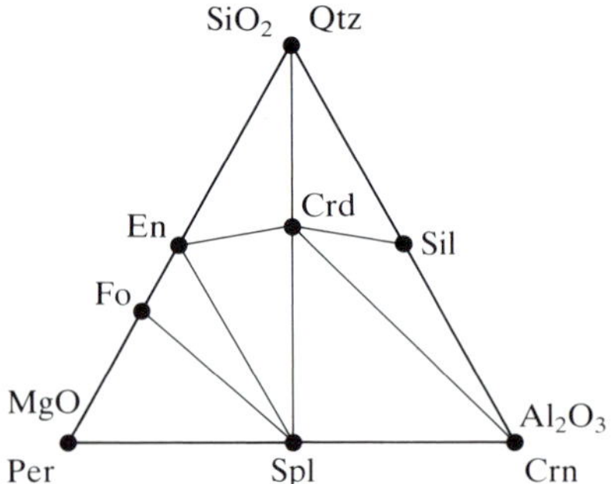

(b) Lower *T* and higher *P* than (a)

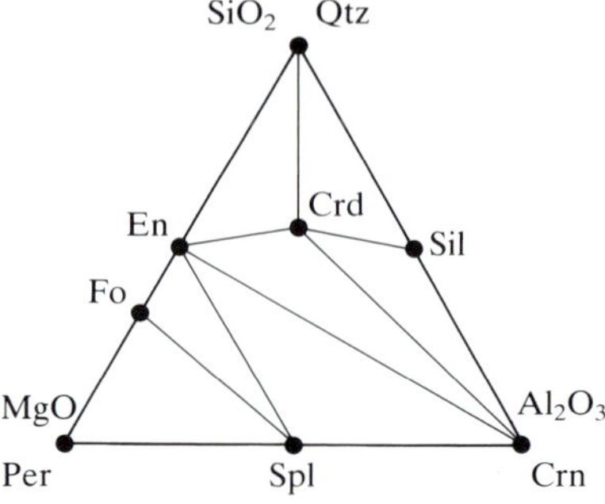

Figure 5.1 Two composition–paragenesis diagrams of the closed system SiO_2–Al_2O_3–MgO for different *P–T* conditions. Per: periclase. For other mineral symbols, see Appendix I.

in one part of a metamorphic region and that of (b) holds in another part. In the transitional state between (a) and (b), the equilibrium of the above equation holds in which all four phases coexist. Since the variance of the system in the transitional state is 1, the equilibrium holds on a line in the metamorphic region, and the line is an isograd. The paragenetic relations of (a) and (b) hold on the opposite sides of the isograd.

5.2.2.2 Diagrams with solid-solution minerals. Most metamorphic minerals are solid solutions showing considerable ranges of composition. Their compositions are represented by areas (fields) in the triangular diagram in Figure 5.2. The triangular diagram has subtriangles to represent three-phase parageneses. However, unlike the case of Figure 5.1, the subtriangles are separated by two-phase fields. Rocks plotting within the area representing each mineral have only one phase.

If the composition of a rock plots within the area representing a mineral, $c = 3$ and $p = 1$, and hence $F = 4$. Out of the four degrees of freedom, two refer to temperature and pressure, whereas the remaining two correspond to the divariant compositional variation illustrated in the diagram. When two phases coexist in a rock, $c = 3$ and $p = 2$, and hence $F = 3$. Thus, the compositional variation in the diagram is univariant. In other words, the composition of each of two coexisting minerals is on a curve in the diagram. These curves delimit the outlines of the composition fields of the solid-solution minerals. In Figure 5.2, points on the outline of a mineral's field are connected by a straight line to

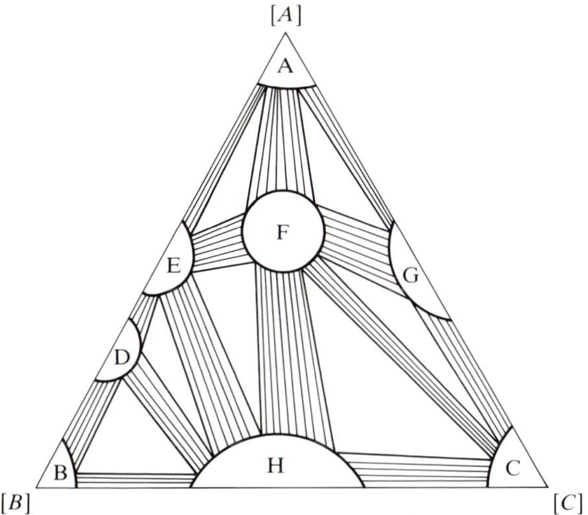

Figure 5.2 Composition–paragenesis diagram of the three-component closed system [*A*]–[*B*]–[*C*] with solid-solution minerals A, B, C, D, E, F, G and H. The same relation holds for a system which has fixed components [*A*], [*B*] and [*C*] and is externally buffered with regard to other components. Note that the phase relations in this diagram are essentially the same as those in Figure 5.1a, except that all minerals show wide ranges of solid solution.

points on the outline of the coexisting mineral's field, in order to show the equilibrium compositional relations. When three phases coexist in a rock, the composition of each mineral is fixed, and the paragenesis is represented by a unique subtriangle in the diagram.

5.2.3 Triangular composition–paragenesis diagrams under externally controlled chemical potentials of certain volatile components

According to Korzhinskii's mineralogical phase rule, the number of minerals coexisting stably in a system with three fixed components under a definite set of external conditions (temperature, pressure and chemical potentials of the externally buffered components) is generally three, two, or one. Let us now look at the paragenetic relations shown in Figure 5.2 from a different point of view. We assume that the diagram shows paragenetic relations in an externally buffered system containing three fixed components [*A*], [*B*] and [*C*]. The triangular diagram has subtriangles that represent three-phase parageneses, and such subtriangles are separated by two-phase fields. Rocks plotting on the area of each mineral have only one phase.

This illustrates that composition–paragenesis diagrams under constant chemical potentials of certain volatile components are very similar to those for the corresponding closed systems, the only differences being that only the fixed components are taken at the apices in the former.

5.2.4 Techniques for reducing the number of components to be expressed in the composition–paragenesis diagram

Ordinary metamorphic rocks are made up of more than 10 components such as SiO_2, TiO_2, Al_2O_3, Fe_2O_3, FeO, MgO, MnO, CaO, Na_2O, K_2O, H_2O and P_2O_5. On the other hand, an ordinary triangular composition–paragenesis diagram can show relationships between the chemical and mineralogical compositions rigorously, only when the rocks have three components if they are closed systems, and three fixed components if they are open systems. Thus, various techniques have been devised to reduce the number of components to be expressed in diagrams.

Limiting the kinds of rocks to be shown in a diagram. An important technique for reducing the number of components for graphical representation is limiting the kinds of rocks to be plotted on the diagram so that we can ignore components that are absent or minor in the mineral assemblages to be discussed. For example, CaO is usually minor in typical pelitic rocks, and so can generally be ignored in diagrams dealing with pelitic rocks alone such as *A'KF* and *AFM* diagrams. Mineral assemblages in siliceous dolomitic limestones may be treated as composed of SiO_2, CaO and MgO together with H_2O and CO_2 to a good approximation.

Indifferent components. (Korzhinskii 1959) If the metamorphic rocks that we deal with contain zircon $ZrSiO_4$, an increase or decrease of the ZrO_2 content of rocks causes only an increase or decrease, respectively, in the amount of zircon, but no changes in the paragenetic relations of the rocks. Therefore, component $ZrSiO_4$ and mineral zircon may be ignored simultaneously. The partial assemblage formed by the remaining minerals has the same variance as the complete assemblage with zircon. Components such as $ZrSiO_4$ in this case are called indifferent.

Externally buffered components. If the chemical potentials of H_2O, CO_2 and any other components are externally controlled at constant values, they need not be shown at an apex of the composition–paragenesis diagram.

Excess components. Most of the commonly used composition–paragenesis diagrams deal only with quartz-bearing metamorphic rocks (so-called rocks with *excess SiO_2*). Under a given *P-T* condition, the presence of quartz maintains a definite chemical potential of SiO_2 in the rock, which is equal to the molar

Gibbs energy of quartz. An increase or a decrease of SiO_2 content of such rocks under a constant temperature and pressure causes only an increase or decrease, respectively, in the amount of quartz, without any change in the paragenetic relations of minerals. So, in the discussion of paragenetic relations of minerals, we may simultaneously disregard the component SiO_2 and the phase quartz without changing the variance of the system. This enables us to reduce by one the number of components to be considered.

More generally, any component that behaves like SiO_2 in this case, is called an **excess component**, and it may be ignored in the composition–paragenesis diagrams. It differs from indifferent components in the respect that it is contained in large amounts in associated minerals such as biotite.

Sillimanite has a nearly constant composition Al_2SiO_5. When we discuss mineral assemblages of sillimanite-bearing rocks only, we may regard Al_2SiO_5 as an excess component. For an analogous reason, other minerals that occur in all rocks under consideration may be ignored, insofar as the minerals have a constant composition.

5.2.5 Effects of quartz in metamorphic rocks

Since quartz is widespread in metapelitic and metapsammitic rocks, SiO_2 is commonly treated as an excess component in composition–paragenesis diagrams. However, quartz-free rocks are not rare, particularly among metabasites and metamorphosed calcareous rocks. In these rocks, the amount of SiO_2 influences the kinds of minerals and the P–T conditions of some reactions. In this kind of case, SiO_2 must be shown at an apex of the composition–paragenesis diagram for the rocks (Fig. 5.1).

A brief comment is given below on what occurs in quartz-free rocks. For example, ordinary sillimanite-bearing metapelites contain quartz. When the SiO_2 content of the rock is reduced, quartz decreases in amount and then disappears. A further decrease of the SiO_2 content of the rock begins to change sillimanite into corundum by the reaction: $Al_2SiO_5 - SiO_2 = Al_2O_3$, as is clear from Figure 5.1. If sillimanite and corundum coexist, the variance of the system is 2, and the state of the system is definite at a given P and T. The chemical potential of SiO_2 in the rock is kept constant at a certain value, which is lower than the molar Gibbs energy of quartz. Ultimately all sillimanite may disappear, and then the chemical potential of SiO_2 becomes still lower and variable. In this case, corundum is a mineral characteristic of a silica-deficient environment.

In an analogous way, spinel may form together with, or instead of, cordierite in quartz-free rocks. With decreasing SiO_2 content of the rock, magnesian orthopyroxene may be changed into olivine, which in turn may be changed into periclase (Fig. 5.1). There are many other minerals that form characteristically in quartz-free rocks.

There are many prograde metamorphic reactions that have quartz among the reactants. In quartz-free rocks, these reactions cannot occur. So the mineral assemblages, analogous to the reactant assemblages of the above reactions, but not including quartz, continue to be stable at temperatures higher than the reaction temperatures in quartz-bearing rocks (e.g. Chatterjee & Johannes 1974). Hence, the progressive sequence of index minerals may differ between quartz-bearing and quartz-free rocks with similar compositions except for their SiO_2 content.

5.2.6 Composition–paragenesis diagrams for metamorphic rocks of various compositional groups

5.2.6.1 Brief history of composition–paragenesis diagrams. The history of composition–paragenesis diagrams began with Eskola's (1915) *ACF* and *A'KF* diagrams. The *ACF* diagram was used for metamorphic rocks of almost all major compositional groups, including pelitic, calcareous and mafic rocks, whereas the *A'KF* diagram was used for metapelitic rocks. Eskola used these diagrams to demonstrate the existence of regular relationships between the chemical and mineralogical compositions of metamorphic rocks exposed in his study area (the Orijärvi area of Finland), because he regarded the existence of such relationships as strong evidence for his view that the metamorphic mineral assemblages represent equilibrium states under the *P–T* conditions that prevailed in the area.

Meantime, a few types of composition–paragenesis diagrams were successfully used for metamorphic rocks with relatively simple chemical compositions (i.e. metamorphic rocks composed of a relatively small number of components). In particular, Bowen (1940) discussed the metamorphism of quartz-bearing dolomitic limestones in response to increasing temperature. He used the tetrahedron SiO_2–CaO–MgO–CO_2 to show the compositions and mineral assemblages of the rocks, and projected the paragenetic relations in the rocks from the CO_2 apex onto the SiO_2–CaO–MgO face. This triangular diagram shows the paragenetic relations either in the presence of pure CO_2 gas (as was assumed by Bowen), or under an externally controlled chemical potential of CO_2.

Since then, there have been many attempts to construct and/or use composition–paragenesis diagrams for metamorphic rocks of various compositional groups. The most successful composition–paragenesis diagram was J. B. Thompson's (1957) *AFM* diagram for metapelites. This diagram could explain the paragenetic relations in metapelites in direct connection to the phase rule. Moreover, thermodynamic analysis of paragenetic relations based on this diagram led to a chemographic classification of metamorphic reactions. These results were important, particularly, because they demonstrated that metapelites were amenable to rigorous thermodynamic treatment in spite of their complicated chemical compositions. Because of its importance, the *AFM* diagram is described and discussed in detail in a later section.

A few other types of composition–paragenesis diagrams for metapelites and metabasites have been used successfully.

Rigorous thermodynamic discussions of composition–paragenesis diagrams were given by Korzhinskii (1950, 1959, 1965) and J. B. Thompson (1957, 1970, 1979, 1982a,b).

5.2.6.2 Composition–paragenesis diagrams and the variances of mineral assemblages. The degrees of success of various attempts to construct composition–paragenesis diagrams depended mainly on the variances of the mineral assemblages that occur in the compositional groups of metamorphic rocks under consideration. If a mineral assemblage with variance F is equilibrated under an externally controlled temperature, pressure and chemical potential of H_2O, the number of independent compositional parameters of the assemblage is $F - 3$. On the other hand, if the number of independent compositional parameters is 2 or smaller, the state of the system can be represented in a two-dimensional composition–paragenesis diagram under constant external conditions. This condition may be written as:

$$F - 3 \leq 2$$

Then,

$$F \leq 5 \tag{5.3}$$

If we neglect minor components, this condition (eq. 5.3) is fulfilled by most mineral assemblages in metapelites, and hence construction of composition–paragenesis diagrams has been very successful for metapelites. On the other hand, this condition (eq. 5.3) is not fulfilled by the majority of mineral assemblages in metabasites. For example, amphibolite composed of hornblende and plagioclase alone is a common and typical metabasite. The number of components is at least seven, and so $F = 7$, at least. Thus, in general, construction of composition–paragenesis diagrams has not been so successful for metabasites as for metapelites. However, composition–paragenesis diagrams have been constructed for certain types of metabasites with a relatively large number of minerals and so with a relatively small variance.

When the mineral assemblage of one metamorphic rock differs from that of another, there are two possible cases: (a) the difference in mineral assemblage may have been caused by an essential difference in bulk-rock chemical composition, and so there can be no balanced chemical reaction between the two assemblages; (b) the two different mineral assemblages can be connected by a balanced chemical reaction. In the latter case, the reaction may be realized by a change in external conditions, e.g. temperature and pressure.

If the two rocks belong to entirely different compositional groups, the presence of an essential difference in bulk-rock chemical composition may be self-evident. When the two rocks belong to the same compositional groups, the difference in chemical composition is relatively small, and it may not be so evident whether the chemical difference is essential or not.

If the metamorphic rocks are composed of only three components, whether the difference in bulk-rock chemical composition is essential or not, can be directly observed by plotting the assemblages in a triangular composition–paragenesis diagram. However, this method does not work for ordinary metamorphic rocks, which are composed of many components, because the composition–paragenesis diagrams for these rocks, even when they are triangular, actually are only projections from some very complicated compositional space onto a plane. To resolve this type of question, Greenwood (1967b), Perry (1967), Reid et al. (1973) and Braun & Stout (1975) devised analytical methods based on linear algebra.

5.3 Eskola's *ACF* and *A'KF* diagrams

5.3.1 Merits and demerits of ACF and A'KF diagrams

Eskola's (1915) two diagrams are composition–paragenesis diagrams that are very convenient for synoptical representation of metamorphic minerals that occur in an area. By glancing over some diagrams of this type, we can notice, recall and compare the minerals that occur in metapelites, metabasites and calc-silicate rocks in various areas which were metamorphosed under different *P–T* conditions. Mainly for this reason, the diagrams are still used, not only to summarize the results of petrographic works, but also for the purpose of definition and characterization of individual metamorphic facies, although it is now clear that they have too many defects for use in rigorous thermodynamic discussions.

Although metamorphic rocks usually contain small amounts of magnetite, hematite, ilmenite, sphene, rutile, apatite and/or sulfides, these accessory minerals are disregarded. TiO_2, P_2O_5 and S, being contained mainly or entirely in them, are disregarded. *ACF* and *A'KF* diagrams are used only for rocks containing quartz.

Eskola disregarded the component H_2O by stating that H_2O is always present in excess during metamorphism. If H_2O is externally buffered, this is justified.

Eskola disregarded Na_2O and the albite component of plagioclase, although this is not justifiable, because plagioclase is a solid solution with variable composition, and moreover the Na_2O contents of hornblende, micas and K-feldspar play an important rôle. He added Fe_2O_3 to Al_2O_3 on the assumption that they play similar rôles in solid-solution minerals. He added MgO and MnO to FeO for an analogous reason. Now we know that the replacement of Al^{3+} by Fe^{3+} and of Fe^{2+} by Mg^{2+} and Mn^{2+} does not justify these groupings of components, since the mutually replaceable atoms have distinctly different properties. This procedure, however, is a necessary evil in order to reduce the number of compositional variables. Thus, Eskola's diagrams cannot rigorously show paragenetic relations thermodynamically. At the same time, however, we accept all his procedures as necessary for constructing a diagram to be used for a very wide composition range of metamorphic rocks.

In this way, Eskola arrived at the following four components: $Al_2O_3+Fe_2O_3$, CaO, FeO+MgO+MnO, and K_2O. In making triangular diagrams, three components must be chosen out of the four. Eskola made two different choices, as follows.

5.3.2 The ACF diagram

Here, Eskola chose the mole proportions of Al_2O_3 + Fe_2O_3, CaO, and FeO + MgO + MnO for the apices of the triangular diagram. Since K_2O is disregarded in this case, K-feldspar cannot be represented in the diagram. As explained above, Na_2O and the albite component of plagioclase are also disregarded. So, the amount of Al_2O_3 used in the diagram is: (the Al_2O_3 content of the rock) − (the Al_2O_3 contained in K-feldspar and the albite component). Since the number of moles of Al_2O_3 in the K-feldspar and the albite is the same as that of (Na_2O + K_2O), we have the following three groups of components:

$$A = Al_2O_3 + Fe_2O_3 - (Na_2O + K_2O)$$
$$C = CaO$$
$$F = FeO + MgO + MnO$$

Therefore, the *ACF* diagram is triangular, having these three groups at the apices calculated on a molar basis (Fig. 5.3). Figure 5.4 shows the positions of common metamorphic minerals and rocks in the *ACF* diagram.

Since the micas and hornblendes contain considerable amounts of K_2O and Na_2O, strictly speaking they cannot be plotted on the *ACF* diagram, although they are customarily shown. Eskola showed calcite at the *C* apex (by ignoring CO_2) in some cases, and disregarded the mineral in others. If CO_2 is externally buffered, this component may be ignored.

The stabilities of CaO-bearing minerals vary with the chemical potential of

CO_2. Hence, the mineral assemblages of metabasites and metamorphosed calcareous rocks show great diversity in response to variation in the chemical potential of CO_2, particularly at low temperatures (§11.1–11.3). These paragenetic changes cannot be treated by the *ACF* diagram.

ACF diagrams in the literature are usually divided into subtriangles, each of which indicates a three-mineral paragenesis, as shown in Figure 5.3. This custom is generally followed in this book. However, as the *ACF* diagram is based on procedures that are not justifiable from a rigorous thermodynamic viewpoint, it is not an exact composition–paragenesis diagram, and cannot always be divided into subtriangles showing parageneses. In many cases, four or five minerals on the diagram occur in the same rock, as for example stated in the captions to Figures 11.3 and 11.4.

The position of a mineral in the diagram is not changed by Fe^{2+}–Mg–Mn and Al–Fe^{3+} substitutions. Solid-solution minerals involving other substitutions appear as lines or areas in the diagram. The most important of these substitutions is the **Tschermak substitution** AlAl↔(Mg,Fe^{2+})Si, which shifts the mineral point parallel to the *AF* side. For this reason, the compositions of muscovite, biotite, chlorite and hornblende appear either as a line segment or an elongated field parallel to the *AF* side. Because of the Tschermak substitution, muscovite may be regarded as approximately a solid solution between Al-muscovite $KAl_3Si_3O_{10}(OH)_2$ and celadonite $K(Mg,Fe)AlSi_4O_{10}(OH)_2$, and biotite may be regarded as approximately a solid solution between the eastonite–siderophyllite series $K(Mg,Fe)_{2.5}Al_2Si_{2.5}O_{10}(OH)_2$ and the phlogopite–annite series $K(Mg,Fe)_3AlSi_3O_{10}(OH)_2$. As the diagram is not exact, very accurate

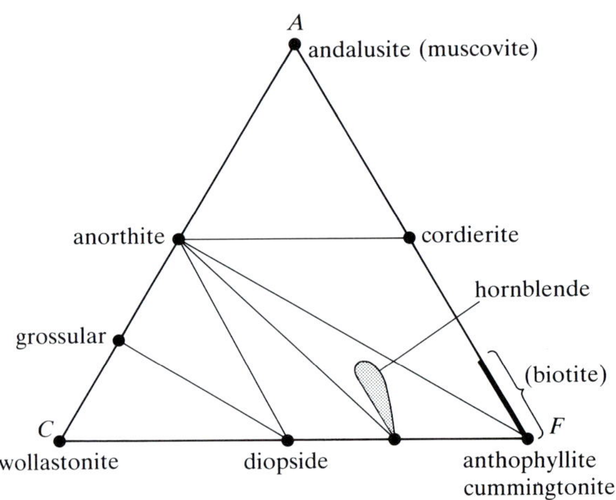

Figure 5.3 *ACF* diagram for the metamorphic rocks in the Orijärvi region, Finland. Based on Eskola (1915).

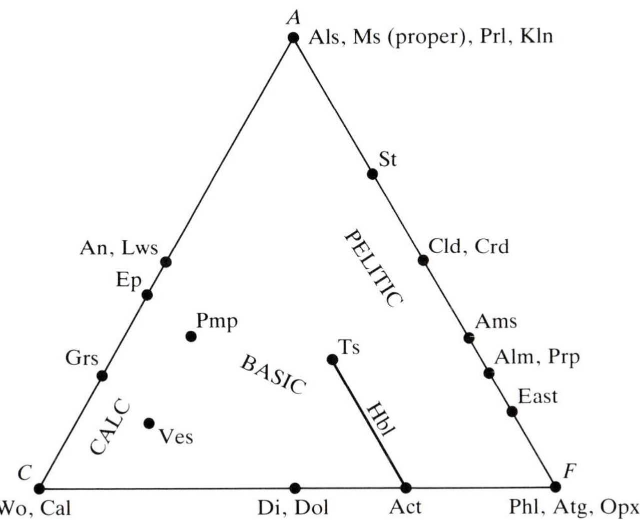

Figure 5.4 Positions of common metamorphic minerals and rocks in the *ACF* diagram. Ves: vesuvianite. For other mineral symbols, see Appendix I. The compositions of chlorite and biotite plot approximately between the points of antigorite (Atg) and amesite (Ams), and between those of eastonite (East) and phlogopite (Phl), respectively.

representation of the solid-solution ranges is usually unnecessary. In this book, many minerals are schematically shown by dots.

The merit of the *ACF* diagram is that all the common metamorphic rocks – including pelitic, psammitic, calcareous and mafic ones – can be plotted to show their approximate mutual chemical and paragenetic relations.

5.3.3 The A'KF diagram for metapelites

Pelitic rocks are usually low in CaO, containing $Al_2O_3 > (Na_2O + K_2O + CaO)$ calculated on a molar basis (Table 1.1). Such rocks are usually said to have *excess* Al_2O_3, because they contain more Al_2O_3 than is necessary to combine with all the Na_2O, K_2O and CaO present to form feldspars. They should plot within the triangle *A*–anorthite–*F* of the *ACF* diagram (Fig. 5.3). In pelitic rocks metamorphosed at medium and high temperatures, the CaO is usually contained mainly in the anorthite component of plagioclase.

Minerals characteristic of metapelites are very low in CaO, and plot very close to the *A*–*F* side. The effects of K_2O on these minerals can be clarified by examining the excess Al_2O_3, that is, the Al_2O_3 remaining after the formation of Na-, K- and Ca-feldspars. Hence, the *A'KF* diagram uses excess Al_2O_3 for the *A'* apex. At low temperatures, epidote and possibly other Ca–Al minerals form in place of the anorthite component. Since epidote and anorthite have somewhat different compositions, this has an effect on other minerals, although in general it may be neglected.

Thus, for metapelitic rocks, we can use an *A'KF* diagram that has the following groups of components on a molar basis at the apices:

$$A' = Al_2O_3 + Fe_2O_3 - (Na_2O + K_2O + CaO)$$
$$K = K_2O$$
$$F = FeO + MgO + MnO$$

An example is given in Figure 5.5.

Eskola named this the ***AKF* diagram**. However, since the meaning of *A* in this case differs from that of *A* in the *ACF* diagram, to avoid possible confusion the term ***A'KF* diagram** is used in this book, after Winkler (1979).

Minerals characteristic of metapelites plot very close to the *A'F* side of the *A'KF* diagram, except for biotite and muscovite. At the *P–T* conditions where both muscovite and biotite are stable, the diagram is separated into two fields by the muscovite–biotite join, as shown in Figure 5.5. This means that virtually K-free minerals plotting on the right-hand side of the join cannot coexist with K-feldspar. Typical metapelites plot on the right-hand side of the join, and so do not contain K-feldspar. With increasing temperature, muscovite begins to decompose by reaction with quartz, and begins to coexist with K-feldspar. This coexistence is related to the K–Na substitution in the two minerals, and so cannot be expressed by the *A'KF* diagram, but can be represented by the Al_2O_3–$NaAlO_2$–$KAlO_2$ diagram (Fig. 5.16b).

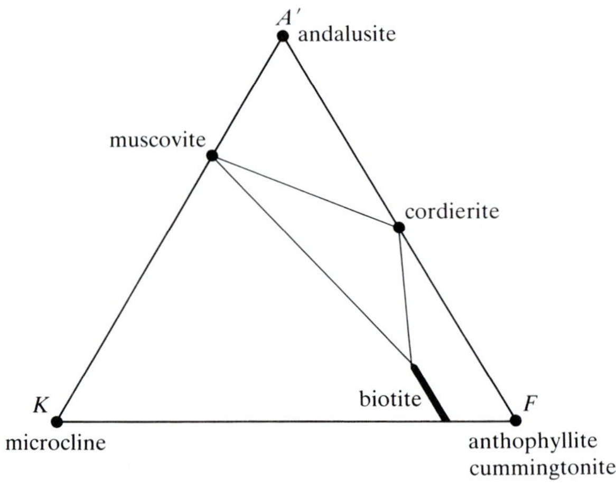

Figure 5.5 *A'KF* diagram for the metamorphic rocks of the Orijärvi region, Finland. Based on Eskola (1914, 1915).

5.4 J. B. Thompson's *AFM* diagram for metapelites

In the *ACF* and *A'KF* diagrams, FeO and MgO are treated jointly as *F*. Among the minerals of metapelites, however, garnet, staurolite and chloritoid have a high FeO/MgO ratio, whereas cordierite has a low one. The FeO/MgO ratio of rocks has a great influence on the mineral assemblage. Hence, triangular diagrams with FeO and MgO at different apices are helpful in paragenetic analysis. J. B. Thompson's (1957) *AFM* diagram for metapelites is a very useful example of this kind of diagram. The thermodynamic relations underlying the *AFM* diagram were elaborately discussed by A. B. Thompson (1976).

5.4.1 Construction of the AFM *diagram*

For simplification, Thompson regarded metapelites as being composed of six components SiO_2, Al_2O_3, FeO, MgO, K_2O and H_2O by neglecting minor ones such as CaO, Na_2O, MnO, Fe_2O_3 and TiO_2. He assumed that the chemical potential of H_2O is externally controlled at a constant value. Therefore, H_2O does not appear in this composition–paragenesis diagram. The whole group of rocks plotted on one diagram was metamorphosed at the same temperature, pressure, and chemical potential of H_2O. His diagram treats quartz-bearing rocks only, and so component SiO_2 and phase quartz are ignored. Thus, only four main components Al_2O_3, FeO, MgO and K_2O (denoted as *A*, *F*, *M* and *K*, respectively) remain in his graphical representation, and the composition of a metapelite is represented by a point in the *AFMK* tetrahedron (Fig. 5.6).

Muscovite is very widespread in metapelites. Thompson confined his discussion to muscovite-bearing rocks, and hence this mineral need not be expressed in the composition–paragenesis diagram. He assumed the muscovite composition to be $KAl_3Si_3O_{10}(OH)_2 = KAl_3O_5.3SiO_2.H_2O$. All the other minerals within the tetrahedron are projected from the muscovite point KAl_3O_5 onto the Al_2O_3–FeO–MgO plane (Fig. 5.6).

Compositions in the subtetrahedron *A*–*F*–*M*–KAl_3O_5 project onto the triangle *AFM*. Except for biotite, the ferromagnesian minerals of metapelites, such as staurolite, cordierite and garnet, lie practically on the *AFM* face of the tetrahedron, just inside the above-mentioned subtetrahedron, and so plot within the triangle *AFM*. On the other hand, minerals in the volume KAl_3O_5–*F*–*M*–$K_2Mg_3O_4$–$K_2Fe_3O_4$ project onto the downward extension of the *AFM* plane beyond the line *FM*. Biotite belongs to this type. The tielines from almandine, cordierite and biotite towards K-feldspar extend downwards away from the *A* apex. In this sense, K-feldspar may be regarded as being situated downwards. The height of the projected point on the diagram is determined by the parameter

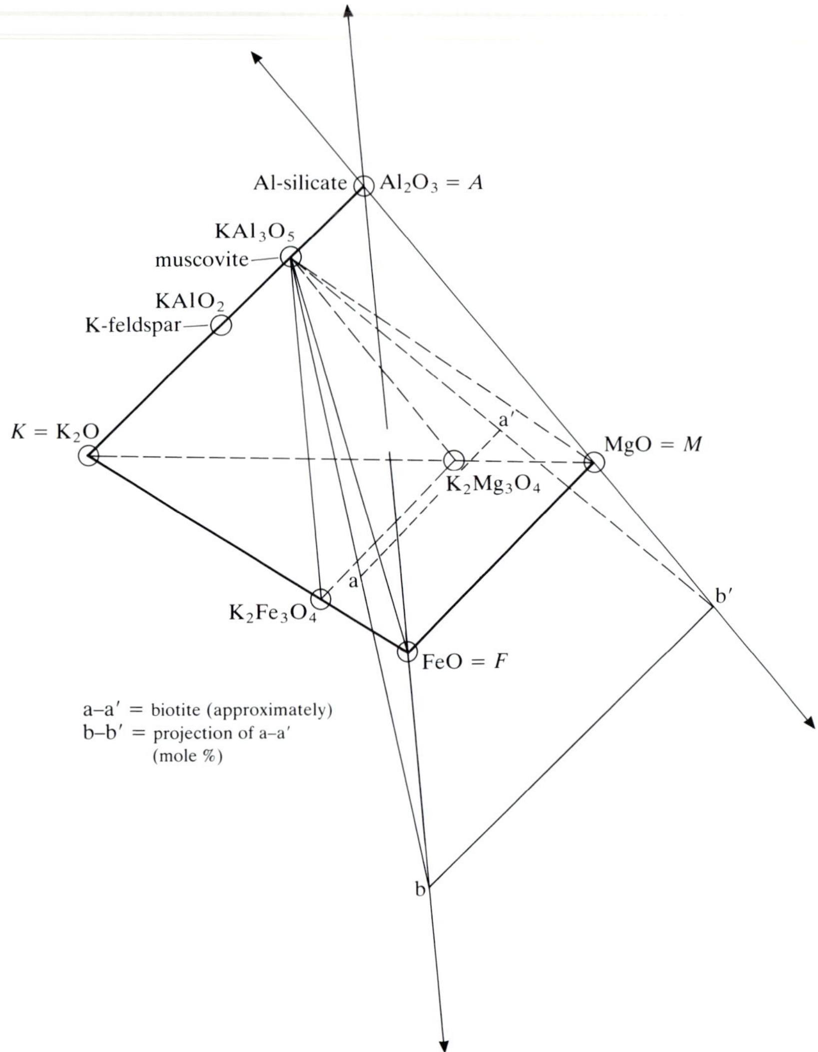

Figure 5.6 Thompson's projection through the idealized muscovite point onto the plane determined by the Al_2O_3, FeO and MgO apices (J. B. Thompson 1957).

$(Al_2O_3 - 3K_2O)/(Al_2O_3 - 3K_2O + FeO + MgO)$ (Fig. 5.7). The height above the *FM* side is positive, and that below it is negative.

Most metapelite minerals are solid solutions, particularly based on Fe–Mg substitution. Such minerals appear as horizontal lines on the diagram. Muscovite, biotite and chlorite show a wide range of Tschermak substitution AlAl→(Fe,Mg)Si. This substitution gives a vertical width to the composition fields of biotite and chlorite (Fig. 5.7).

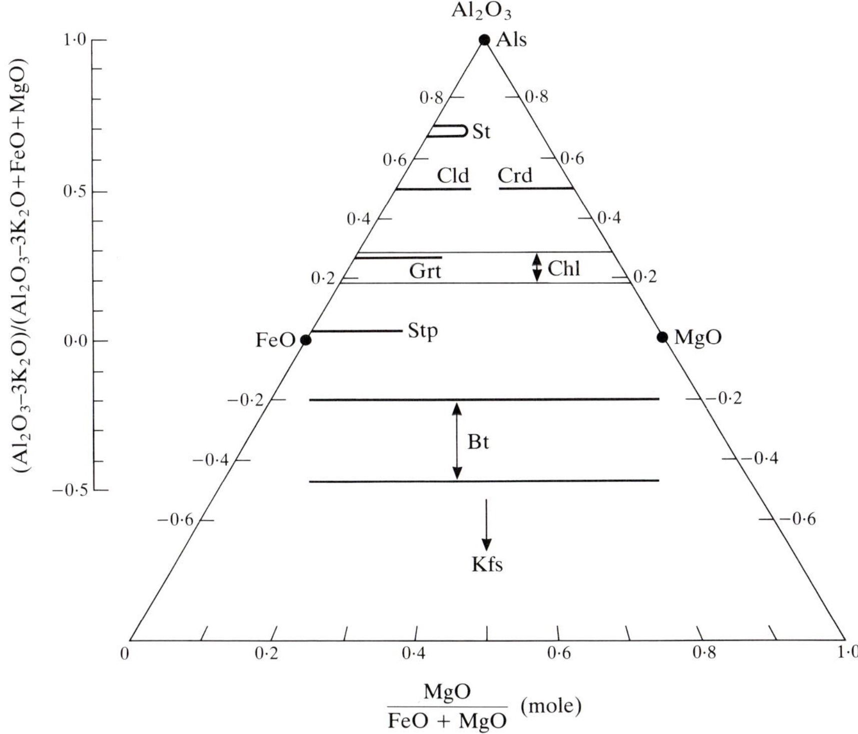

Figure 5.7 Positions of various metapelite minerals in Thompson's *AFM* diagram.

AFM *diagram projected from the K-feldspar point.* Muscovite is present and K-feldspar is absent in most metapelites in the ordinary temperature range of regional metamorphism. At higher temperatures (i.e. in the granulite facies), however, K-feldspar forms at the expense of muscovite. Since such high-temperature metapelites contain K-feldspar and quartz, we can use a modified *AFM* diagram in which all other minerals are projected from the K-feldspar point onto the Al_2O_3–FeO–MgO plane of the Al_2O_3–FeO–MgO–K_2O tetrahedron (Fig. 10.3).

In the following section, only the ordinary type of *AFM* diagram is discussed, although analogous relations hold for the modified type as well.

5.4.2 *Paragenetic relations in the* AFM *diagram*

5.4.2.1 Fe–Mg distribution and topologic relations. The topologic relations of mineral assemblages in the *AFM* diagram are largely controlled by the Fe–Mg distribution between coexisting ferromagnesian minerals. In ordinary cases, the MgO/FeO ratio of coexisting ferromagnesian minerals in metapelites increases in the following order: garnet, staurolite, chloritoid, biotite, chlorite, cordierite (A. B. Thompson 1976: 405).

In some cases, a reversal of the MgO/FeO order has been observed. Thus, in the staurolite–chloritoid pair, chloritoid has a higher or lower MgO/FeO ratio than the coexisting staurolite, when the MgO/(FeO + MgO) ratio of the pair is higher or lower than 0.09, respectively (Grambling 1983). In other words, the distribution coefficient (MgO/FeO in staurolite)/(MgO/FeO in chloritoid) varies with the average MgO/FeO ratio of the pair and passes unity at about MgO/(FeO + MgO) = 0.09. Similar reversals of the Fe–Mg partition have been suggested for the olivine–orthopyroxene and the clinopyroxene–orthopyroxene pairs (Grover & Orville 1969, Kretz 1981).

The MgO/(FeO + MgO) ratios of garnet, staurolite and chloritoid are very small in ordinary cases. When we discuss the paragenetic relations in rocks with large MgO/(FeO + MgO) ratios, the small differences in the MgO/(FeO + MgO) ratio between these minerals do not influence the topologic relations of the *AFM* diagram. So, for simplicity the MgO/(FeO + MgO) ratio of all these minerals may well be regarded as zero.

5.4.2.2 Three-, two- and one-phase fields. Metapelites have been simplified as being composed of six components: Al_2O_3, FeO, MgO, K_2O (or KAl_3O_5), SiO_2 and H_2O. Only H_2O is externally buffered. From Korzhinskii's mineralogical phase rule (eq. 5.2), the maximum number of stably coexisting phases is five. Since two of them are always quartz and muscovite, which do not appear in the diagram, the mineral assemblages to be represented in the diagram contain three, two or one phase(s). The diagram is divided into fields, each showing parageneses of three, two or one phase(s), as illustrated in Figure 5.8.

A three-phase field is a triangle with the three coexisting minerals at the apices. The diagram of Figure 5.8 shows four such three-phase fields, e.g. the sillimanite–staurolite–biotite triangle and the almandine–biotite–K-feldspar triangle extending downward. Such a three-phase field persists over a range of temperatures in prograde as well as in progressive metamorphism under given pressures and chemical potentials of H_2O. In this case, the compositions of solid-solution minerals at the apices are specific at a given temperature, and change with temperature. In other words, the paragenesis triangle changes its shape and position with temperature.

Three-phase fields are separated by two-phase fields, e.g. the almandine–biotite field. At a specific temperature, pressure and chemical potential of H_2O, the mineral assemblage of the two-phase field (i.e. the two phases + muscovite + quartz) has one compositional parameter that is variable with the bulk-

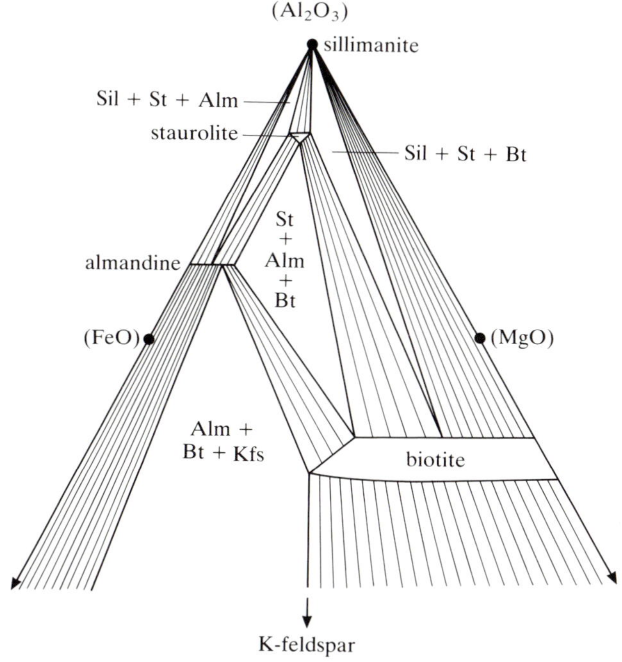

Figure 5.8 *AFM* diagram showing the paragenetic relations observed in metapelites in the lower sillimanite–muscovite zone of west-central New Hampshire (J. B. Thompson 1957).

rock chemical composition. For example, the FeO/MgO ratio of either of the two phases may be chosen as this compositional parameter. If the FeO/MgO ratio of one of the two coexisting phases is arbitrarily specified, that of the other is automatically fixed by the equilibrium conditions. The composition relations between the two coexisting phases may be shown by connecting line segments in the diagram.

A one-phase field, e.g. the biotite or almandine field, is surrounded by the lines that represent the compositions of the mineral in the adjacent two-phase fields.

5.4.2.3 Continuous and discontinuous reactions. We have already defined continuous and discontinuous reactions in §2.3.2. Under a specific pressure and chemical potential of H_2O, the equilibrium of a continuous reaction holds over a range of temperatures, whereas that of a discontinuous reaction holds at a specific temperature. The difference lies in the variance of the relevant systems, and is of vital importance in understanding the nature of isograds. This classification was first introduced by J. B. Thompson in relation to his *AFM* diagram (J. B. Thompson 1957, J. B. Thompson & Norton 1968), and then elaborated by A. B. Thompson (1976).

In the presence of a pure H_2O fluid, a continuous reaction proceeds over a range of temperatures under a given pressure, whereas a discontinuous reaction takes place at a specific temperature under a given pressure. This means that a discontinuous reaction represents a univariant equilibrium and holds on a univariant curve in *P–T* space.

5.4.3 *Two-phase fields in the* AFM *diagram*

Variance. A rock plotting in a two-phase field has four phases (the 2 phases + quartz + muscovite) in the absence of an aqueous fluid in the six-component system; thus the variance is 4. At a given temperature, pressure and chemical potential of H_2O, one compositional parameter is still variable with bulk-rock chemical composition. The compositional parameter may well be the FeO/MgO ratio, the $Al_2O_3/(FeO + MgO)$ ratio, and so on. If an aqueous fluid is present, the system has five phases, and so a variance of 3. At a given temperature and pressure, one compositional parameter is still variable, just as in the preceding case.

Reactions. In a two-phase field, if both of the two phase are Fe–Mg solid solutions, an Fe–Mg exchange reaction occurs between them, and there is no net-transfer reaction. Prograde chemical changes may occur in the compositions of the two minerals because of a change in the equilibrium constant of the exchange reaction. However, the compositions of the minerals also vary with bulk-rock composition.

If the two phases on the diagram and muscovite show both Fe–Mg and Tschermak substitution, there is one net-transfer reaction in addition to exchange reactions (see Miyashiro & Shido 1985: Table 5). If the two phases are chlorite and biotite, the reaction is represented by:

$$\underset{\text{chlorite}}{Mg_6Si_4O_{10}(OH)_8} + 3\,\underset{\text{muscovite}}{KAl_3Si_3O_{10}(OH)_2}$$
$$= 3\,\underset{\text{biotite}}{KMg_3AlSi_3O_{10}(OH)_2} + 3\,Al_2Mg_{-1}Si_{-1} + 7\,\underset{\text{quartz}}{SiO_2} + 4\,H_2O \qquad (5.4)$$

This reaction proceeds continuously over a range of temperature.

5.4.4 *Three-phase fields in the* AFM *diagram and the corresponding continuous reactions*

5.4.4.1 Variance. In the absence of an aqueous fluid, or in the presence of an impure fluid, the variance of the mineral assemblage of a three-phase field (with quartz and muscovite) is 3. At a given temperature, pressure and chemical potential of H_2O, the state of the system is specific. In other words, the compositions of all solid-solution minerals are functions of temperature, pressure and chemical potential of H_2O, and are independent of bulk-rock composition.

In the presence of a pure H_2O fluid, the variance of a three-phase field assemblage + the fluid phase is 2. The compositions of the minerals are functions of temperature and pressure alone. Therefore, the compositional variation of a solid-solution mineral in a specific three-phase field can be shown by a series of isopleths on a *P–T* diagram (e.g. Fig. 10.7).

5.4.4.2 If solid-solution minerals show only Fe–Mg substitution, each paragenesis subtriangle corresponds to one continuous reaction. If we assume that solid-solution minerals show only Fe–Mg substitution, any three-phase field in the *AFM* diagram corresponds to a continuous reaction (net-transfer reaction). The *AFM* diagram of Figure 5.9a, for example, shows the kyanite–staurolite–chlorite and staurolite–biotite–chlorite triangles. Since each assemblage is accompanied by muscovite and quartz, then the following continuous reactions (5.5) and (5.6) occur in the above two assemblages, respectively.

$$\underset{\text{kyanite}}{55\,Al_2SiO_5} + \underset{\text{chlorite}}{4\,(Fe,Mg)_7Al_4Si_4O_{15}(OH)_{12}}$$
$$= \underset{\text{staurolite}}{14\,(Fe,Mg)_2Al_9Si_4O_{23}(OH)} + \underset{\text{quartz}}{15\,SiO_2} + 17\,H_2O \qquad (5.5)$$

$$\underset{\text{chlorite}}{31\,(Fe,Mg)_7Al_4Si_4O_{15}(OH)_{12}} + \underset{\text{muscovite}}{55\,KAl_3Si_3O_{10}(OH)_2}$$
$$= \underset{\text{staurolite}}{26\,(Fe,Mg)_2Al_9Si_4O_{23}(OH)} + \underset{\text{biotite}}{55\,K(Fe,Mg)_3AlSi_3O_{10}(OH)_2}$$
$$+ \underset{\text{quartz}}{20\,SiO_2} + 173\,H_2O \qquad (5.6)$$

Note that, in the mineral formulas used here, FeO and MgO are lumped together as if they were one component (Fe,Mg)O.

More generally the paragenesis triangle *ABC* corresponds to the following equation:

$$aA + bB + cC + s\,SiO_2 + m\,KAl_3Si_3O_{10}(OH)_2 + n\,H_2O = 0 \qquad (5.7)$$

Here, *A*, *B* and *C* represent mineral formulas in which FeO and MgO are lumped together, and *a*, *b*, *c*, *s*, *m* and *n* are stoichiometric coefficients. Since any one of these coefficients may be taken as 1, we have to determine the re-

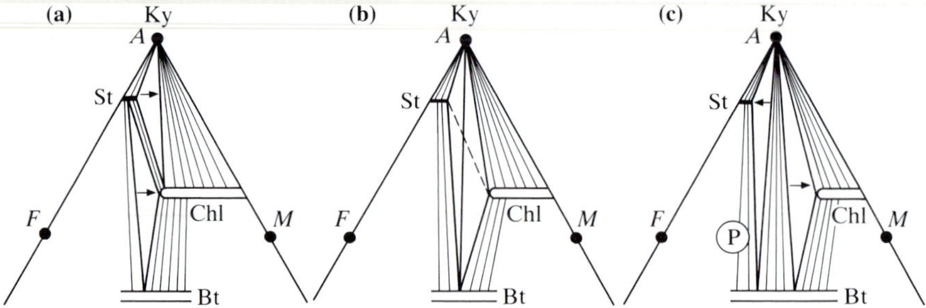

Figure 5.9 *AFM* diagrams showing the changes of paragenetic relations among kyanite, staurolite, biotite and chlorite with increasing temperature in the order of (a) → (b) → (c). Diagram (b) shows the tieline-switching reaction: staurolite + chlorite → kyanite + biotite, that is, reaction (5.11). Arrows indicate the direction of movement of paragenesis triangles with increasing temperature.

maining five coefficients from five equations balancing the mole number of each of the five components: SiO_2, Al_2O_3, (Fe, Mg)O, K_2O and H_2O. The coefficients will have positive and negative signs, depending on whether the terms are on the reactants or products side.

Actually each of the above reactions (5.5) and (5.6) must be regarded as a concise expression of two reactions: one for the Fe and the other for the Mg end-members (J. B. Thompson & Norton 1968). For example, reaction (5.5) represents the following two reactions:

$$55\,Al_2SiO_5+4\,Fe_7Al_4Si_4O_{15}(OH)_{12} = 14\,Fe_2Al_9Si_4O_{23}(OH)+15\,SiO_2+17\,H_2O \tag{5.8}$$

$$55\,Al_2SiO_5+4\,Mg_7Al_4Si_4O_{15}(OH)_{12} = 14\,Mg_2Al_9Si_4O_{23}(OH)+15\,SiO_2+17\,H_2O \tag{5.9}$$

Subtracting reaction (5.9) from reaction (5.8), we obtain the following Fe–Mg exchange reaction between chlorite and staurolite:

$$\begin{aligned} &2\,Fe_7Al_4Si_4O_{15}(OH)_{12} + 7\,Mg_2Al_9Si_4O_{23}(OH) \\ &= 2\,Mg_7Al_4Si_4O_{15}(OH)_{12} + 7\,Fe_2Al_9Si_4O_{23}(OH) \end{aligned} \tag{5.10}$$

Of the three reactions (5.8)–(5.10), any two are independent. Hence, concise expressions like (5.5) and (5.6) in which FeO and MgO are lumped together implicitly involve an Fe–Mg exchange reaction.

When an intergranular fluid of H_2O composition is present, the equilibria of reactions (5.8) and (5.9) are univariant, whereas those of reaction (5.5) are divariant. On a *P–T* diagram, equilibria of reactions (5.8) and (5.9) will hold on a univariant curve, whereas those of reaction (5.5) will hold in the field between the two curves.

The side that has liberated H_2O represents higher temperature. Chlorite is a low-temperature mineral with a very large amount of (OH) group in its structure. In reactions (5.5) and (5.6), the decomposition of chlorite dominates the amount of dehydration in the reactions. Both muscovite and biotite contain

2 (OH) per K. If muscovite and biotite occur on the opposite sides of an equation and they are the only K_2O-bearing minerals, as in reaction (5.6), the terms of muscovite and biotite must have the same coefficients, and so the amount of (OH) in muscovite is the same as that in biotite. In this case, muscovite and biotite do not contribute to the dehydration of the reaction.

Note that, in reaction (5.5), the only *AFM* phase on the right-hand side, staurolite, has a value of coordinate "*A*" intermediate between those of kyanite and chlorite of the left-hand side. For this reason the equation is balanced. Analogous relations hold for all such equations.

Chlorite, biotite and staurolite show a considerable extent of Tschermak substitution, and so mineral formulas with considerably different Al_2O_3/(FeO + MgO) ratios might well be used for them. In reactions (5.5) and (5.6), the mineral formulas given by J. B. Thompson & Norton (1968) were used. If we use formulas with different Al_2O_3/(Fe,Mg)O, we thereby obtain different coefficients.

5.4.4.3 Progressive change in Mg/Fe ratio of minerals by continuous reactions. The progress of continuous reactions changes the chemical compositions of relevant solid-solution minerals.

Reactions (5.5) and (5.6) are balanced only because FeO and MgO are lumped together as if they were a single component (Fe,Mg)O. In these assemblages, chlorite has a higher MgO/FeO ratio than the associated staurolite and biotite. Therefore, if the compositions of coexisting minerals are entered in the equations, the right-hand (products) side of each equation has a lower MgO/FeO ratio than the left-hand (reactants) side.

The progress of reaction (5.5) decreases the amount of chlorite and increases the amount of staurolite which has a smaller MgO/FeO ratio than the associated chlorite. In spite of the increase of the staurolite/chlorite ratio, the bulk-rock MgO/FeO ratio is kept constant by increase of the MgO/FeO ratio in both staurolite and chlorite. (Staurolite and chlorite change their MgO/FeO ratios sympathetically because of the exchange reaction between them.) This compositional adjustment may be remembered as a simple rule that, as a reaction advances, the MgO/FeO ratio of the product approaches the average MgO/FeO ratio of the assemblage (i.e. the bulk-rock MgO/FeO ratio).

The progressive changes in the proportion and the MgO/FeO ratio of coexisting ferromagnesian minerals can be visualized by pseudo-binary diagrams like Figure 5.10, in which the abscissa and ordinate represent the MgO/FeO ratio and temperature, respectively (A. B. Thompson 1976). The abscissa may be regarded as a projection from the *A* apex onto an arbitrary Fe–Mg line in the *AFM* diagram.

In the lower half of Figure 5.10, the continuous reaction loop, denoted as

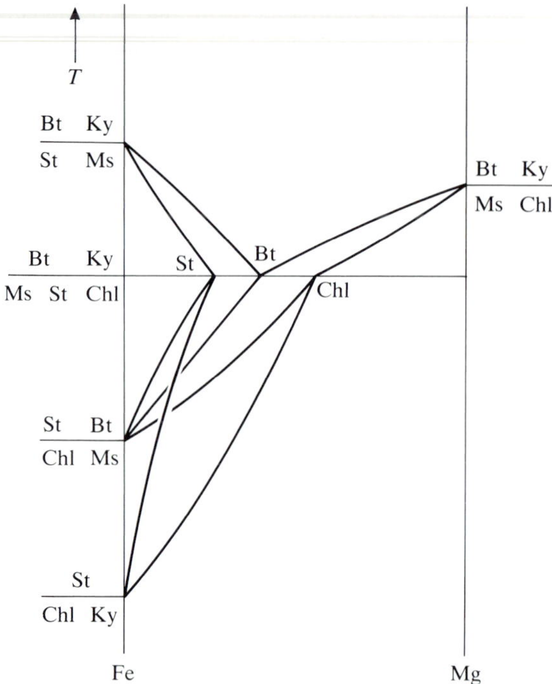

Figure 5.10 The changing MgO/FeO ratios of coexisting minerals, corresponding to the (a) → (b) → (c) of Figure 5.9. The horizontal line in the middle corresponds to the tieline-switching in Figure 5.9b. The symbol St/Chl Ky, for example, means the continuous reaction: chlorite + kyanite → staurolite, that is, reaction (5.5). After A. B. Thompson (1976).

St/Chl Ky, indicates the MgO/FeO ratios of coexisting chlorite and staurolite in reaction (5.5). With increasing temperature, both of them become higher in MgO/FeO. At MgO/FeO = 0, the number of components, and so the variance, decrease by 1, and the reaction takes place at a specific temperature. At temperatures below this loop, the rock does not contain staurolite.

The other loop in the lower half, denoted as St Bt/Chl Ms, indicates the MgO/FeO ratios of coexisting chlorite, staurolite and biotite in reaction (5.6), all of which also increase with increasing temperature.

5.4.4.4 If solid-solution minerals show both Fe–Mg and Tschermak substitution, each paragenesis triangle corresponds to two continuous reactions. If we assume that ferromagnesian minerals and muscovite show both Fe–Mg and Tschermak substitution, each paragenesis triangle corresponds to two continuous reactions (net-transfer reactions). The compositions of all solid-solution minerals are functions of temperature, pressure and chemical potential of H_2O. For example, in the triangle biotite–chlorite–staurolite in Figure 5.9a, both reactions (5.4) and (5.6) hold. For detailed discussions, see Miyashiro & Shido (1985).

5.4.5 Four-phase assemblages in the AFM *diagram and the corresponding discontinuous reactions*

In the absence of an aqueous fluid, the assemblage of four phases on the *AFM* diagram has a variance of 2, and so is stable only at a specific temperature under a given pressure and chemical potential of H_2O. Hence, it represents a discontinuous reaction. In the presence of a pure H_2O fluid, the variance of such a system is 1, and so is stable only at a specific temperature under a given pressure. Thus, the system is univariant, and the reaction is discontinuous.

Discontinuous reactions may be classified into tieline switching reactions and terminal reactions. A **tieline-switching reaction** causes a discontinuous change in the topology of the *AFM* diagram by topological change of tielines connecting coexisting minerals, whereas a **terminal reaction** causes the appearance of a new phase or the disappearance of an existing phase in the *AFM* diagram.

5.4.5.1 Tieline-switching reactions. The following is a tieline-switching reaction, which takes place at the kyanite isograd in many metapelite regions.

$$\text{staurolite + chlorite + muscovite + quartz = kyanite + biotite + } H_2O \qquad (5.11)$$

This equilibrium holds only at a specific temperature under a given pressure and chemical potential of H_2O. At equilibrium, all minerals in the assemblage have definite compositions (including definite MgO/FeO ratios of ferromagnesian minerals). The equilibrium state is illustrated in Figure 5.9b.

At temperatures lower than the reaction temperature of reaction (5.11), the two paragenesis triangles – kyanite–chlorite–staurolite and chlorite–staurolite–biotite – are separated by a two-phase field staurolite–chlorite, as in Figure 5.9a. The width of this field becomes narrower with increasing temperature, and it vanishes when the reaction temperature of reaction (5.11) is reached. The tieline switches from staurolite–chlorite to kyanite–biotite at this temperature (Fig. 5.9b). With further increase of temperature, two new paragenesis triangles form: staurolite–biotite–kyanite and chlorite–biotite–kyanite (Fig. 5.9c). The corresponding continuous reactions are:

$$\text{6 staurolite + 4 muscovite + 7 quartz = 31 kyanite + 4 biotite + 3 } H_2O \qquad (5.12)$$

$$\text{3 chlorite + 7 muscovite + quartz = 13 kyanite + 7 biotite + 18 } H_2O \qquad (5.13)$$

With the progress of the reactions, the MgO/FeO ratios of ferromagnesian minerals of the former triangle decrease, whereas those of the latter increase. Hence, the two-phase field kyanite–biotite separating the two paragenesis triangles becomes wider with increasing temperature.

In Figure 5.10, the equilibrium state for equation (5.11) is shown on the horizontal line in the middle of the diagram. On the low-temperature side of it, the MgO/FeO ratios of staurolites in the two assemblages differ but approach each other with increasing temperature and finally coincide at the tieline-switching temperature. With further increase of temperature, two new loops form corresponding to the new continuous reactions (5.12) and (5.13).

Reaction (5.11) includes seven terms. Taking the coefficient of any one term as 1, the remaining six coefficients can be determined from six balancing equations with respect to SiO_2, Al_2O_3, FeO, MgO, K_2O and H_2O. Thus, in this case FeO and MgO must be treated as different components. Therefore, to make the balancing calculation, we must know the MgO/FeO ratios of coexisting staurolite, chlorite and biotite in the assemblage by chemical analysis. For this reason the coefficients are not shown in reaction (5.11). Generally, such coefficients are not integers.

5.4.5.2 Terminal reactions. A tieline-switching reaction is represented by an equation having two *AFM* phases on each side. On the other hand, a terminal reaction is represented by an equation that has one *AFM* phase on the one side and three *AFM* phases on the other. Two examples of terminal reactions are shown below:

$$\text{chlorite} + \text{muscovite} + \text{quartz} = \text{andalusite} + \text{biotite} + \text{cordierite} + H_2O \qquad (5.14)$$

$$\text{staurolite} + \text{muscovite} + \text{quartz} = \text{sillimanite} + \text{biotite} + \text{almandine} + H_2O \qquad (5.15)$$

By reactions (5.14) and (5.15), with increasing temperature, chlorite and staurolite, respectively, finally disappear from the *AFM* diagram, and instead assemblages of three *AFM* minerals are produced (e.g. Carmichael 1970, Burnell & Rutherford 1984). The chlorite and staurolite that disappear must, therefore, be located in the *AFM* diagram within the triangle of the three *AFM* phases produced (Fig. 5.11a,b). Coefficients are not written in these equations for the same reason as in reaction (5.11).

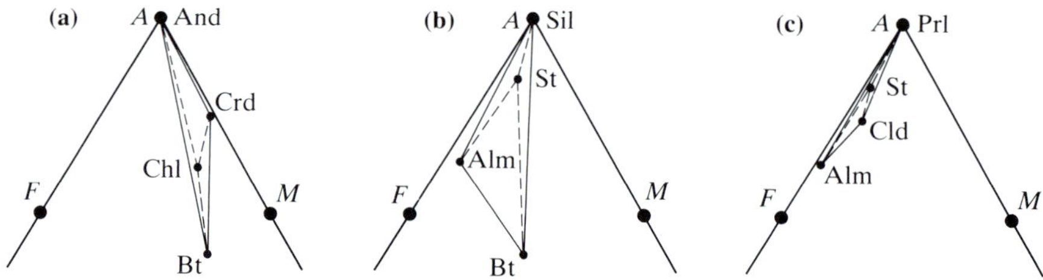

Figure 5.11 Parts of *AFM* diagrams showing the terminal reactions representing the final disappearance of (a) chlorite and (b) staurolite, and (c) the first appearance of staurolite in the *AFM* diagram. Based on A. B. Thompson (1976) and Zen (1981).

The progressive changes in paragenetic relations near the discontinuous reaction (5.15) are represented diagrammatically in Figure 5.12. At temperatures lower than the discontinuous reaction point, there are three paragenesis triangles, each of which has staurolite at an apex and which are separated by three two-phase fields (Fig. 5.12a). Each of the triangles corresponds to a continuous reaction. The MgO/FeO ratios of staurolites in the three triangles approach one another and coincide at the terminal reaction temperature, at which the mineral

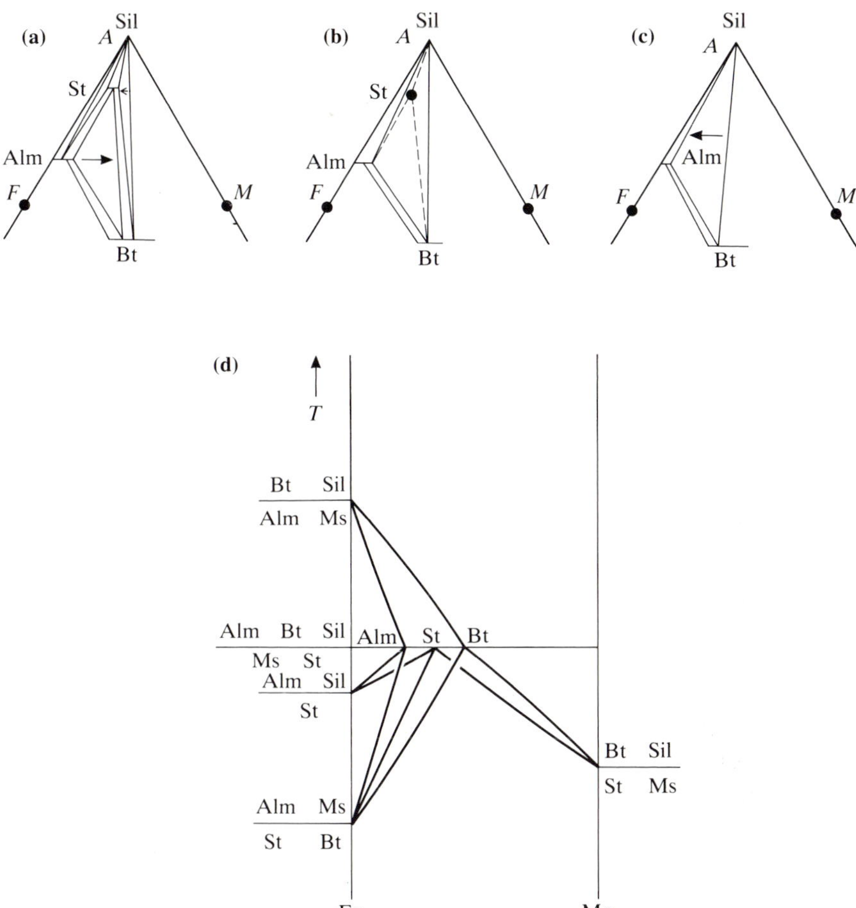

Figure 5.12 Diagrams (a), (b) and (c) indicate the changing paragenetic relations with increasing temperature near the temperature of the terminal reaction that causes the disappearance of staurolite (reaction 5.15). Diagram (b) and the horizontal line in the middle of (d) correspond to the terminal reaction. Arrows indicate the direction of movement of paragenesis triangles with increasing temperature. Based on A. B. Thompson (1976).

disappears (Fig. 5.12b). With further increase of temperature, the sillimanite–almandine–biotite triangle remains with its corresponding reaction continuing (Fig. 5.12c). The changes in the MgO/FeO ratios of ferromagnesian minerals are shown in Figure 5.12d.

Zen (1981) proposed that the following terminal reaction could cause the appearance of staurolite in the *AFM* diagram (Fig. 5.11c):

$$\text{chloritoid} + \text{almandine} + \text{pyrophyllite} = \text{staurolite} + \text{quartz} + H_2O \qquad (5.16)$$

It would be extremely rare for the staurolite-in isograd of a progressive metamorphic region actually to represent reaction (5.16), because this reaction occurs only in unusually FeO-rich bulk chemical compositions. In ordinary metamorphic regions, staurolite begins to occur in metapelites of more common compositions at a temperature higher than that of reaction (5.16) by a tieline-switching reaction such as that of Figure 10.1c.

Note that some minerals appear in the *AFM* diagram by discontinuous reactions within subsystems. For example, kyanite can appear by the discontinuous reaction:

$$\text{pyrophyllite} = \text{kyanite} + 3\ \text{quartz} + H_2O$$

in the Al_2O_3–SiO_2–H_2O system, and sillimanite usually appears by the discontinuous reaction: kyanite = sillimanite in the Al_2SiO_5 system.

5.4.6 Isograds based on continuous and discontinuous reactions

Most common metapelites plot in the horizontal zone lying between the chlorite and the biotite fields in the *AFM* diagram, and also over a limited range of MgO/FeO ratios. We assume that the pelitic rocks of a metamorphic region plot in the composition field P in Figure 5.9c, and thus the rocks contain staurolite and biotite. The staurolite–biotite side of the kyanite–staurolite–biotite triangle moves to the left with increasing temperature, and therefore kyanite begins to form at the temperature in which the staurolite–biotite side touches the composition field P. This temperature is that of the kyanite isograd in this region. Therefore, the temperature of the kyanite isograd based on continuous reaction of this type depends on the position of the composition field P.

Chinner (1965) found that the classical kyanite isograd of Barrow (1893) is based on the above continuous reaction, and the MgO/(FeO + MgO) ratio of the biotite coexisting with kyanite and staurolite at the isograd in Barrow's original area is about 0.50. By using the garnet–biotite geothermometer, Baker (1985) found that the biotite MgO/(FeO + MgO) value of 0.50 corresponds to about 630°C. If the composition range of metapelites exposed in the area were lower in MgO/(FeO + MgO) ratio, kyanite would begin to appear at a higher temperature.

On the other hand, the kyanite isograd based on the discontinuous reaction

(5.11) is independent of the bulk chemical composition of the rocks exposed in the area, and can be determined by observation of any rocks plotting in the composition field surrounded by the four relevant AFM minerals. In a part of the Grenville province, Carmichael (1970) drew a kyanite isograd based on this discontinuous reaction (5.11).

J. B. Thompson (1957) considered discontinuous reactions to be more suitable than continuous reactions as a basis for isograds, because the temperature of the former are independent of the bulk chemical compositions of the rocks exposed in the metamorphic regions.

5.4.7 *Effects of solid solution of muscovite*

5.4.7.1 K–Na substitution in muscovite in high-temperature metapelites. In the AFM diagram, muscovite and K-feldspar are treated as their K end-members. This muscovite with the idealized composition decomposes at high temperature by the reaction:

muscovite + quartz = K-feldspar + sillimanite + H_2O

which is a univariant reaction in the presence of pure H_2O fluid. So, at temperatures below this reaction point, we can use the AFM diagram, whereas at higher temperatures we can use a modified AFM diagram in which metapelite minerals are projected from the K-feldspar point instead of the muscovite point (§5.4.1).

In reality, however, muscovite and K-feldspar in metapelites are K–Na solid solutions. The decomposition of muscovite solid solution begins to occur at a temperature lower than the above reaction point, and continues over a wide temperature range, until finally muscovite comes to have the K end-member composition, and disappears by the above reaction. In this temperature range, the four minerals muscovite, K-feldspar, sillimanite and quartz can coexist. Moreover, since metapelites usually contain some CaO, the five minerals muscovite, K-feldspar, sillimanite, quartz, and plagioclase (with some CaO) could coexist over a wide temperature range (e.g. Evans & Guidotti 1966). These relations cannot be shown in the AFM diagram.

For such problems, J. B. Thompson & A. B. Thompson (1976) proposed the use of the Al_2O_3–$NaAlO_2$–$KAlO_2$ diagram, which can deal with the K–Na substitution in the white micas and alkali feldspars (§5.6).

5.4.7.2 Tschermak substitution in muscovite and its relation to the AFM diagram. In the 1950s when the use of the AFM diagram was proposed, muscovite in metapelites was generally considered to be a K–Na solid solution, and so naturally it was simplified to its K end-member. However, Lambert (1959) showed that muscovites in low-temperature metamorphic rocks are commonly phengites, which are derived from the simplified muscovite of the *AFM* diagram, here called Al-muscovite $KAl_3Si_3O_{10}(OH)_2$, by application of Tschermak substitution AlAl→(Fe + Mg)Si. Ernst (1963b) then showed that the extent of the above substitution in muscovite decreases with increasing metamorphic temperature, and that this is related to the formation of biotite.

This has raised two problems in the use of *AFM* diagrams. First, although muscovite is the projection point of the *AFM* diagram, its composition is highly variable. Secondly, the formation of biotite cannot be analyzed by use of the *AFM* diagram, because it is essentially related to the compositional variation of muscovite.

Apart from the K–Na substitution, muscovite, biotite and chlorite show a wide range of Tschermak as well as Fe–Mg substitutions. In this respect, each of the three minerals may approximately be regarded as being composed of three components. In the *AFMK* tetrahedron, their possible chemical variations occur on three planes which are identical or parallel to the *AFM* plane, as illustrated in Figure 5.13.

In metapelites, a muscovite with variable composition coexists with an assemblage of biotite and minerals which lie virtually on the *AFM* plane. The composition of muscovite is controlled by the chemical potentials of Al_2O_3, FeO and MgO, which in turn are constrained by the mineral assemblage of the rock.

If the rock has three *AFM* phases, the state of the system is definite at a definite pressure, temperature, and chemical potential of H_2O, and so the chemical potentials of Al_2O_3, FeO and MgO have specific values. In other words, under a specific pressure, temperature, and chemical potential of H_2O, the muscovite coexisting with three *AFM* phases is represented by a specific point in a triangular **muscovite composition diagram** (e.g. Fig. 5.14b). Under a specific pressure, temperature, and chemical potential of H_2O, a muscovite coexisting with two *AFM* phases has one compositional parameter which varies with the bulk-rock chemical composition, and so is represented by a line on the triangular muscovite composition diagram. A muscovite coexisting with only one *AFM* phase has two compositional parameters which vary with the bulk-rock chemical composition, and so is represented by an area in a muscovite composition diagram (J. B. Thompson 1979).

This relationship is illustrated in Figure 5.14, which shows how the mineral assemblages in an *AFM* diagram under a specific pressure, temperature, and chemical potential of H_2O correspond to the chemical variations of the muscovite coexisting with them. Variations in external conditions can change the paragenetic relations in the *AFM* diagram, and so change the composition diagram of muscovite.

If the *AFM* phases show higher MgO/FeO ratios, the system tends to have a

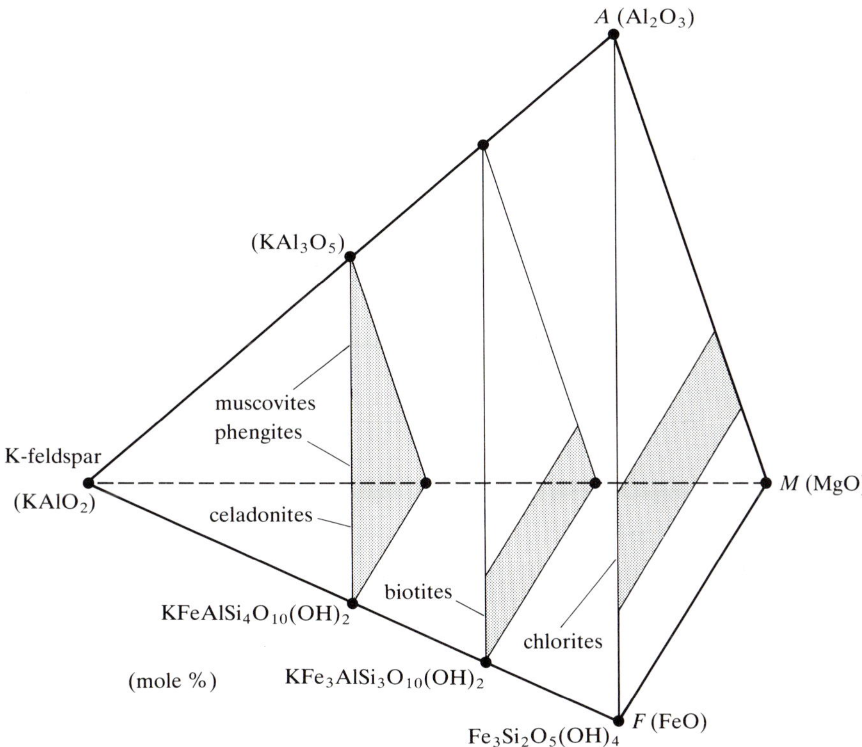

Figure 5.13 Micas and chlorites in the *A–F–M*–K-feldspar tetrahedron within the *AFMK* tetrahedron. The compositions of micas and chlorites lie within the stippled area in planes within or parallel to the *AFM* plane (J. B. Thompson 1979).

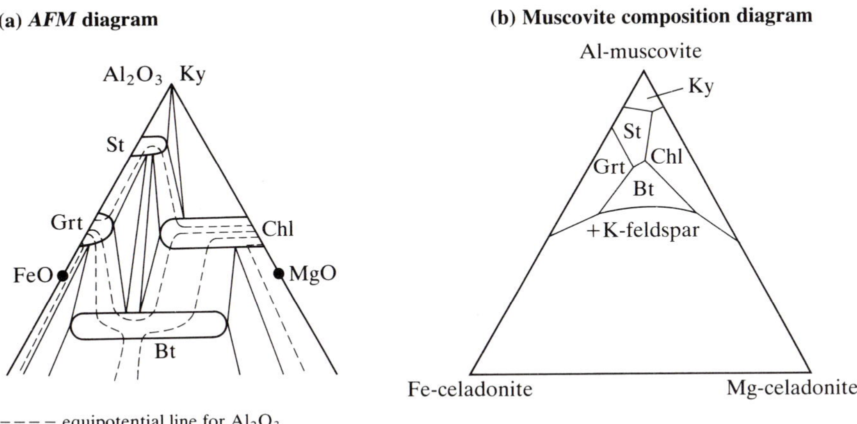

Figure 5.14 Schematic *AFM* diagram and corresponding muscovite composition diagram for the staurolite zone of the Barrovian sequence. The composition fields of most minerals in the *AFM* diagram are arbitrarily enlarged for convenience of illustration (Miyashiro & Shido 1985).

higher chemical potential of MgO, and so the coexisting muscovite should also tend to have a higher MgO/FeO ratio. In an analogous way, if the *AFM* phases show higher $Al_2O_3/(FeO + MgO)$ ratios, the system tends to have a higher chemical potential of Al_2O_3, and so the coexisting muscovite should also tend to have a higher $Al_2O_3/(FeO + MgO)$ ratio.

In the *AFM* diagram, each point corresponds to a specific value of the chemical potential of Al_2O_3. In any one-phase field, the chemical potential changes continuously with composition. In any two-phase field, two coexisting phases have the same value of the chemical potential. Any three-phase field has a specific value of the chemical potential. All such relations can be visualized by drawing Korzhinskii's (1959) equipotential lines on the *AFM* diagram. A corresponding quantitative muscovite composition diagram can be constructed from it (Miyashiro & Shido 1985).

In the *AFMK* tetrahedron of Figure 5.13, the mineral assemblages can be shown by connecting a muscovite composition with the compositions of coexisting biotite and virtually K_2O-free *AFM* phases. J. B. Thompson (1979) states that the biotite and virtually K_2O-free *AFM* phases and the tielines between them form a well defined surface within the tetrahedron, that is wholly visible from the idealized muscovite (Al-muscovite) point KAl_3O_5, and so this surface can be projected from the idealized muscovite point onto the *AFM* plane without overlap. In this meaning, an *AFM* diagram can be constructed even for low-temperature metapelites in which the muscovite is phengitic.

Although the *AFM* diagram can be constructed for low-temperature metapelites as well, it cannot express the relation of the Tschermak substitution in muscovite to the biotite-forming reaction in metapelites. Such relations and reactions can be treated to some extent by combination of an Eskola *A'KF* diagram with a muscovite composition diagram.

5.5 Applicability of the *AFM* diagram, and the effect of MnO and local variation of the chemical potential of H_2O

5.5.1 Introduction

Many students have used the *AFM* diagram in many metamorphic regions. It has been found that the diagram was generally useful for understanding mineral parageneses and progressive mineralogical changes (e.g. Albee 1965a, Guidotti 1970, Carmichael 1970). This was thought to mean that the underlying assumptions are usually approximately justified. The *AFM* diagram is based on two major assumptions: (a) that the mineralogical compositions of metapelitic rocks can be simplified as belonging to the six-component system SiO_2–Al_2O_3–FeO–MgO–K_2O–H_2O, and (b) that, in a group of metapelites under discussion, not only temperature and pressure but also the chemical potential of H_2O is externally controlled and kept at a constant value.

Some authors, however, found various types of discrepancies compared with their expectations from the diagrams. Particularly disturbing were crossing tie-lines and the common coexistence of four and even five *AFM* phases in the same rocks. In many cases the discrepancies appear to be due to the presence of minor components that are ignored in the diagram. Another important cause for discrepancies is local variation in the chemical potential of H_2O. These factors are discussed below.

5.5.2 *Effects of minor components*

The *AFM* diagram is based on the simplification that metapelites belong to the six-component system SiO_2–Al_2O_3–FeO–MgO–K_2O–H_2O. Actual metapelites usually contain variable quantities of CaO and MnO. Relatively high contents of these components in metapelites, particularly of MnO, tend to promote the formation of garnet. Thus, relatively MnO-rich rocks may begin to form garnet at a lower temperature than MnO-poor rocks (e.g. Miyashiro 1953), and may produce garnet in mineral assemblages that would not have garnet in the simplified six-component system. In other words, such garnet violates the paragenetic rules of the six-component system, thus causing, for example, widespread occurrence of four coexisting *AFM* phases over a range of metamorphic grades (e.g. Chinner 1965, Zen 1981). The MnO of the rock is very highly concentrated in garnet. Ilmenite is the only other common metapelite mineral that could contain a considerable amount of MnO, up to a few per cent by weight (e.g. Zen 1981). The role of CaO is more complicated, since it can be contained in some other minerals in metapelites such as calcite, epidote and plagioclase.

Albee (1968, 1972) discovered that staurolite concentrates ZnO. Hence, relatively high contents of ZnO will promote the formation of staurolite in metapelites. Some biotites and chlorites contain considerable amounts of TiO_2 and Fe_2O_3, respectively. These components must also have some effects on the stability of the minerals.

5.5.3 *Symmes & Ferry's (1992) calculation of the effects of MnO*

A higher MnO content in metapelites tends to decrease the temperature of the first appearance of garnet, and tends to produce garnet-bearing four-phase assemblages on the diagram, as stated above. Although these observations were widely known, it was still generally held until recently that the essential features of the observed phase relations in metapelites could be explained by the above six-component system, and that a relatively high content of MnO only modified the relations. However, recent thermodynamic calculations by Symmes & Ferry (1992) have made it clear that the effects of MnO are more crucial than was expected.

On the basis of Berman's (1988) internally consistent thermodynamic data set, Symmes & Ferry (1992) calculated *AFM* diagrams for typical MnO-free metapelitic compositions at a pressure of 5 kbar at 510°, 540° and 570°C. The calculated *AFM* diagrams show paragenetic relations that are very different from those known in natural metapelites. In the calculated *AFM* diagrams of the MnO-free system, garnet occurs only at relatively high temperatures in rocks with very high FeO/MgO ratios, whereas in natural metapelites, garnet is widespread in rocks of middle and high grades, not only with high but also with intermediate FeO/MgO ratios. In the calculated *AFM* diagrams of the MnO-free system, garnet does not coexist with chlorite or with kyanite, whereas in natural metapelites garnet commonly coexists with chlorite and with kyanite in the middle grades. These differences are so critical for understanding metapelitic mineral assemblages that the idealized representation of metapelitic rocks as belonging to the six-component system SiO_2–Al_2O_3–FeO–MgO–K_2O–H_2O is not justifiable.

Metapelites usually contain 0.05–0.54 wt % MnO. If the mole ratios of MnO/(FeO + MgO + MnO) are expressed by X_{MnO}, the "normal" range of X_{MnO} of metapelites is 0.01–0.04. Then, Symmes & Ferry (1992) calculated mineralogical compositions of metapelites with "normal" MnO contents. The results of these calculations show good agreement with observed mineral assemblages of natural metapelites. Garnet appears in metapelites of a very wide range of FeO/MgO ratios. The parageneses garnet + chlorite and garnet + kyanite appear. A small amount of MnO content is thus shown to be essential for the formation of observed mineral assemblages of natural metapelites.

Their calculations show that an increase in the X_{MnO} of pelitic rocks decreases the temperature of the first appearance of garnet in prograde metamorphism. With increasing temperature, garnet begins to occur at about 510°C for $X_{MnO} = 0.01$, and at about 430°C for $X_{MnO} = 0.04$ in a biotite-zone metapelite of average composition under a pressure ranging from 3 to 7 kbar.

Discontinuous reactions in the six-component system become continuous reactions with the addition of MnO to the system. However, these reactions proceed, and are completed, over a very narrow temperature range. Hence, Symmes & Ferry use the terms **gradual** and **abrupt** reactions for MnO-bearing systems in place of continuous and discontinuous reactions in the six-component systems.

Garnets in middle-grade metamorphic rocks are usually strongly zoned with an MnO-rich core. However, this zoning does not affect the temperature of the first appearance of garnet with increasing temperature.

5.5.4 Local variation of the chemical potential of H_2O and its effect on the mineral assemblages

The *AFM* diagram is based on the assumption of uniformity in not only temperature and pressure but also the chemical potential of H_2O in the metamorphic area under consideration. Even in the early days of the application of the *AFM* diagram, some authors questioned the uniformity of the chemical potential of H_2O, when they found discrepancies between the expectations from the diagram and observed mineral assemblages (e.g. Evans & Guidotti 1966, Guidotti 1970, 1974). More recently, reliable descriptions showing the existence of local variation in the chemical potential of H_2O in metapelites have been published (Rumble 1978, Dickenson 1988, Grambling 1990).

The equilibria of dehydration reactions are controlled by pressure, temperature, and the chemical potential of H_2O. So far in this chapter, we have considered the effect of increasing temperature on these equilibria under a constant pressure and chemical potential of H_2O. However, we can consider the effect of variation of chemical potential of H_2O in an analogous way. Decrease of the chemical potential of H_2O should cause the progress of dehydration, and so its effect will usually be analogous to that of increasing temperature. In Figure 5.9, for example, successive decreases in the chemical potential of H_2O cause a continuous reaction (a), and then a tieline-switching reaction (b), and further a continuous reaction (c). Dickenson (1988) has found that metapelites in different layers within a single outcrop show compositional variations of solid-solution minerals and variations of mineral assemblages, just as expected from a variation of the chemical potential of H_2O.

If H_2O is regarded as a fixed component, a composition–paragenesis diagram showing H_2O at one of the apices may be used (Rumble 1978, Dickenson 1988).

5.6 Thompson & Thompson's Al_2O_3–$NaAlO_2$–$KAlO_2$ diagram for metapelites

J. B. Thompson & A. B. Thompson (1976) have shown that paragenetic relations of Fe,Mg-free minerals in metapelites can be successfully analyzed by means of the Al_2O_3–$NaAlO_2$–$KAlO_2$ diagram (Fig. 5.15). This diagram is that part of the triangular diagram Al_2O_3–Na_2O–K_2O which satisfies the condition $Al_2O_3 > (Na_2O + K_2O)$. Since metapelites commonly contain both micas and alkali feldspars showing a K–Na substitution, the paragenetic relations among these minerals can be analyzed by use of this diagram, but not the *AFM* diagram. This diagram represents a partial system of metapelitic composition, which has five components: Al_2O_3, $NaAlO_2$, $KAlO_2$, SiO_2 and H_2O. We assume that the system is externally buffered with regard to H_2O. According to Korzhinskii's mineralogical phase rule (eq. 5.2), the maximum number of coexisting phases is four. We consider only quartz-bearing metapelites, and therefore do not show quartz in the diagram. So the diagram is divided into three-, two- and one-phase fields, just as the *AFM* diagram is. Any equilibrium assemblage of three minerals on the diagram has a variance of 3, and so is in a specific state under a specific pressure, temperature, and chemical potential of H_2O. Any paragenesis triangle corresponds to a continuous reaction involving K–Na solid-solution minerals. A four-phase assemblage on the diagram corresponds to a discontinuous reaction. Thus, we can use this diagram in an way analogous to the *AFM* diagram.

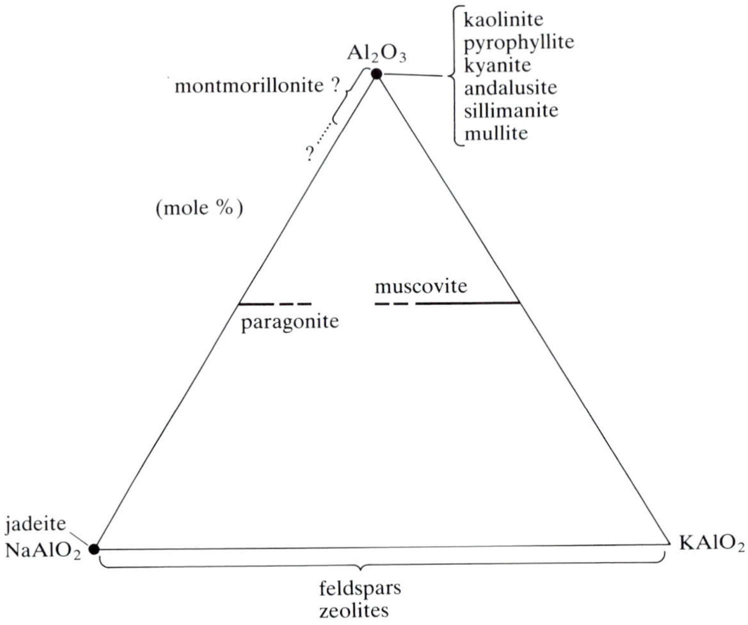

Figure 5.15 Positions of various minerals in the Al_2O_3–$NaAlO_2$–$KAlO_2$ diagram. After J. B. Thompson & A. B. Thompson (1976).

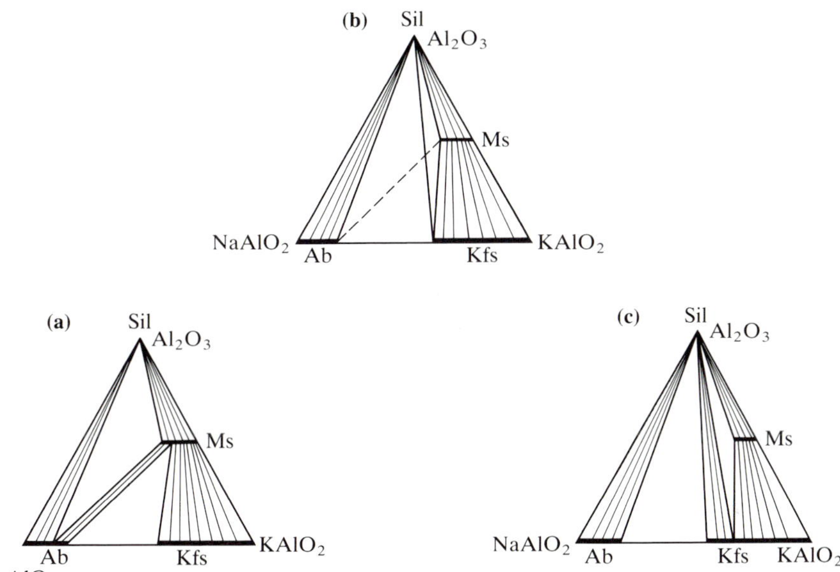

Figure 5.16 Schematic Al_2O_3–$NaAlO_2$–$KAlO_2$ diagrams showing a tieline-switching reaction: muscovite + albite + quartz = sillimanite + K-feldspar + H_2O with increasing temperature. See text.

In the construction of the diagrams it has been assumed that the K/Na ratio is always greater in mica than in the associated alkali feldspar. The Tschermak substitution in muscovite is ignored.

Figure 5.16 gives examples of Al_2O_3–$NaAlO_3$–$KAlO_3$ diagrams schematically showing the progressive changes of paragenetic relations of pelitic rocks in the transition from the sillimanite–muscovite to the K-feldspar–sillimanite zone (Figs 8.4, 8.5). In diagram (a) of Figure 5.16 (sillimanite–muscovite zone), there are two three-phase assemblages: muscovite + albite + sillimanite and muscovite + albite + K-feldspar, which is separated by a two-phase field: muscovite + albite. With increasing temperature, the two-phase field decreases in width and finally disappears by a tieline switching from muscovite + albite to sillimanite + K-feldspar (diagram b). Then, with further increase of temperature, the K-feldspar + sillimanite field becomes wider, as shown in diagram (c) (K-feldspar–sillimanite zone). Since actual pelitic rocks usually contain small amounts of CaO, they have CaO-bearing plagioclase in place of albite. So the four-phase assemblage of diagram (b) occurs in a zone of considerable width in progressive metamorphic regions.

Pelitic rocks usually contain some CaO, which may be contained in paragonite, plagioclase, jadeite and analcime. The high contents of CaO in these minerals may allow them to form assemblages that are not stable for CaO-free sodic minerals.

5.7 Composition–paragenesis diagrams for metabasites

Many authors used *ACF* diagrams to express the relationships between the chemical and mineralogical compositions of metabasites. However, the *ACF* diagram is based on a thermodynamically unjustifiable procedure, and cannot be an exact composition–paragenesis diagram (§5.3).

Metabasites are usually mainly composed of at least eight components: SiO_2, Al_2O_3, Fe_2O_3, FeO, MgO, CaO, Na_2O and H_2O. In addition, MnO, K_2O, TiO_2 and CO_2 play an important rôle in the formation of additional minor minerals. Thus, the number of components to be considered tends to be greater than that in metapelites. On the other hand, many metabasites have a relatively small number of phases, e.g. typical amphibolite which is composed essentially of only two minerals, hornblende and plagioclase. So, many mineral assemblages of metabasites show a relatively high variance. Reduction of the number of components necessary for graphical representation is difficult.

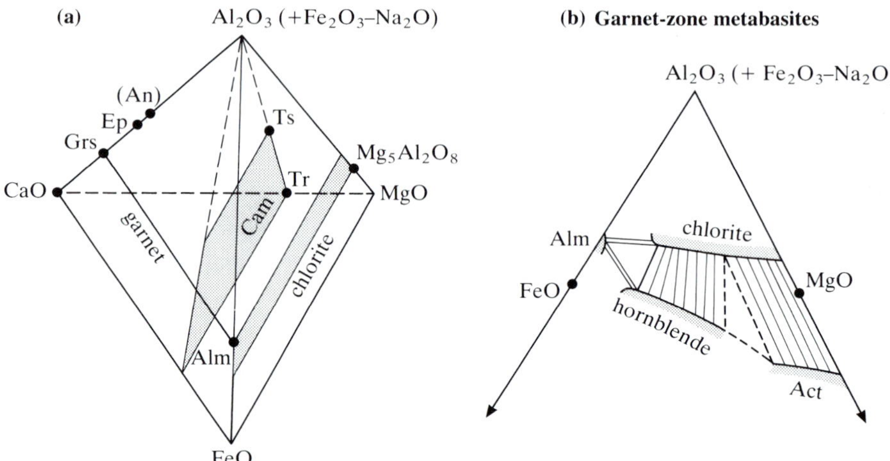

Figure 5.17 Harte & Graham's (1975) diagram. (a) $CaO–Al_2O_3–FeO–MgO$ tetrahedron, showing molar compositions of relevant metabasite minerals. The composition planes of Ca-amphiboles ("Cam", including tremolite, actinolite and hornblende) and chlorite are stippled. The subtraction of Na_2O from Al_2O_3 is for adjustment to albite just as in the *ACF* diagram. (b) Epidote projection of quartz-, plagioclase- and epidote-bearing metabasite assemblages in the garnet zone of the southern Alps of New Zealand.

However, many low-temperature metabasites contain a relatively large number of phases, so that effective reduction of the number of components needed for graphical representation is possible. Two examples of composition–paragenesis diagrams for such low-temperature metabasites is reviewed in the following pages.

5.7.1 Harte & Graham's (1975) diagram for low-temperature metabasites

Metabasites crystallized at low temperatures (in the greenschist facies and the transitional state between it and the amphibolite facies) commonly contain epidote, albite and quartz. In these rocks, Na_2O is almost entirely contained in the albite. SiO_2 is an excess component, and the chemical potential of H_2O may be assumed to be externally controlled. Therefore, Harte & Graham (1975) dismissed Na_2O, SiO_2 and H_2O from the diagrammatic representation, and treated metabasite approximately as composed of the four components Al_2O_3, FeO, MgO and CaO. They used a tetrahedron with the above four components at its apices (Fig. 5.17a).

If the Fe_2O_3 in epidote is combined with Al_2O_3, epidote plots on the Al_2O_3–CaO edge of the tetrahedron. Epidote (including all epidote-group minerals) is present in almost all of these low-grade metabasites. So Harte & Graham treated only epidote-bearing metabasites, and projected the paragenetic relations within this tetrahedron from the epidote point onto the extended Al_2O_3–FeO–MgO face. In this way, they obtained a triangular composition–paragenesis diagram Al_2O_3(+ Fe_2O_3)–FeO–MgO (Fig. 5.17b). This diagram can show the progressive changes in the paragenetic relations between chlorite, Ca-amphibole and garnet in metabasites at low temperatures.

In metabasites metamorphosed at higher temperatures (i.e. in the amphibolite facies), the plagioclase becomes more calcic than albite, and the Na_2O content of Ca-amphibole increases and may plays an essential rôle in the formation of the mineral. Hence, dismissing Na_2O may not be justified. Moreover, epidote usually disappears and so the projection from the epidote point becomes impossible. MnO must influence the occurrence and paragenetic relations of garnet.

5.7.2 Laird's (1980) diagrams for low-temperature metabasites

In metamorphism at low temperatures and a wide range of pressures (in the greenschist, low amphibolite and blueschist facies), metabasites commonly show the mineral assemblage: amphibole (actinolite, hornblende or glaucophane) + chlorite + epidote + plagioclase + quartz + Ti phase (sphene, ilmenite or rutile)

± Fe^{3+}oxide (magnetite or hematite) ± carbonate (calcite or dolomite) ± K mica (biotite or potassic white mica). Laird (1980) termed this mineral assemblage the "common assemblage", and discussed the changes of paragenetic relations of minerals in this assemblage with temperature and pressure. Some of her metabasite samples from Vermont contained stilpnomelane, garnet, paragonite or omphacite in addition to the common assemblage.

Twelve oxides are necessary to describe the "common assemblage": SiO_2, Al_2O_3, Fe_2O_3, TiO_2, FeO, MgO, MnO, CaO, Na_2O, K_2O, H_2O and CO_2. To reduce the number of components for graphical representation, the following procedure was adopted. SiO_2 and TiO_2 may be dismissed because quartz and one major Ti phase occur. K_2O and CO_2 may also be eliminated, because either they are minor components and no major K_2O- or CO_2-bearing mineral occurs, or they are major components and just one K-mica and carbonate occur in each sample. MnO is minor in all phases except in garnet cores, and may thus be ignored. The chemical potential of H_2O is assumed to be externally controlled, and the fugacity of O_2 is buffered by the Fe^{3+}oxide + quartz + Fe silicate assemblage. It follows that six oxide components: Al_2O_3, Fe_2O_3, FeO, MgO, CaO and Na_2O, must be considered in composition–paragenesis diagrams, which show coexisting amphibole, chlorite, epidote and plagioclase. The common assemblage has a variance of 4, and is univariant at a constant temperature, pressure, and chemical potential of H_2O. To show paragenetic relations of minerals, Laird used a tetrahedron with Na_2O, (Al_2O_3 + Fe_2O_3), CaO and (FeO + MgO + MnO) at the four apices. To plot analytical data, she used projections of minerals onto faces of this tetrahedron.

Most metabasites contain amphibole as a major constituent mineral. Amphibole is an exceptionally complicated solid-solution mineral, showing many types of atomic substitutions. As a result, such rocks have many possible reactions that involve the components of the amphibole. J. B. Thompson et al. (1982) discussed the reactions in low-temperature metabasites and described a procedure to give a set of independent reactions.

6 Buffering of intergranular fluid and infiltration of externally derived fluid

6.1 Internal and external buffering

The composition of intergranular fluid in a rock undergoing metamorphism is controlled by two factors: reactions with the minerals of the rock concerned and influence acting from outside the rock. Both of these factors operate together in most cases. We consider the first factor in §6.1.1, and the second factor in §6.1.2.

6.1.1 Internal buffering of fluid composition

In many cases of regional metamorphism, the composition of an intergranular fluid appears to be controlled by reaction with the mineral assemblage of the rock surrounding it. The control of the fluid composition and hence of the chemical potentials and fugacities of components in the fluid by reaction with the mineral assemblage of the host rock is called **internal buffering**, in contrast to external buffering which is discussed in the next section.

Internal buffering of an intergranular fluid takes place through chemical reactions that involve components of the fluid, changing its composition until a state of equilibrium is achieved. In other words, the chemical potentials and fugacities of components that participate in the reactions are controlled by the equilibrium conditions. When there are no reactions, there is no internal buffering. When buffering operates, the stringency, or the preciseness, of control of the fluid composition tends to decrease with increasing variance of the system. Only stringent types of buffering are buffering in the proper sense. Very stringent buffering occurs when the reacting systems have a variance of 1 or 2.

If a system of rock + intergranular fluid is closed, the fluid composition will usually be internally buffered. Even when part of the fluid that is being produced leaks out of the system, internal buffering is still possible. Since the mineral assemblage can vary from rock to rock, the composition of the internally buffered fluid can also vary from place to place in a metamorphic rock mass. Equilibria usually hold within the range of a few millimeters in middle-grade metamorphism and possibly of a few centimeters in high-grade metamorphism. Within these ranges, internally buffered fluid will have a virtually uniform composition.

The fact that mineral assemblages of metamorphic rocks vary from rock to rock, and incompatible assemblages occur commonly in adjacent rocks, indicates that the chemical potentials of SiO_2, Al_2O_3, FeO, MgO, and so on vary from rock to rock, being internally buffered by the mineral assemblages of individual rocks (e.g. Fig. 5.14a). Moreover, petrological studies over the past 30 years have shown that, in regional metamorphism, the intergranular fluid is usually internally buffered with regard to O_2, H_2 and S in rocks where reactions involving these components take place. On the other hand, the chemical potentials of H_2O and CO_2 may be buffered either internally or externally, according to circumstances.

The concept of internal buffering with respect to volatile components began with James & Howland's (1955) study of the internal buffering of O_2 fugacity in metamorphosed iron formations. The idea of buffering with respect to O_2 was supported and developed by J. B. Thompson (1957, 1972), Albee (1965a), and others. On the other hand, J. B. Thompson (1955, 1957, 1970) used an assumption of externally buffered chemical potentials for H_2O and CO_2 in his theory of mineral parageneses. Other authors recognized the effectiveness of internal buffering with regard to H_2O and CO_2 in order to explain deviations of mineral assemblages that they observed from Thompson's theory (e.g. Guidotti 1970, 1974, Ferry 1983a, Dickenson 1988). Greenwood (1975) and Ferry (e.g. 1983a), in particular, contributed to the formulation of a general theory of buffering with regard to H_2O and CO_2.

6.1.2 External buffering of fluid composition

There may be metamorphic areas where the chemical composition of the intergranular fluid is kept virtually constant by influences from outside the area, and hence outside the individual rocks of the area. Such a mechanism of external control of the fluid composition is called **external buffering**. It could take place, for example, by pervasive and active infiltration of an externally derived fluid of a constant composition into individual rocks.

Since internal buffering works through reactions between mineral assemblages and intergranular fluids, it does not work where there is no reaction. Hence, where there is no reaction, external buffering can work relatively effectively.

Even in rocks undergoing dehydration and decarbonation reactions, approximate external buffering may hold with regard to H_2O and CO_2 under certain circumstances.

If a rock is initially not in equilibrium with the externally buffered intergranular fluid, minerals in the rock may react with the fluid, and thereby the mineral assemblage and the compositions of solid-solution minerals may change until the rock reaches equilibrium with the fluid. In this way, the externally buffered components of intergranular fluid tend to reach a state of equilibrium with the minerals in the rock.

The idea of external buffering on a small scale combined with larger-scale variation of the chemical potentials of many rock-forming components was the basis of Ramberg's (1944b, 1945, 1948, 1949, 1952) grand hypothesis of crustal-scale dry granitization. Korzhinskii (1936, 1950, 1959, 1965) accepted the idea of external buffering for some components, and J. B. Thompson (1955, 1957, 1970) emphasized external buffering with regard to H_2O and CO_2 in metamorphism. These two authors formulated a rigorous theory of parageneses on this basis.

6.1.2.1 External buffering and composition–paragenesis diagrams. In a regional metamorphic complex, the temperature and pressure of metamorphism (i.e. the thermal peak) are virtually constant within a small area, say, a few hundred meters across. If the chemical potential and fugacity of a certain component are virtually uniform in all rocks through the area, we may think that external buffering is working through the area. The composition–paragenesis diagrams described in §5.3–5.7 show variations of mineral assemblages in individual rocks exposed in such a small area that crystallization occurs under specific pressure, temperature and externally buffered chemical potentials.

The externally buffered component can enter the rocks to cause reaction with the pre-existing minerals, or leave the rocks if released by some reaction within the rocks so as to maintain its chemical potential at a uniform value throughout the area. In other words, external buffering works only in *open systems*. Korzhinskii (1950, 1959) and J. B. Thompson (1955) termed an externally buffered component a "mobile" (or "perfectly mobile") component. This name, however, tends to cause a misunderstanding. Whether a component is "mobile" or not, is defined by the degree of achievement of uniformity in the chemical potential throughout an area. It therefore depends on the rate of migration of the component *relative to the amount* of the component needed to reach uniformity of the chemical potential by changing the mineralogical composition of rocks throughout the area.

6.2 Internal buffering with regard to O_2

6.2.1 Variation of O_2 fugacity from bed to bed in regional metamorphism

It has been well established that the fugacity of O_2 is usually internally buffered by the mineral assemblage of each rock, and that it varies from bed to bed as well as from formation to formation (James & Howland 1955, J. B. Thompson 1957, 1972, Eugster 1959, Chinner 1960). In many metamorphic regions, hematite-bearing and magnetite-bearing rocks are closely associated at the same metamorphic grade, and some rocks contain both hematite and magnetite. Some highly oxidized rocks show mineralogical compositions very different from that of associated ordinary (i.e. less highly oxidized) rocks.

Since the fugacity of H_2 is connected to the fugacity of O_2 by equation (3.21) under a given H_2O fugacity and P–T condition, the variation of the fugacity of O_2 from bed to bed means a variation of the fugacity of H_2 at the same distances. Hence, the fugacity of H_2 is also internally buffered.

The concentration of O_2 is very low (Fig. 3.6a), and so the gradient of concentration of O_2 must also be very small. Hence, flow of fluid and diffusion in fluid can transport only a very tiny amount of O_2, such that the oxidation or reduction caused by the transport is not enough to result in the disappearance of an existing mineral or the appearance of a new mineral in O_2 buffer assemblages. Therefore, rocks tend to preserve their original oxidation state during metamorphism. Fresh igneous materials are commonly in fairly strongly reduced states, whereas weathered or some hydrothermally altered igneous materials show strongly oxidized states. Such differences in the original oxidation state tend to be preserved during metamorphism and result in different mineral assemblages. The same argument applies to H_2.

An analogous situation holds for sulfur. The concentrations of S-bearing molecular species in intergranular fluids are usually very low (§6.5.3), and therefore the chemical potentials of these components in rocks are usually internally buffered during metamorphism.

6.2.2 Mechanism of progressive reduction of iron

As stated above, the long-range transport of oxygen and sulfur should not occur during metamorphism. However, there are other observations which do show progressive reduction of iron in metamorphism. In nearly unmetamorphosed or very low-grade metamorphic rocks, the occurrence of hematite is relatively common, whereas in garnet or higher-grade zones, magnetite is relatively common. It has been observed in some regional metamorphic complexes that the Fe_2O_3

contents and the Fe_2O_3/FeO ratios of the rocks tend to decrease progressively with increasing metamorphic grade in metapelites as well as in metabasites (e.g. Table 1.1; Miyashiro 1958). In an analogous way, the occurrence of pyrite is common in low-grade rocks, whereas the occurrence of pyrrhotite is common in high-grade rocks. What are the mechanisms for this reduction of iron?

In graphite-bearing metapelites, this reduction is probably caused by graphite or a related carbonaceous substance (Miyashiro 1964) through reactions such as:

$$C + 2\,Fe_2O_3 = 4\,FeO + CO_2 \tag{6.1}$$

The CO_2 on the right-hand side will diffuse and flow much more easily than O_2 because of its much higher concentration in fluids. The Fe_2O_3 in the above equation may be hematite, or may be ferric iron in other minerals. In the latter case, some other minerals are involved as in the following (J. B. Thompson 1972):

$$\text{graphite} + 2\,\text{magnetite} + \text{muscovite} + 3\,\text{quartz} = \text{biotite} + \text{almandine} + CO_2 \tag{6.2}$$

In an analogous way, graphite can cause reduction of iron in sulfide, for instance:

$$\begin{aligned} &3\,\text{graphite} + 6\,\text{pyrite} + 2\,\text{biotite} + 2\,\text{sillimanite} \\ &= 12\,\text{pyrrhotite} + 2\,\text{muscovite} + 2\,\text{quartz} + 3\,CO_2 \end{aligned} \tag{6.3}$$

CO_2 produced by such reactions may flow for a long distance.

In addition, in some cases the progressive mineralogical change from hematite to magnetite (as well as from pyrite to pyrrhotite) may take place by adjustment of associated minerals alone with no change in the bulk-rock Fe_2O_3 content or reduction of iron (J. B. Thompson 1972). The following are possible examples of such reactions.

$$3\,\text{hematite} + \text{biotite} + \text{sillimanite} = 3\,\text{magnetite} + \text{muscovite} + \text{quartz} \tag{6.4}$$

$$5\,\text{hematite} + \text{chlorite} = 5\,\text{magnetite} + 2\,\text{chloritoid} + 2\,\text{quartz} + 4\,H_2O \tag{6.5}$$

$$5\,\text{pyrite} + 4\,\text{chlorite} = 10\,\text{pyrrhotite} + 8\,\text{chloritoid} + 8\,\text{quartz} + 5\,\text{magnetite} + 16\,H_2O \tag{6.6}$$

6.3 Prograde reactions under external and internal buffering with regard to H_2O and CO_2

6.3.1 Internal buffering with regard to H_2O and CO_2

Internal buffering of fluids with regard to H_2O and CO_2 operates through equilibria of reactions involving these components. For such internal buffering to operate, the rock must contain both reactants and products of a dehydration–decarbonation reaction.

Thus, if a rock undergoing metamorphism contains both minerals *B* and *D* in any of the reactions (2)–(6) in Figure 3.5, the variance of the system is 2, and the intergranular fluid in equilibrium has a composition on the corresponding isobaric univariant curve in the diagram. At a given pressure and temperature, the CO_2/H_2O ratio of the fluid is specific. With increasing temperature, the reaction proceeds, producing and/or consuming H_2O and/or CO_2. This changes the composition of the intergranular fluid along the curve toward the composition of the fluid that is being produced, until either one of the reactant minerals is used up or the intergranular fluid composition reaches an isobaric invariant point. If one of the reactant minerals is used up, the variance of the system increases by 1 and so buffering discontinues. If the intergranular fluid composition reaches an isobaric invariant point, the variance decreases by 1, and so a more stringent type of buffering begins, as is illustrated later for the model system $MgO–H_2O–CO_2$.

The chemical potentials of H_2O and CO_2 appear to show a considerable variation from place to place even within one outcrop in a regional metamorphic complex. These local variations appear commonly to be caused mainly by variations of the CO_2/H_2O ratio in intergranular fluids.

6.3.2 *External and internal buffering with regard to H_2O and CO_2 in the system $MgO–H_2O–CO_2$*

As a simple example to illustrate buffering with regard to H_2O and CO_2, we consider prograde metamorphism of a magnesite rock that contains a small amount of $H_2O + CO_2$ fluid between mineral grains, after Greenwood (1975). Figure 6.1 shows the relevant phase relations at 2 kbar. We assume that the mole fraction of CO_2 in the initial fluid is very small, say 0.02 as indicated by point E in the diagram.

6.3.2.1 Prograde reactions under external buffering. It is assumed that the fluid is externally buffered and keeps its mole fraction of CO_2 equal to 0.02. Temperature increases without any reaction, until point G is reached, where the following reaction begins:

$$\underset{\text{magnesite}}{MgCO_3} + H_2O = \underset{\text{brucite}}{Mg(OH)_2} + CO_2 \tag{6.7}$$

The variance of this system is 2, and so the equilibrium holds on the isobaric univariant curve (6.7) in Figure 6.1. This reaction consumes H_2O and produces CO_2. To keep the composition of the fluid constant, the CO_2 produced by the reaction is removed from the rock, and an appropriate amount of H_2O migrates from the surroundings into the rock. The reaction continues until all the magnesite present is used up, producing a pure brucite rock. Thus, the variance becomes 3, and so the temperature begins to increase once more, until point N

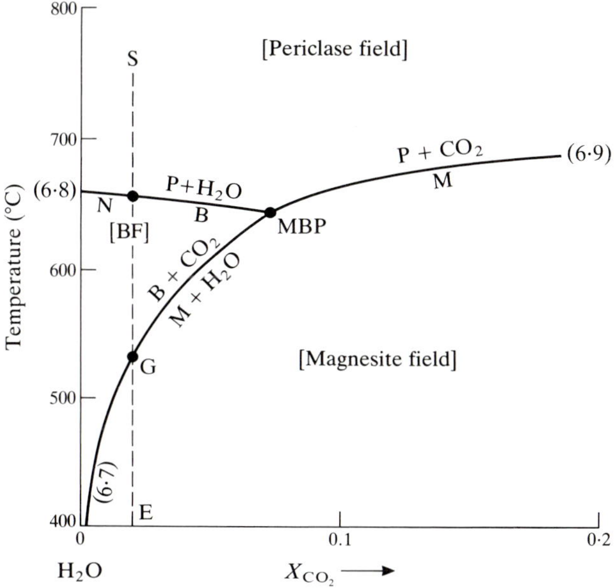

Figure 6.1 H_2O-rich part of the *T–X* diagram at 2 kbar for the system MgO–H_2O–CO_2. The abscissa (*X*) represents the mole fraction of CO_2 in H_2O–CO_2 mixtures. M: magnesite; B: brucite; P: periclase. Bf indicates the brucite field. The equilibrium curves (6.7), (6.8) and (6.9) refer to the reactions with these numbers in the text. The initial fluid composition is indicated by point E (mole fraction of CO_2 = 0.02). For G, N and S, see text. After Greenwood (1975).

is reached. At this point, the following reaction takes place.

$$\underset{\text{brucite}}{Mg(OH)_2} = \underset{\text{periclase}}{MgO} + H_2O \qquad (6.8)$$

After brucite has been used up by this reaction, the temperature begins to increase again towards point S.

In a progressive metamorphic region where the mole fraction of CO_2 in intergranular fluid is externally buffered, we can distinguish three zones made up of magnesite, brucite and periclase, in order of increasing temperature. These represent isobaric divariant assemblages. Isobaric univariant assemblages at G and at N occur on isograds separating the three zones.

6.3.2.2 Prograde reactions under internal buffering. Greenwood (1975) discussed the prograde *T–X–t* paths of internally buffered fluid in the MgO–H_2O–CO_2 system. In this case, the consumption of H_2O and addition of CO_2 at point G by reaction (6.7) causes an increase of the mole fraction of CO_2 in the fluid. The fluid composition moves along curve (6.7), until it reaches the isobaric invariant point MBP. How much brucite is produced in the rock between points G and MBP depends on the amount of intergranular fluid present, which in turn depends on the porosity of the rock. If the porosity is very low, as is the case with ordinary regional metamorphic rocks, the amount of brucite produced is very small, and may be overlooked under the microscope. In this case, the fact that buffering occurs will also be overlooked.

At point MBP, the curve (6.7) intersects two other equilibrium curves, as shown in Figure 6.1; one is curve (6.8) and the other is the equilibrium curve for the following reaction:

$$\overset{\text{magnesite}}{MgCO_3} = \overset{\text{periclase}}{MgO} + CO_2 \tag{6.9}$$

At the invariant point MBP, four phases: magnesite + brucite + periclase + fluid coexist. The three curves radiating from the point represent three-phase assemblages, each of which is derived by removal of one solid phase out of the four.

The reaction which takes place at this invariant point may be expressed by the following general equation with coefficients *a*, *b*, *c* and *m*:

$$a \text{ magnesite} + b \text{ brucite} + c \text{ periclase} + m\,H_2O + CO_2 = 0 \tag{6.10}$$

where m represents the mole ratio H_2O/CO_2 in this reaction. The value of m must be equal to the H_2O/CO_2 ratio at the invariant point, that is 12.4 (corresponding to a mole fraction of CO_2 of 0.0745 as shown in Fig. 6.1), so that the constancy of the mole fraction of CO_2 is maintained. By balancing the equation with respect to MgO, H_2O and CO_2, we can eliminate *a, b* and *c*, thus obtaining the following relation (with $m = 12.4$):

$$m(-\text{ brucite} + \text{periclase} + H_2O) + (-\text{ magnesite} + \text{periclase} + CO_2) = 0 \tag{6.11}$$

The reaction at the invariant point may be regarded as a linear combination of the two reactions (6.8) and (6.9). The system at the invariant point has three system components and five phase components (§3.1). So there are only two independent stoichiometric relations among the phase components. In the above discussion, reactions (6.8) and (6.9) were used as two such independent relations. We could use any other set of two out of the three reactions (6.7), (6.8) and (6.9). The reaction at this invariant point will continue until either magnesite or brucite is used up.

Which of magnesite or brucite is used up at the invariant point MBP depends on the amount of these two minerals present. In the course of prograde metamorphism, the amount of brucite produced is generally small. Moreover, the reaction at the invariant point consumes 12.4 times as much brucite as magnesite. Hence, brucite will usually be used up. Thereafter, the temperature will begin to rise again, and the reaction will proceed along curve (6.9), causing a progressive increase of the mole fraction of CO_2 in the fluid, until magnesite is used up.

Under some other situations, it may happen that magnesite is used up at the invariant point MBP. In this case, the fluid composition thereafter will change along curve (6.8) with progressive decrease of the mole fraction of CO_2.

The most important characteristics of the *T–X–t* paths of internally buffered fluids may be summarized as given below.

On an **isobaric univariant curve**, the reaction generally produces a fluid whose composition differs from the existing intergranular fluid. So, progress of the reaction changes the fluid composition towards an isobaric invariant point. When the amount of the intergranular fluid is very small, a very small amount of reaction causes a large change of fluid composition. In this case, the amount of mineral produced by reaction along the isobaric univariant line may be so small that it tends to be overlooked under the microscope. In a progressive metamorphic region, a buffered isobaric univariant assemblage of this kind occurs over a zone.

At an **isobaric invariant point**, the fluid that evolves has the same composition as the existing intergranular fluid. At such a point, an existing mineral disappears, and a considerable amount of new mineral may form abruptly, which may be easily recognized under the microscope, and may serve as an isograd.

Greenwood (1975) has given a quantitative expression for the compositional change of internally buffered fluid by various types of reactions that involve H_2O and/or CO_2 under the implicit assumption that the system of rock + intergranular fluid is closed. With progress of decarbonation and dehydration, the volume of the fluid+rock usually increases. Therefore, the closed system assumption is unrealistic in real metamorphism. Part of the fluid produced will usually leak out of the system so that the fluid/rock ratio is kept at a low value. A more realistic model, which permits leaking and infiltration of fluids, was discussed by Symmes & Ferry (1991), and is reviewed in §10.2.2.

Baker et al. (1991) made thermodynamic calculations on some internally buffered reactions.

6.4 Buffering with regard to H_2O and CO_2 in graphite-bearing pelitic metamorphic rocks

6.4.1 Homogeneous equilibria in intergranular fluids

In an intergranular fluid mainly or largely composed of H_2O and CO_2, the following reactions occur:

$$H_2O = H_2 + \tfrac{1}{2}O_2 \qquad (6.12)$$

$$CO_2 = CO + \tfrac{1}{2}O_2 \qquad (6.13)$$

$$CO_2 + 2\,H_2O = CH_4 + 2\,O_2 \qquad (6.14)$$

For each of these reactions, an equilibrium condition like the following holds: $\Delta G° = -RT \ln K$ (§3.3). Hence, the fluid always contains not only H_2O and CO_2 but also varied amounts of H_2, O_2, CO and CH_4.

If the fluid contains no elements other than C, H and O, and if the reactions under consideration do not involve any solid, it is a three-component system C–H–O with only one phase (so far as it is not unmixed into an H_2O-rich and a CO_2-rich fluid phase at low temperatures), and so the variance is 4. At a specific pressure and temperature, two variables related to the composition of the fluid must be specified in order to fix the state of the system.

If the fluid is reacting with solid phases, the equilibrium conditions for the reactions concerned provide additional constraints upon the fluid composition. One solid of this type that may control the fluid composition is graphite. In addition, solid phases participating in dehydration and decarbonation equilibria are also important in this regard.

6.4.2 *Graphite and related substances*

Unmetamorphosed pelitic sediments usually contain organic substances, which are gradually transformed into graphite through intermediate steps with increasing metamorphism. In the process, they gradually lose H, N and O, and increase their C content. Well crystallized graphite usually forms in the amphibolite facies and higher temperatures (Itaya 1981, Buseck & Huang 1985). A small amount of graphite or a related substance commonly occurs in metamorphosed limestones as well, and it is usually also regarded as of biogenic origin.

In some cases, graphite in rocks may be oxidized and may completely disappear, producing CO_2. In other cases, graphite is precipitated from fluid containing CO_2 and/or CH_4 (e.g. Rumble et al. 1989). The carbon isotope ratio may give a clue to the origin of graphite.

6.4.3 *Fluids in equilibrium with graphite*

6.4.3.1 Equilibria and the reducing effect. If we assume, for simplicity, that the carbonaceous substances are graphite, the following equilibria will be approached between the graphite and components in the fluid:

$$C + O_2 = CO_2 \tag{6.15}$$

$$C + \tfrac{1}{2}O_2 = CO \tag{6.16}$$

$$C + 2\,H_2 = CH_4 \tag{6.17}$$

For the first, for example, the equilibrium condition may be as follows:

$$\Delta G°(1\text{bar}, T) + (P-1)\Delta V_s° = -RT \ln K = -RT \ln f_2/f_1 \tag{6.18}$$

where f_1 and f_2 denote the fugacities of O_2 and CO_2, respectively. Thus, at a constant pressure and temperature, the fugacity of O_2 varies in proportion to that of CO_2.

On the other hand, in rocks undergoing metamorphism, the partial pressure of CO_2 cannot be much higher than the lithostatic pressure. So the presence of graphite imposes an upper limit on the fugacity of O_2 in crustal rocks. The upper limits of the fugacity of O_2 for a series of pressures are shown as curves labeled "stability limit of graphite" in Figure 6.2.

For the ordinary range of pressure in regional metamorphism, the upper limits of the fugacity of O_2 lie in the magnetite field of Figure 3.6a. This means that graphite has a reducing effect. Although the fugacity of CO_2 may vary from rock to rock, the amount of variation is usually relatively small. So the relation in equation (6.18) leads to the achievement of a relatively uniform distribution of O_2 fugacity in pelitic metamorphic rocks, and so to their relatively regular progressive mineralogical changes (§3.6.3).

The thermodynamic and petrological significance of graphite in metamorphic rocks was first discussed by Miyashiro (1964) and French (1966), based on an assumption of ideal behavior of the relevant gases. Ohmoto & Kerrick (1977) and Frost (1979) made more elaborate and extensive calculations, taking the non-ideal behavior of gases into consideration.

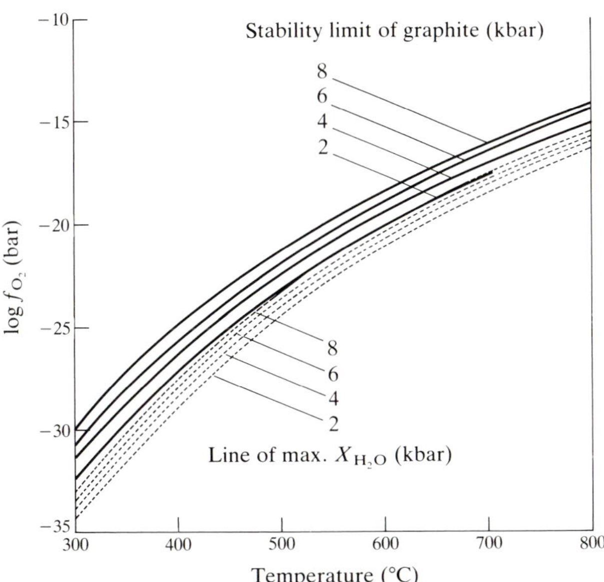

Figure 6.2 Solid lines show the stability limits of graphite, and dashed lines represent the lines on which the mole fraction of H_2O reaches the maximum value possible under the prevailing pressure and temperature for fluids of the C–H–O system, at 2, 4, 6 and 8 kbar (Ohmoto & Kerrick 1977).

6.4.3.2 Variance and buffering in graphite-bearing rocks. A fluid coexisting with graphite belongs to the three-component system C–H–O with two phases, and so the variance is 3. At a specific pressure and temperature, the state of the system is fixed by choosing only one variable related to the composition of the fluid. The system can be externally buffered with regard to H_2O or CO_2, but not to both.

In a rock of this kind with a variance of 3, the composition of an intergranular fluid is specific, if the chemical potential of H_2O is externally buffered at a given pressure and temperature, or if the O_2 fugacity is internally controlled at a given value at a given pressure and temperature. The latter case is discussed quantitatively later (Fig. 6.3).

If a reaction involving H_2O and/or CO_2 occurs, and if all relevant solid phases have fixed compositions, a certain relation holds among P, T, and mole fractions of H_2O and/or CO_2, as exemplified by equation (3.16). This decreases the variance of the system by 1. Therefore, if such a reaction occurs in a graphite-bearing rock with an intergranular fluid of the C–H–O system, the variance of the rock is 2. Such a reaction occurs at a specific temperature, if the rock is externally buffered with regard to H_2O (or CO_2) under a given pressure, or if the O_2 fugacity of the rock is internally controlled at a given value under a given pressure. The latter case is discussed later (Fig. 6.5).

6.5 Compositions of internally buffered fluids in pelitic metamorphic rocks

6.5.1 Calculated composition of fluids of the C–H–O system in rocks with or without graphite

When a fluid of the C–H–O system is present by itself (i.e. not reacting with solids of the surrounding rock) at a specific pressure and temperature, we must specify two variables related to the composition in order to fix the composition of the fluid, as stated above. As the two variables, Frost (1979) chose the fugacity of O_2 and the mole fraction of C relative to H_2 (denoted as X_c). X_c is 1.0 in pure CO_2 and in pure CO, and is 0.333 in pure CH_4. In a fluid made up of these species, $X_c = (X \text{ of } CO_2) + (X \text{ of } CO) + 0.333(X \text{ of } CH_4)$.

Some of the results of his calculation of equilibrium compositions of C–H–O fluids are shown in the four diagrams of Figure 6.3. In each diagram, the curve A–A′ shows the composition of the fluids in equilibrium with graphite at 1 kbar and 700 °C. At a constant pressure and temperature, this system is univariant. The compositions of the fluids that are not associated with graphite plot in the divariant field on the left-upper side of the curve A–A′. (The field on the right-lower side of the curve A–A′ represents the bulk composition of coexisting graphite and fluid.) Point A, which is situated at $X_C = 1$ on the curve A–A′, represents the highest possible O_2 fugacity in graphite-bearing rocks at the pressure and temperature concerned. In other words, O_2 fugacities higher than this are realized only in graphite-free rocks. At point A, the fluid is mainly composed of CO_2 accompanied by a very small amount of CO.

A comparison of the four diagrams of Figure 6.3 reveals that for the values of O_2 fugacity near point A or higher, the fluids are approximately binary mixtures of H_2O and CO_2, and the mole fractions of other components such as CH_4 and H_2 are negligibly small. As the value of the O_2 fugacity decreases, the mole fraction of CO_2 decreases, whereas those of CH_4 and H_2 increase, and eventually fluids are composed almost entirely of $H_2O + CH_4 + H_2$. This relation holds for both graphite-bearing and graphite-free rocks.

The curve indicating the fluid composition in graphite-bearing rocks such as A–A′ of Figure 6.3 moves to higher O_2 values with increasing pressure and temperature. Hence, the high-O_2 fugacity limit for the presence of graphite as represented by point A in Figure 6.3 becomes higher with increasing pressure and temperature, as shown in Figure 6.2. The O_2 fugacity values of the limiting curve lie within the stability field of magnetite of Figure 3.6a for the pressure range of 2–8 kbar in the temperature range of 400°–800°C. In this *P–T* range, hematite is stable only in graphite-free rocks.

On the curve A–A′ in Figure 6.3a, the mole fraction of H_2O in the fluid coexisting with graphite reaches the maximum value at the point marked with a star. At this point, the mole fraction of CO_2 is equal to that of CH_4. At values of O_2 fugacity higher than this point, the fluids are approximately binary mixtures of H_2O and CO_2, whereas at lower O_2 fugacities, fluids are mainly composed of H_2O and CH_4.

The point showing the maximum mole fraction of H_2O in graphite-bearing rocks, as indicated by a star in Figure 6.3a, moves upwards (to higher O_2 fugacity values) with increasing pressure and temperature, as shown in Figure 6.2. The O_2 fugacity value of this point also lies usually within the stability field of magnetite and is close to the line of QFM buffer at 400°–500°C shown in Figure 3.6a (Ohmoto & Kerrick 1977).

Ohmoto & Kerrick have shown that the maximum value of the mole fraction of H_2O in fluids coexisting with graphite, as realized at the point marked by the star in Figure 6.3a, decreases with decreasing pressure and increasing temperature, as shown in Figure 6.4. It is about 0.9 at the *P–T* condition of the triple point of andalusite, kyanite and sillimanite (500°C and 3.76 kbar), and is about 0.7 at 800°C at the same pressure. So the assumption of pure H_2O composition is not a good approximation in metamorphism at low pressure and high temperature.

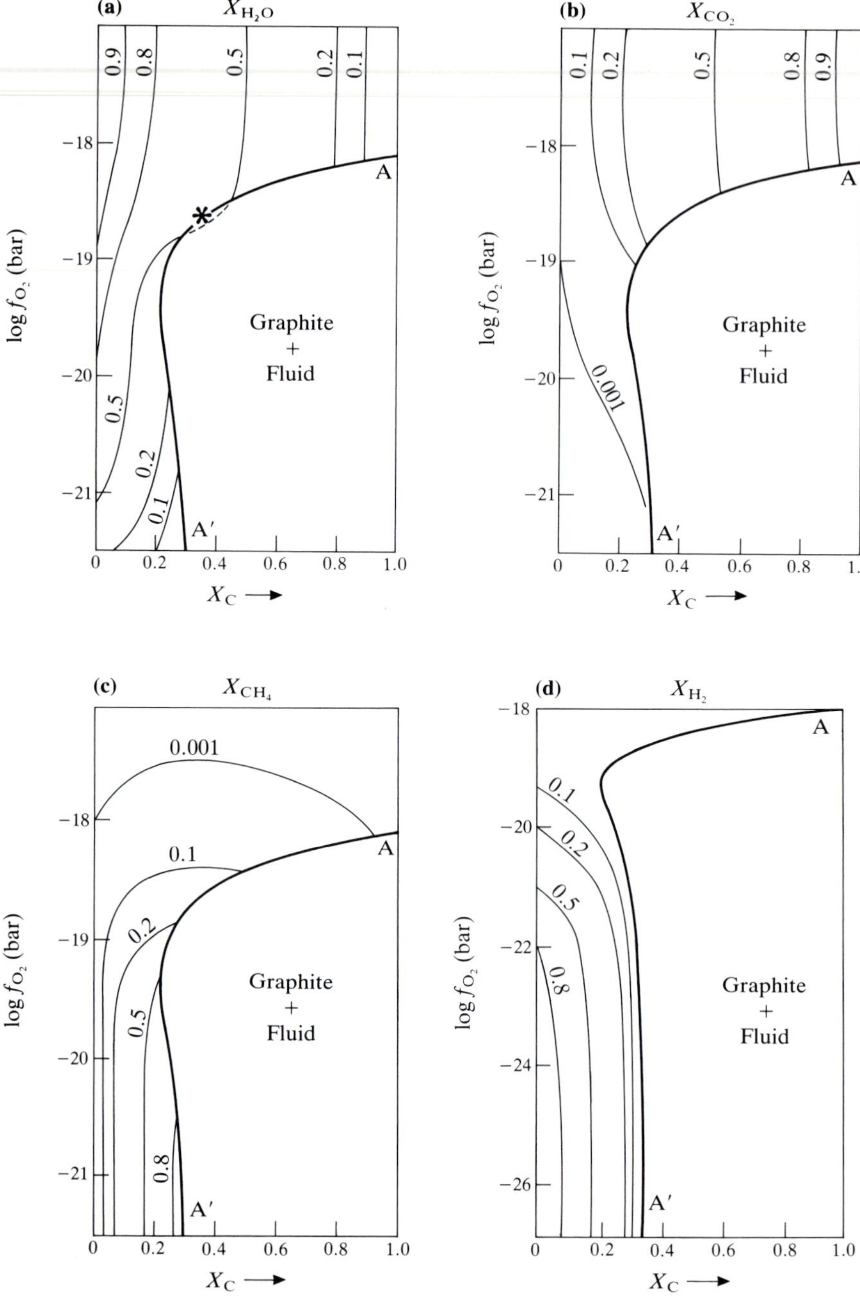

Figure 6.3 Mole fraction of (a) H_2O, (b) CO_2, (c) CH_4, and (d) H_2 in fluids of the C–H–O system at 1 kbar and 700°C (Frost 1979). In diagram (a), the star indicates that composition of the fluid equilibrated with graphite which shows the maximum value of mole fraction of H_2O possible at the prevailing pressure and temperature.

In the presence of graphite, the mole fractions of H_2 and CO exceed 0.1 only at pressures below 2 kbar and temperatures above 800°C.

6.5.2 *Dehydration and decarbonation reactions in graphite-bearing rocks*

6.5.2.1 Variance of the system. If a dehydration reaction is taking place in a graphite-bearing rock, and if the fluid is assumed to belong to the C–H–O system, the variance is 2. At a given pressure and temperature, the fluid has a specific composition. In a T-X or a T versus O_2 fugacity diagram at a fixed pressure, such an equilibrium holds on a curve. If two dehydration–decarbonation reactions are taking place simultaneously in a graphite-bearing rock, the variance is 1. So the equilibrium occurs at an isobaric invariant point (Ohmoto & Kerrick 1977).

If the solid phases involved in dehydration or decarbonation reactions are solid solutions, the activity of H_2O or CO_2, respectively, varies not only with pressure and temperature but also with the composition of the solid phases.

6.5.2.2 Equilibrium curves of dehydration and decarbonation reactions in graphite-bearing rocks. We consider a simple dehydration reaction: $B = D + H_2O$, where solid phases B and D have fixed compositions. The hydrous phase B will be stable, if at all, at relatively high values of H_2O fugacity, and hence at relatively high values of the mole fraction of H_2O at a given pressure and temperature. The mole fraction of H_2O reaches a maximum value at an O_2 fugacity slightly lower than the high-O_2 fugacity limit for the presence of graphite at a given pressure and temperature (Fig. 6.3a), and along a line on the lower side of the stability limit (high-O_2 fugacity limit line for the presence) of graphite in a temperature versus O_2 fugacity diagram like Figure 6.2. So hydrous mineral B must be stable, if at all, in a field along the line of maximum mole fraction of H_2O, and the stability field will usually terminate on the high-temperature side, as shown by curve 1 in Figure 6.5. Hydrous mineral B is stable within a parabola-like curve.

Now, let us consider a simple decarbonation reaction: B + calcite = D + CO_2, where solid phases B and D have fixed compositions. At a constant pressure and temperature, the mole fraction of CO_2 increases rapidly with increasing O_2 fugacity, approaching 1 at the stability limit of graphite in Figures 6.3b and 6.5. The calcite-bearing side is stable, if at all, in a field close to the stability limit of graphite. So, the equilibrium curve for the decarbonation reaction must have a positive slope and must terminate where it reaches the stability limit of graphite, as shown by curve 2 of Figure 6.5.

In a combined dehydration–decarbonation reaction: B + calcite $= D + H_2O +$

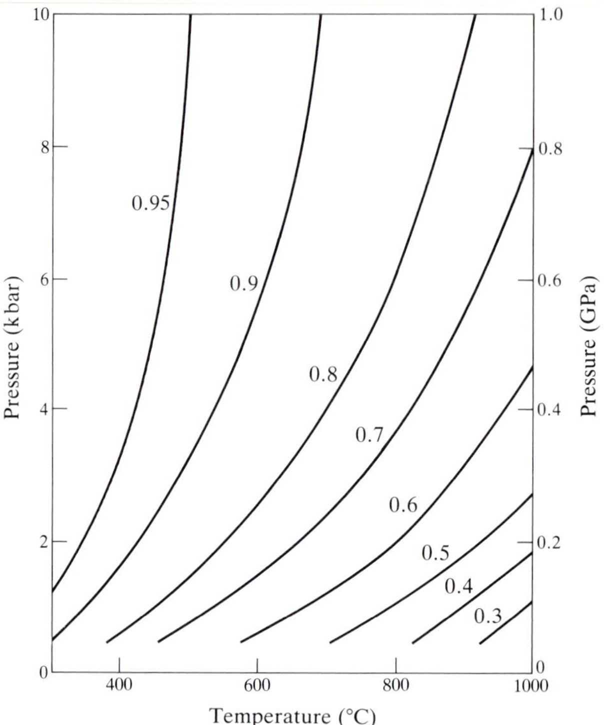

Figure 6.4 Maximum possible values of the mole fraction of H_2O in C–H–O fluids equilibrated with graphite as a function of pressure and temperature (Ohmoto & Kerrick 1977, see also Poulson & Ohmoto 1989: Fig. 4).

CO_2, the minerals *B* and *D* are assumed to have fixed compositions. The equilibrium curve of this type of reaction has a parabola-like shape, as shown by curve 3 in Figure 6.5, because the left-hand side is stable only under relatively high values of the mole fractions of both H_2O and CO_2. The axial line of the parabola-like curve is somewhat displaced upward from the line of the maximum mole fraction of H_2O.

Figure 6.5 shows, in addition, a melting curve in the presence of a fluid phase as curve 4. The melting is expressed by: $B + H_2O$ = melt. The melting temperature decreases with increasing mole fraction of H_2O, and so in this diagram the equilibrium curve has a parabola-like shape concave towards higher temperatures with an axial line coincident with the line of maximum mole fraction of H_2O.

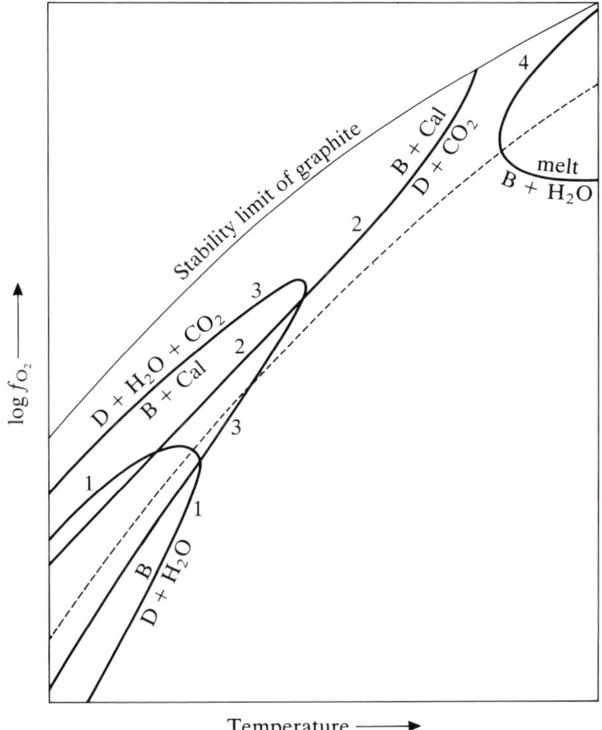

Figure 6.5 Schematic diagram showing equilibrium curves of dehydration and decarbonation reactions in graphite-bearing metamorphic rocks at a constant pressure (Ohmoto & Kerrick 1977). The dashed line indicates the line of maximum mole fraction of H_2O in the fluid. B and D represent minerals with fixed compositions.

6.5.2.3 Temperature–O_2 fugacity–time paths of fluids in the prograde metamorphism of graphite-bearing pelitic rocks. Figure 6.6 shows dehydration equilibrium curves on the temperature–O_2 fugacity diagram for three different K–Na–Al hydrous silicate assemblages in graphite-bearing rocks at 4 kbar fluid pressure. Ohmoto & Kerrick (1977) discussed paths of fluids on this type of diagram caused by increase in the temperature of metamorphism.

An intergranular fluid in a rock showing the assemblage graphite + pyrophyllite + kyanite + quartz has a composition represented by a point on the equilibrium curve (1) in Figure 6.6. With increase of temperature, the dehydration reaction (1) of Figure 6.6 takes place and the evolved H_2O increases the mole fraction of H_2O in the fluid. So, the composition of the fluid moves along the curve to higher temperatures and towards the line of maximum mole fraction of H_2O. If the rock contains much pyrophyllite, the amount of H_2O evolved by the

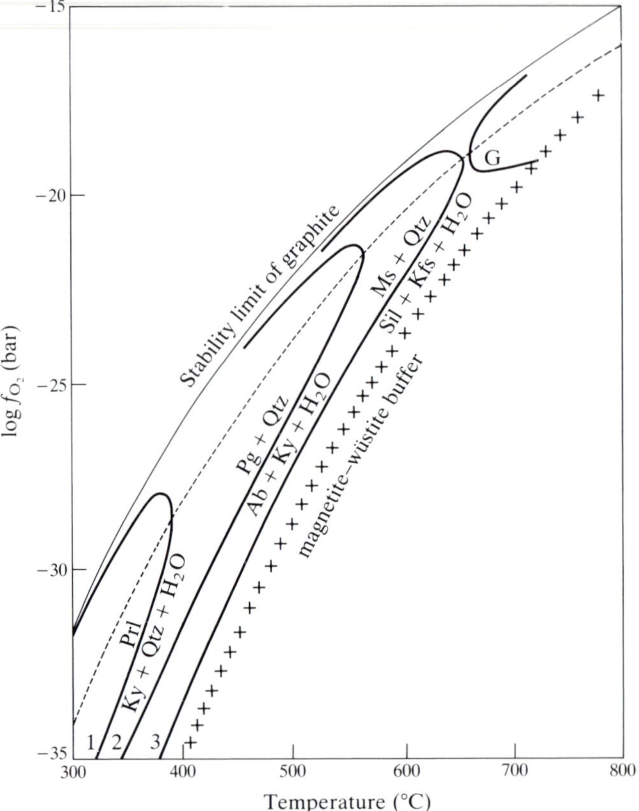

Figure 6.6 Equilibrium curves of the following dehydration reactions in graphite-bearing pelitic metamorphic rocks at 4 kbar fluid pressure (Ohmoto & Kerrick 1977). (1) pyrophyllite = kyanite + quartz + H_2O, (2) paragonite + quartz = albite + kyanite + H_2O, (3) muscovite + quartz = sillimanite + K-feldspar + H_2O. The dashed line indicates the line of maximum mole fraction of H_2O. The line G is the minimum melting curve of granite.

reaction is large relative to the amount of fluid that was initially present in the rock, and so the fluid composition reaches the line of maximum mole fraction of H_2O before all the pyrophyllite is used up. Then, the remaining pyrophyllite is used up by reaction (1) at this point. The H_2O evolved at this point is consumed by the reaction: $2C + 2H_2O \rightarrow CO_2 + CH_4$, so that the mole fraction of H_2O remains constant. Since the line of maximum mole fraction of H_2O agrees with the line where the mole fraction of CO_2 is equal to that of CH_4, as stated before, the above reaction does not change the fluid composition away from the line of maximum mole fraction of H_2O.

If the rock initially contains a relatively small amount of pyrophyllite, the mineral may be used up before the fluid composition reaches the line of maximum mole fraction of H_2O.

In either case, if pyrophyllite is used up, the fluid composition leaves the equilibrium curve of reaction (1), and goes into an isobaric divariant field on the high-temperature side of it with increase of temperature, until it reaches the next equilibrium curve for a dehydration reaction that can occur for the bulk chemical composition of the rock. In the isobaric divariant field between the two dehydration curves, no dehydration reaction occurs, and the mole fraction of H_2O continues to decrease with increasing temperature by the reaction: $2\,C + 2\,H_2O \rightarrow CO_2 + CH_4$. If the fluid composition had reached the line of maximum mole fraction of H_2O by the preceding reaction (1), it continues to advance along the line in the divariant field, because the above reaction produces CO_2 and CH_4 in equal amounts.

6.5.3 *Sulfides and sulfur-bearing molecular species in fluids*

Metamorphic rocks commonly contain small amounts of pyrrhotite and other sulfides, which react with the fluid to produce H_2S, S_2, and other sulfur-bearing molecular species, though the mole fractions of these are usually smaller than 0.01 in ordinary metamorphic conditions. The reactions are, for example,

$$\underset{\text{pyrrhotite}}{FeS} + \tfrac{1}{2}S_2 = \underset{\text{pyrite}}{FeS_2} \qquad (6.19)$$

$$\tfrac{1}{2}S_2 + O_2 = SO_2 \qquad (6.20)$$

$$\tfrac{1}{2}S_2 + H_2O = H_2S + \tfrac{1}{2}O_2 \qquad (6.21)$$

$$4\,S_2 = S_8 \qquad (6.22)$$

As illustrated in Figure 6.7, pyrrhotite is stable at an intermediate range of fugacity of S_2 between the stability fields of native iron and pyrite at the same pressure and temperature. Pyrrhotite $Fe_{1-x}S$ shows a variable degree of deficiency of Fe, with x ranging from 0 to about 0.12. It shows $x = 0$ when it coexists with native iron, and shows the largest value of x possible under the prevailing pressure and temperature when it coexists with pyrite.

Iron oxides commonly coexist with iron sulfides. Their stability relations with respect to the fugacities of O_2 and S_2 are also illustrated in Figure 6.7. Although the fugacity of S_2 may be used as an indicator of the sulfidization condition of the system, the mole fraction of S_2 is usually much smaller than those of H_2S and SO_2 (Eugster & Skippen 1967, Froese 1977).

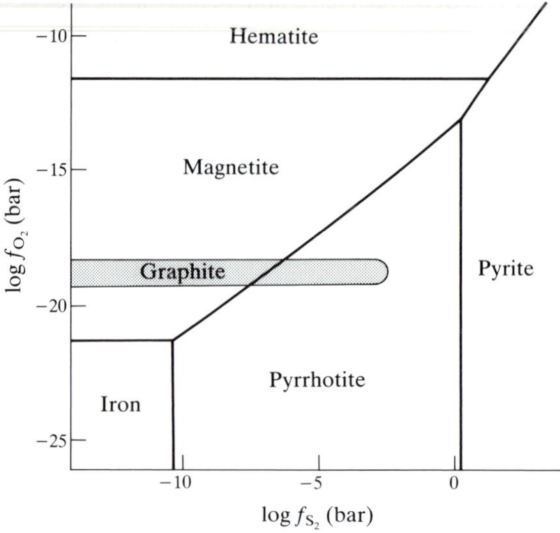

Figure 6.7 Stability fields of oxides and sulfides of iron with respect to the fugacities of O_2 and S_2 under a total pressure of 2 kbar and an H_2O pressure of 1 kbar at 700°C (Froese 1977; 1981: Fig. 18). The stability field of graphite is stippled.

6.6 Buffering with regard to H_2O and CO_2 in metabasic and calcareous rocks

6.6.1 Metabasites mixed with metasediments

Many pelitic metamorphic regions contain small masses of metabasites. If the protoliths of the metabasites were unaltered igneous rocks, the mineral changes in the low-grade stage must have required a supply of a large amount of H_2O. As an empirical fact we know that the necessary amount of H_2O is usually supplied to metabasites. This can be ascribed to the easy migration of a large amount of H_2O due to the presence of intergranular fluid mainly composed of the component. The intergranular fluid may be produced and maintained by the dehydration reactions in the surrounding pelitic rocks.

Metabasites, particularly in low-grade metamorphic zones, are susceptible to the CO_2/H_2O ratio of the intergranular fluid. Greenschists, for example, contain Ca-rich minerals such as actinolite and epidote, which may react with fluid with precipitation of calcite and/or dolomite (ankerite), if the concentration of CO_2 increases in the fluid. Because of such reactions, infiltration of CO_2-bearing fluid into metabasites at low temperature can easily be detected.

Graham et al. (1983) have found, in a greenschist-facies area of the Dalradian in the Scottish Highlands, that the interiors of some metamorphosed dolerite and gabbro sills show mineral assemblages containing Ca-amphibole (actinolite or hornblende) and epidote, but free of carbonate, whereas margins of the sills show mineral assemblages free of amphibole (and possibly also of epidote), but containing calcite and/or ankerite (§11.3.3). This suggests that a hydrous fluid containing some CO_2 infiltrated from the surrounding metasediments into the metabasite sill during metamorphism, and caused the mineral changes. The $^{13}C/^{12}C$ ratio of carbon in carbonate minerals has been found to be very low, suggesting that the CO_2 in infiltrating fluid might have been generated mainly by oxidation of biogenic organic matter or graphite in metasediments.

6.6.2 *Prograde metamorphism in carbonate rock bodies and in regions of mixed pelitic and carbonate rocks*

In the case of a very *large carbonate rock body*, fluids from the surrounding pelitic rocks may not be able to infiltrate the body. If reactions involving H_2O and/or CO_2 occur in the carbonate rock body, the fluids in it are internally buffered and they change their composition along the equilibrium curves of the reactions with increasing temperature. A beautiful example of a case of this type in a large dolomite mass intruded by a granodiorite in Montana was described by Rice (1977). Because of decarbonation reactions, intergranular fluids in the mass were high in CO_2 with a mole per cent of CO_2 equal to, or greater than, 75.

In prograde metamorphism of a heterogeneous complex composed of *interbedded pelitic and carbonate rocks*, diffusion in, and mixing of, fluids are very important. Since each bed has its own bulk-rock chemical and mineralogical compositions, fluids produced by prograde reactions may differ in CO_2/H_2O ratio from bed to bed. In most dehydration and decarbonation reactions, the total volume of solid phases decreases. This may make the flow, infiltration and mixing of fluids occur relatively easily. Nevertheless, heterogeneity of fluid composition between beds may remain.

In the case of relatively thin layers (say a few meters wide) of impure limestones occurring in a dominantly pelitic metamorphic region, the inner part of the limestone layers may develop internally buffered CO_2-rich fluids as long as a decarbonation reaction is in progress. On the other hand, the margins of the layers may show lower concentration of CO_2 because of infiltration of fluids or diffusion of H_2O from the surrounding pelitic rocks. An example of a case of this kind from Connecticut was described by Hewitt (1973). When a reactant is exhausted in a limestone layer, the reaction is terminated, and the production of

CO_2 stops. Then, mixing of, and diffusion in, fluids decrease the differences in CO_2/H_2O ratio between the limestone and the surrounding pelitic rocks, until a further increase of temperature begins a new reaction to produce CO_2 in the limestone.

Since the middle 1970s, John M. Ferry has published a series of ingenious investigations on the regional metamorphism of thin layers of argillaceous carbonate rocks, interbedded with pelitic and other rocks in the Augusta–Waterville area in southern Maine. Since the carbonate rocks are impure, and layers of carbonate rocks are usually only several centimeters thick, infiltration of externally derived H_2O-rich fluids into the carbonate rocks as well as into associated pelitic ones was so intensive and pervasive at all stages of prograde metamorphism that the intergranular fluids always kept H_2O-rich compositions, usually below 25 mole % CO_2. Since his results are important for understanding the effects of infiltration, they are reviewed in detail in the next section.

Carmichael (1970) described a typical case of *externally buffered progressive metamorphism* in the Whetstone Lake area, Canada. There, calcareous and other types of metamorphic rocks occur mixed, and the CO_2/H_2O ratio of intergranular fluid was locally uniform and varied gradually on a regional scale.

6.7 Ferry's research on the fluid flow and infiltration during regional metamorphism in Maine

Ferry has done monumental work in the Augusta–Waterville area in Maine, which has provided strong evidence for the existence of fluid flow in regional metamorphism and has clarified relationships between buffering, infiltration and fluid flow.

6.7.1 Zonal mapping

The Acadian (Devonian) regional metamorphic complex extending from Augusta to Waterville in Maine (Fig. 6.8) is made up of pelitic and psammitic rocks and quartzite interbedded with a subordinate amount of argillaceous carbonate rocks. Compositional layering is on a scale of 2–15 cm.

6.7.1.1 Pelitic rocks. In the pelitic rocks, the grade of metamorphism increases generally southward and towards associated syn-metamorphic granitoid bodies. With respect to the progressive mineralogical changes in pelitic rocks, the western half of the area has been divided by biotite, garnet, staurolite(–cordierite)–andalusite, and sillimanite isograds in order of increasing temperature, as shown in Figure 6.8. These isograds show relatively smooth shapes, lying generally in an east–west direction. The regional metamorphism occurred at a relatively low pressure, belonging to the andalusite–sillimanite series (§8.1). Ferry (1980, 1983a, 1986) has estimated the pressure at 3.5 kbar, and the temperatures of the biotite, garnet and sillimanite zones at 400 °C, 460 °C and 550 °C, respectively.

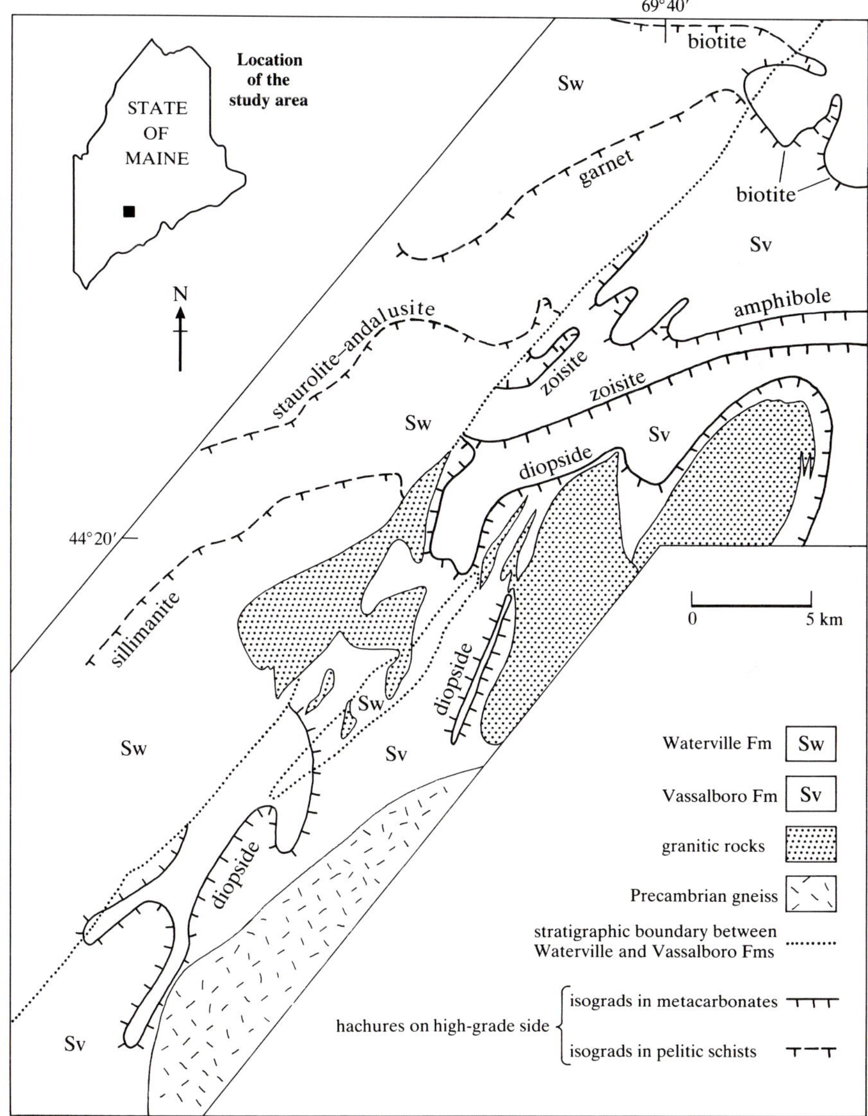

Figure 6.8 Progressive metamorphic zones for argillaceous carbonate rocks and for pelitic schists in the Augusta–Waterville area, southern Maine (Ferry 1983a).

The wide Acadian metamorphic region on the west side of this area is one of the most fully documented pelitic metamorphic regions in the world (§9.4.3). There, Guidotti, Holdaway and their co-workers distinguished two successive regional metamorphic events, named M_2 and M_3, both in the Devonian, and considered that the presently observed mineral assemblages were formed mainly by M_3, while cordierite and andalusite are relics formed by M_2 (Holdaway et al. 1982). Holdaway et al. (1988) recognized a further M_5 metamorphism in the early Carboniferous (§9.4.3). Ferry considers that his zones are concerned with products of $M_2 + M_3$ (with no distinction) or M_3.

6.7.1.2 Argillaceous carbonate rocks. In the progressive metamorphism of argillaceous carbonate rock layers in this area, the lowest-grade rocks are mainly composed of ankerite, quartz, albite and muscovite. Ankerite is used up by reaction with muscovite, etc., with resultant production of calcite. Then, calcite decreases by successive decarbonation reactions with increasing temperature, but nevertheless persists to the highest grade.

In the *Vassalboro Formation*, which is exposed in the eastern half of this area, Ferry (1983a,b) mapped isograds for the progressive metamorphism of argillaceous carbonate rocks, as shown in Figure 6.8. They are characterized by the first appearance of the following minerals in order of increasing temperature: (1) biotite, (2) amphibole, (3) zoisite, (4) diopside. The metamorphic area on the low-temperature side of the biotite isograd was called the ankerite zone. The mineral assemblages in these zones are summarized in Table 6.1. This sequence of progressive changes is caused by successive decarbonation reactions, which involve either dehydration or hydration, and consume or liberate quartz.

Initially, Ferry (1976a,b) had defined these and two more isograds as reaction isograds in the CaO–MgO–SiO_2–Al_2O_3–K_2O–H_2O–CO_2 system. So, for example, the biotite isograd was called the biotite–chlorite isograd at that time. Later Ferry (1983a,b) found that the actual reactions were much more complicated, and so redefined four of them as tentative isograds, and then assigned each of them an overall reaction.

The shapes of isograds in the argillaceous carbonate rocks are very irregular, forming isolated concentric patterns in some places. At first, Ferry (1976a) ascribed the irregularity to variations in the CO_2/H_2O ratio in intergranular fluids. Later, Ferry (1983a) came to consider it to be due to an irregular distribution of fluid flows or different bulk-rock compositions.

In the *Waterville Formation*, which is exposed in the western half of the area, carbonate rocks showing different mineral assemblages indicating different extents of progress of decarbonation, commonly occur in different parts within single outcrops (Ferry 1987).

Table 6.1 Progressive metamorphism of argillaceous carbonate rocks in the Vassalboro Formation in the Augusta–Waterville area, Main (Ferry 1983a, etc.).

Zone	Mineral assemblage
Ankerite	ankerite +Qtz + Ab + Ms±Cal±Chl (accessory: ± pyrite ± Ilm ± graphitic material
Biotite	*Lower-grade part:* Bt + ankerite + Qtz + Ab + Ms + Cal + Chl *Higher-grade part:* Bt + Qtz + oligoclase/labradorite + Cal + Chl ± Ms (no ankerite) (accessory: ± pyrite ± Ilm ± Spn ± graphitic material)
Amphibole	calcic amph. + Qtz + andesine/labradorite + Cal + Bt ± Chl (accessory: ± Ilm ± Spn ± graphitic material)
Zoisite	zoisite + calcic amph. + Qtz + andesine/anorthite + Cal ± Bt ± microcline (accessory: ± Ms ± scapolite ± Grt ± Spn ±graphite)
Diopside	diopside + zoisite + calcic amph . + Cal + Qtz + andesine/ anorthite ± Bt ± microcline (accessory: ± scapolite ± Grt ± Spn ± graphite)

Note: Rocks in any zone may contain small amounts of pyrrhotite, apatite and tourmaline, in addition to the above.

6.7.2 Composition of intergranular fluids in argillaceous carbonate rocks

Ferry (1976a, 1983a, 1987) calculated the compositions of intergranular fluids in the argillaceous carbonate rocks at the thermal peak from various observed mineral assemblages. The results have shown that the fluids were always approximately mixtures of H_2O and CO_2 alone with a low CO_2/H_2O ratio.

The metamorphosed carbonate rocks in the biotite and higher zones in the Vassalboro Formation contain sphene and not rutile, and so the following equilibrium gives the upper limit for the fugacity and mole fraction of CO_2:

$$\overset{\text{sphene}}{CaTiSiO_5} + CO_2 = \overset{\text{rutile}}{TiO_2} + \overset{\text{calcite}}{CaCO_3} + \overset{\text{quartz}}{SiO_2} \qquad (6.23)$$

The assemblage zoisite + plagioclase + calcite occurs in rocks of the zoisite and diopside zones in the Vassalboro Formation. The assemblage indicates equilibrium for the following stoichiometric relation among phase components:

$$2\,\overset{\text{zoisite}}{Ca_2Al_3Si_3O_{12}(OH)} + CO_2 = 3\,\overset{\text{plagioclase}}{CaAl_2Si_2O_8} + \overset{\text{calcite}}{CaCO_3} + H_2O \qquad (6.24)$$

Ferry (1983a) concluded that the fluids in argillaceous carbonate rocks were aqueous, with a mole fraction of CO_2 always smaller than 0.3, and commonly smaller than 0.15.

In argillaceous carbonate rocks in the Waterville Formation which is exposed to the west of the Vassalboro Formation, the calculated compositions of intergranular fluids are similarly hydrous, with a mole fraction of CO_2 in a range of 0.07–0.24 (Ferry 1987).

6.7.3 Infiltration of H_2O-rich fluids

6.7.3.1 Infiltration of aqueous fluids into argillaceous carbonate rocks as a driving force of prograde reactions. Ferry's investigation of prograde reactions in argillaceous carbonate rocks has shown that the progress of reactions produced CO_2-rich fluids with a mole fraction of CO_2 greater than 0.5. On the other hand, the compositions of intergranular fluids, calculated from the mineralogical compositions of rocks, were H_2O-rich with a mole fraction of CO_2 smaller than 0.3, as stated above. To resolve this paradox, Ferry (1976a,b, 1980, 1983a) assumed that there had been infiltration of the carbonate rocks by externally derived highly H_2O-rich fluids. A large amount of highly H_2O-rich fluids flowed into the rocks, and mixed with the CO_2-rich fluids that were produced by the prograde reactions so that the intergranular fluids maintained H_2O-rich compositions. The infiltrating fluids may have been derived from the adjacent and underlying pelitic rocks undergoing metamorphism, or from crystallizing plutons in the neighborhood, or from the upper mantle, or from the hydrosphere, or a combination of these.

Infiltration of H_2O-rich fluids that have compositions not in equilibrium with the mineral assemblages causes reactions. The CO_2 produced by the reactions is removed from the rocks by the flowing fluids so that reactions continue. When one of the reactants is used up, the relevant reaction ends. Then, as the fluid is no longer buffered by the mineral assemblage, it becomes more H_2O-rich by simple mixing of infiltrating fluid, until the next prograde reaction begins. Such a course of compositional change of intergranular fluid is schematically illustrated by a dotted line in Figure 6.9. Thus, infiltration of H_2O-rich fluid is a driving force of metamorphic reactions. It may work in combination with rising temperature in many cases, or by itself in others.

Fluid flows appear to be pervasive generally, causing reactions in most, or all, rocks in the volume through which they pass, although some rock layers are probably more permeable than others. The infiltration and mixing of externally driven aqueous fluid occurs at all times during the course of prograde metamorphism, thus maintaining the composition of intergranular fluids at a low mole fraction of CO_2.

The effect of infiltrating fluid is particularly essential in the case of zoisite-forming reactions. Zoisite in the zoisite zone appears to form by a reaction on

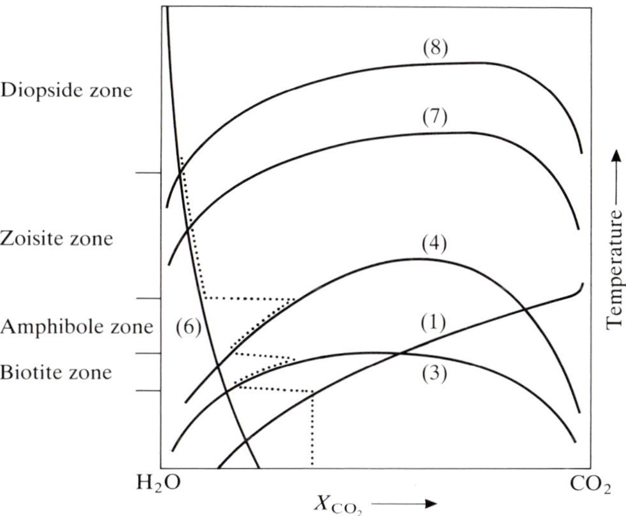

Figure 6.9 Isobaric *T–X* (mole fraction of CO_2) diagram illustrating schematic equilibrium curves for prograde reactions in argillaceous carbonate rocks of the Vassalboro Formation (Ferry 1983a). A dotted line shows a possible course of the compositional change of intergranular fluid in prograde metamorphism. Prograde reactions occur in the order of (1), (3), (4), (6), (7) and (8). Zoisite is stable on the left side of curve (6).

curve (6) in Figure 6.9, such as:

$$\begin{aligned} &1.31\,\text{anorthite} + 0.07\,\text{Fe component of amphibole} \\ &+ 0.59\,\text{Fe component of calcite} + 0.01\,\text{HCl} + 0.59\,H_2O \\ &= 1.00\,\text{zoisite} + 0.002\,\text{sphene} + 0.10\,\text{quartz} \\ &\quad + 0.01\,\text{NaCl} + 0.59\,CO_2 + 0.16\,H_2 \end{aligned} \tag{6.25}$$

The reaction produces zoisite when the mole fraction of H_2O is increased in the fluid. If temperature increases at a constant composition of fluid, such a reaction should proceed in the reverse direction (Fig. 6.9). In some other metamorphic areas in the world, indeed, it is known that reactions like the above proceed in the reverse direction with increasing metamorphic grade. So the prograde formation of zoisite in this area is a result of, and evidence for, infiltration of metamorphic rocks by aqueous fluids (Ferry 1976a).

6.7.3.2 Fluid/rock ratios in argillaceous carbonate rocks. Ferry (1980, 1983a) attempted to calculate the amount of fluid involved in progressive reactions on the assumption that the externally derived fluid had pure H_2O composition. This assumption gives the minimum possible amount of infiltrating aqueous fluids. The amount of H_2O required to account for the mineralogical change in metamorphosed carbonate rocks is (the amount of H_2O for change of reactant minerals to product minerals) + (the amount of H_2O added to keep the composition of the fluid at the equilibrium composition). When the numbers of moles of H_2O and CO_2 generated by the reaction per liter of rock are denoted by n_1 and n_2, respectively, and when the number of moles of H_2O added is denoted by m_1, he calculated the value of m_1 from the relation:

$$\text{mole fraction of } CO_2 \text{ in intergranular fluid} = n_2/(n_2 + n_1 + m_1) \tag{6.26}$$

This equation assumes that "batch mixing" (single-stage mixing) has occurred.

Ferry expressed the results as a volumetric fluid/rock ratio. The value of fluid/rock ratio calculated in this way is a function of temperature, bulk-rock composition and the mole fraction of CO_2 in intergranular fluid at a constant pressure. Rocks with suitable bulk compositions to produce more CO_2 give higher fluid/rock ratios. Rock layers that are more permeable to externally derived aqueous fluids probably tend to show lower mole fractions of CO_2, and greater values of m_1 in the above equation, and greater fluid/rock ratios.

As Ferry (1983a) took into consideration only the reactions in the biotite and higher-grade zones, his results represent the amount of fluid required in prograde changes after the rocks passed the mineralogical state of the biotite isograd. In other words, the fluid/rock ratio at the biotite isograd is taken as zero. The calculated values of fluid/rock ratio generally tend to increase with metamorphic grade, and spatially toward the associated syn-metamorphic granitoid intrusions. In high-grade zones adjacent to granitoid intrusions, the values are higher than 2.1. In other words, the total amount of aqueous fluids that flowed through was more than twice the volume of the rock.

A similar calculation that takes the fluid/rock ratio at the zoisite isograd as zero, gives values mostly below 0.5 (Ferry 1980, 1983a). This means that most of the infiltrating fluid flowed through at metamorphic grades lower than that of the zoisite isograd during an early stage of prograde metamorphism.

The observed increase of fluid/rock ratio toward the syn-metamorphic granitoid intrusions suggests the possibility that the main part of the fluid may have come from crystallizing magmas of the intrusions. In addition to the vicinity of granitoid intrusions, however, the fluid/rock ratio shows high values in a regional-scale NE-SW-trending zone of the metamorphic region. This zone may represent a channelway of high fluid flux, although it may be a zone showing suitable bulk-rock compositions to give high fluid/rock ratios.

6.7.3.3 Comparison with associated pelitic metamorphic rocks of the biotite zone. For a comparison, we can look at the result of Ferry's (1984) investigation of fluid flow and fluid/rock ratio at the biotite isograd of associated pelitic rocks in the Waterville Formation. Almost all the metapelites in the chlorite zone (i.e. the zone on the low-temperature side of the biotite isograd) contain more than 5 vol. % of carbonates: ankerite + siderite, or less commonly ankerite + calcite. Other main constituent minerals are muscovite, chlorite, quartz, and albite, usually accompanied by small amounts of rutile, graphitic material, pyrite and pyrrhotite. On the other hand, no pelites in the biotite zone contain ankerite or siderite. These two minerals and rutile are completely used up by reaction with muscovite, quartz and other minerals, leading to the formation of biotite, plagioclase, ilmenite, etc. Ferry derived overall mineral reactions, which vary to some extent from rock to rock, depending on their mineralogical composition.

Ferry calculated the compositions of intergranular fluids that were equilibrated with observed mineral assemblages from the following relations among phase components:

$$\text{dolomite} + 2\,\text{magnesite} + \text{muscovite} + 2\text{quartz} = \text{phlogopite} + \text{anorthite} + 4CO_2 \quad (6.27)$$

$$3\,\text{dolomite} + \text{muscovite} + 2\,\text{quartz} = \text{phlogopite} + \text{anorthite} + 2\,\text{calcite} + 4\,CO_2 \quad (6.28)$$

Estimating the *P-T* condition of the biotite isograd at 3.5kbar and 400°C, Ferry found that the fluids were essentially mixtures of H_2O and CO_2 with 98–96 mole % H_2O and 2–4 mole % CO_2 under these conditions. The sum of the mole fractions of H_2, CH_4, CO and H_2S was less than 0.01.

Thus, the fluids equilibrated with the mineral assemblages were highly hydrous in pelitic rocks as well. On the other hand, biotite-forming reactions liberated CO_2 together with much smaller amounts of H_2O and H_2S. Therefore, the highly hydrous intergranular fluids must have formed by the mixing of externally derived highly hydrous fluids, just as in the case of carbonate rocks.

The minimum volumetric ratio of fluid/rock was calculated under the assumption that the infiltrating fluid was pure H_2O, as in the case of carbonate rocks, yielding values of the order of 1 or 2.

Ferry (1984) calculated the mole fraction of CO_2 from equilibria for the two stoichiometric relations (6.27) and (6.28). Wood & Graham (1986) showed that the equilibria are very sensitive to temperature. They considered that Ferry's value of temperature (400°C) for the biotite isograd was about 40° too low, so that revision for this error leads to the lowering of the fluid/rock ratio by an order of magnitude to about 0.1 or 0.2, and that such a small amount of external H_2O could be derived from nearby carbonate-free pelitic rocks or granitoid bodies and would not require a regional-scale flow. However, Ferry (1986) did not accept this criticism of his value of temperature. This discussion has highlighted the need for very accurate values of temperature and other factors for reliable calculation of the composition of fluids and fluid/rock ratios.

6.7.3.4 Comparison with the Acadian metamorphic region in New Hampshire and Vermont. To confirm the validity of his idea of flow and infiltration of fluids in other parts of the Acadian regional metamorphic complex, Ferry (1988) made a similar investigation in an area along the state boundary between New Hampshire and Vermont, about 150 km west of the Augusta–Waterville area.

Ferry examined pelitic, psammitic and calcareous metamorphic rocks, as well as metaigneous rocks in the chlorite and the biotite zones. All rocks in the chlorite zone contain carbonate (ankerite, siderite and/or calcite). Equivalent rocks in the biotite zone and higher grades are either, devoid of carbonate, or contain very small amounts (except for limestones). Thus, in the transition from the chlorite zone to the biotite zone, various reactions involving decarbonation occurred, some of which produced biotite. In spite of the liberation of CO_2, the calculated composition of the intergranular fluids in the biotite zone is generally H_2O-rich with a mole fraction of CO_2 in a range of 0.04–0.07. This may be considered to be due to the infiltration of externally derived H_2O-rich fluid. The calculated fluid/rock ratios are 0–0.2 in limestones, 0.2–1 in metapsammitic rocks and 1–3 metapelitic and metaigneous rocks. Thus, except for limestones, the fluid/rock ratios are significantly higher than the likely rock porosity during metamorphism, and so infiltration of an externally derived H_2O-rich fluid occurred. The fluid must have driven decarbonation reactions.

The data suggest that there is a significant difference in permeability between rock types during metamorphism. Fluid flow appears to have occurred preferentially through metapelitic and metaigneous rocks.

6.8 Formation of anomalously high-grade metamorphic areas by flow of a hot fluid in New Hampshire

If fluid flows are focused into narrow zones within the crust, the value of fluid flux increases, and so the fluid flows may cause a marked increase of metamorphic grade in a relatively small area (§2.6.3). A well documented example of an apparent case of this kind is reviewed below.

There is a Paleozoic regional-scale metamorphic region belonging to the lower sillimanite–muscovite zone in central New Hampshire. (This region straddles the boundary between regions of the low and medium-*P/T* ratio facies series.) Chamberlain & Lyons (1983) and Chamberlain & Rumble (1988) discovered the presence of 10 small anomalously high-grade metamorphic areas within the region. The anomalous areas are irregular in shape and a few to several kilometers across. Their metamorphic grades are higher than that of the surrounding region, being as high as the cordierite–garnet–K-feldspar (sillimanite) zone (high granulite facies) in the central part of anomalous areas (cf. Fig. 10.3). Temperature measurements using the garnet–biotite Fe–Mg exchange geothermometer have confirmed that the anomalous areas really equilibrated at higher temperatures than the surrounding region. Temperature increases rapidly over a distance of a few kilometers from 550°C in the surrounding region to 700°C in the anomalous areas.

Chamberlain & Rumble (1988) ascribed the formation of the anomalously high-grade areas to intense heating during metamorphism by flows of high-temperature fluids that were focused through narrow zones with a high fracture permeability. Chamberlain & Rumble discovered an abundant occurrence of graphite–quartz veins with minor rutile and ilmenite within the anomalous areas, whereas such veins are relatively rare outside. The veins are considered to represent channelways through which hot fluids flowed during metamorphism. The mineral assemblage of the vein suggests that the flowing fluids were mainly composed of H_2O with $CH_4 > CO_2$ and trivial amounts of CO and H_2. This composition is similar to that of fluid that would be released by metamorphic devolatilization of pelitic rocks containing organic substances.

The regional-scale sillimanite–muscovite zone region (outside the anomalous areas) appears to have undergone a considerable extent of oxygen isotope homogenization during metamorphism. The homogenization is considered to have occurred by exchange reactions between metasediments and pervasively infiltrating fluids during metamorphism. However, rocks within the anomalous areas appear to have undergone further oxygen isotope exchange reactions with infiltrating fluids flowing through the fractures, which are now represented by the network of graphite–quartz veins.

PART II

Metamorphic facies and metamorphic belts

7 The concept and system of metamorphic facies

7.1 Eskola's concept and practical system of metamorphic facies

7.1.1 General remarks

Eskola (1920) proposed the concept and practical system of metamorphic facies in an attempt to classify metamorphic rocks on the basis of the conditions of their formation. Since he did not have a clear idea of the rôle of the chemical potential of H_2O and CO_2, he considered only temperature and pressure as the external conditions controlling metamorphism.

As a preliminary, Eskola (1915) had devised the *ACF* and *A'KF* diagrams (§5.3). So, he constructed *ACF* diagrams for some well investigated metamorphic areas, throughout each of which temperature and pressure might be regarded as virtually constant. He observed that *ACF* diagrams constructed for some of the areas were similar to one another. He ascribed the similarities to similarities in the *P–T* conditions of metamorphic crystallization. However, *ACF* diagrams showed marked differences between other areas. He ascribed the differences to differences in the *P–T* conditions of metamorphism. In this way he could classify metamorphic areas throughout the world into a number of groups, each of which he characterized by similarities in *ACF* diagrams, and hence regarded as having formed in a certain definite range of *P–T* conditions. All rocks of a group were said to belong to a **metamorphic facies**. Thus, each metamorphic facies was regarded as representing a certain definite range of *P–T* conditions. The whole range of *P–T* conditions of metamorphism could thus be divided into a number of subranges, which represent individual metamorphic facies.

After this general outline of Eskola's idea of metamorphic facies, we shall examine his idea more closely.

7.1.2 Eskola's concept and implicit assumptions

Eskola's original definition (Eskola 1915: 115) is quoted below:

> . . . a metamorphic facies includes rocks which . . . may be supposed to have been metamorphosed under identical conditions. As belonging to a certain facies we regard rocks which, if having an identical chemical composition, are composed of the same minerals. . . . every facies may include all possible chemical and genetical varieties. One and the same facies may be found in widely different parts of the world, while in neighbouring localities different facies may be met with.

Thus, each metamorphic facies is a category in a classification of metamorphic rocks, and is supposed to correspond to a definite *P–T* range, because it is characterized by a particular regular relationship between the chemical and mineralogical compositions of metamorphic rocks, as represented by *ACF* and *A'KF* diagrams. It is not necessary to know the numerical values of the *P–T* range, since the boundaries between metamorphic facies are defined by critical mineral assemblages. This early definition of Eskola's is preferable to his later, excessively positivistic and obscure definition (Eskola 1939). However, even the early definition (Eskola 1915) is defective in the respect that it does not explicitly mention two underlying assumptions, the acceptability of which became serious problems during the later development of the concept of metamorphic facies. The assumptions are that (a) spatially closely associated metamorphic rocks are regarded as having been metamorphosed (crystallized) under the same external conditions (*P–T* conditions) unless there is positive evidence against it, and (b) the external conditions controlling metamorphic mineral formation are temperature and pressure alone.

The basis of metamorphic facies lies in the empirical discovery that we can choose a small metamorphic area, in which the metamorphic rocks show a definite regular relationship between their chemical and mineralogical compositions. In different cases, the size of the area (or space) differs in which such a definite regular relationship holds. It may be tens of kilometers or only tens of meters across. It should also be noted here that what are called the "*P–T* conditions" of metamorphism means those at or near the thermal peak.

7.1.3 Eskola's practical system of facies classification: its merit and technical difficulties

7.1.3.1 System of facies classification based on metabasites and its merit of simplicity. Eskola (1920) proposed the definitions of five metamorphic facies, named **greenschist, amphibolite, hornfels, sanidinite and eclogite facies**, each

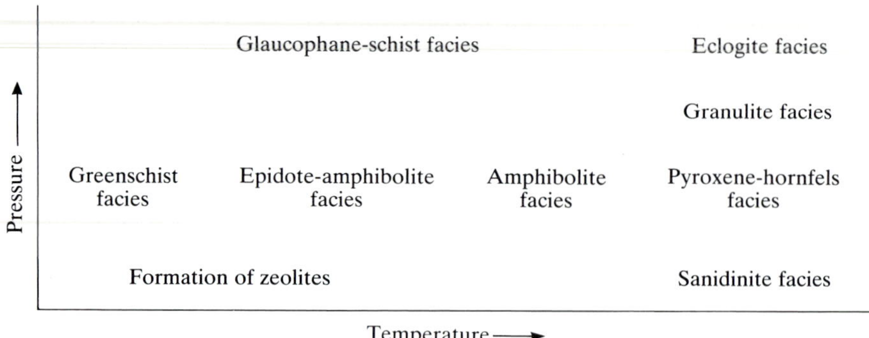

Figure 7.1 Metamorphic facies and their temperature–pressure relations after Eskola (1939).

of which he characterized by distinctive *ACF* and *A'KF* diagrams. In his final systematic account, Eskola (1939) added the **epidote-amphibolite, glaucophane-schist** and **granulite facies**, and changed the name of the hornfels facies into the **pyroxene–hornfels facies** (Fig. 7.1). A later author coined the name of **blueschist facies** to replace the name glaucophane-schist facies, with no change in the meaning. Although this was a violation of Eskola's priority, the name blueschist facies has become so popular in the literature that it is reluctantly accepted in this book.

At the time when Eskola proposed his metamorphic facies, Grubenmann's depth-zone classification of metamorphic rocks was widely used throughout the world (Appendix 3.1.1). Grubenmann (1904–06) assumed the existence of three depth-zones. However, there was no reason for assuming that the number of depth-zones is three. Moreover, the criteria of assigning individual metamorphic rocks into the three depth-zones were not well founded. So Eskola (1920, 1939) intended to replace Grubenmann's classification with his classification of metamorphic rocks based on metamorphic facies, called the **facies classification of metamorphic rocks**. Although Eskola criticized Grubenmann's depth-zones, he still implicitly considered that the number of metamorphic facies must be relatively limited – say, less than 10 – for a classification of rocks to be convenient for use.

Metamorphic rocks show a very great number of mineralogical changes in total, in response to variations of *P–T* conditions, because rocks with different bulk chemical compositions show different mineralogical changes with *P–T* conditions. In order to get a small number of metamorphic facies, Eskola had to ignore most of, and chose only a small number of, the known mineralogical changes for the definition of individual metamorphic facies.

Eskola defined the above-named individual metamorphic facies, such as the greenschist and amphibolite facies, actually on the basis of the principal mineralogical changes in metabasites. In other words, he ignored mineralogical changes in metapelites, which are as sensitive as, or even more sensitive than, metabasites to variation of P–T conditions. This was probably a happy choice, because his facies classification was intended to cover all metamorphic rocks, including eclogites and glaucophane schists, in which mineralogical changes are more remarkable in metabasites than in metapelites.

A great merit of metamorphic facies lies in the simplicity of its use and the resulting convenience. If we can find only a few metabasites in a small area and determine their mineral assemblage(s) under the microscope, we can usually assign all the associated metamorphic rocks of the area to a particular metamorphic facies.

It is important to emphasize that metamorphic facies were originally defined on the basis of the relationships between the chemical and mineralogical compositions of a group of spatially associated rocks, and not on their geologic settings. Certain zones of regional metamorphism show virtually the same relationship between the chemical and mineralogical compositions as certain zones of contact metamorphism. All these zones belong to the same metamorphic facies and should have the same facies name, regardless of the difference in their geologic settings. Turner (1968, 1981) proposed a series of new facies names for contact metamorphism, distinct from those of regional metamorphism, in spite of their close similarity in the relationship between chemical and mineralogical compositions. This is a clear violation of the original concept of metamorphic facies. These names should be abandoned.

7.1.3.2 Technical difficulties. Metamorphic facies in Eskola's original form (or in somewhat revised versions) are based mainly on the mineralogical variations in metabasites. Although this was a wise choice for Eskola's broad classificatory purpose, it makes theoretical analysis of the mineralogical changes with P–T conditions very difficult. The relationships of mineralogical composition to chemical composition and P–T conditions in metabasites have not been well resolved, because metabasites usually show high degrees of freedom (§5.2.6). The chemical compositions of amphiboles in metabasites usually show a complicated variation with varying compositions of the host-rocks even under constant P–T conditions. It is difficult in metabasites to distinguish the effects of P–T conditions from those of bulk-rock chemical composition. This is serious because the metamorphic facies are intended to represent P–T conditions only.

Eskola's ACF and A'KF diagrams cannot show rigorously the relationship between the chemical and mineralogical compositions of metamorphic rocks (§5.3). This difficulty may be considerably relieved by use of improved composition–paragenesis diagrams, although no complete diagrams are available for this purpose.

Table 7.1 The system of metamorphic facies used in this book.

P/T ratio type classification	Metamorphic facies	Critical mineral assemblage in metabasites
Low- and medium-P/T types (see Ch. 11)	Zeolite facies	Zeolite+quartz
	Prehnite–pumpellyite facies	Prehnite and/or pumpellyite+quartz (without zeolite+quartz)
	Greenschist facies	Chlorite+epidote+albite+quartz (without Prh & Pum)
	Amphibolite facies	Hornblende+plagioclase (andesine or more calcic) +quartz
	Granulite facies	Orthopyroxene+Ca-clinopyroxene+garnet+plagioclase+quartz
Characteristics of the high-P/T type (see Ch. 12)	Lawsonite–albite–chlorite facies	Lawsonite+albite+quartz (without glaucophanic amph. & jadeitic pyrox.)
	Blueschist facies	Glaucophanic amphibole (absence of eclogites)
	Eclogite facies	Omphacite+garnet (high in Mg+Fe and Ca)+quartz
Characteristics of contact metamorphism (see Ch. 13)	Pyroxene–hornfels facies	Orthopyroxene+Ca-clinopyroxene+plagioclase +quartz (without garnet)
	Sanidinite facies	Presence of sanidine, high plagioclase, tridymite, pigeonite and/or glass

Note:

1. For the P/T ratio types, see Ch. 8.
2. The zeolite, prehnite–pumpellyite and greenschist facies occur in some areas of the high-P/T type as well, but are not characteristic (diagnostic) of this type (see Fig. 8.3).
3. The so-called albite–epidote amphibolite facies is regarded as a part of the transitional state between the greenschist and amphibolite facies. There is also a wide transitional state between the amphibolite and granulite facies. This transitional state is included in the granulite facies.

7.2 Later changes in the concept and system of metamorphic facies

7.2.1 Coombs's addition of three new metamorphic facies

Eskola in his final version (1939) used eight metamorphic facies, which are greenschist, epidote–amphibolite, amphibolite, granulite, glaucophane–schist, eclogite, pyroxene–hornfels and sanidinite facies (Fig. 7.1). Among these, the epidote-amphibolite facies is not used as a general facies name in this book, because of its essentially transitional nature and the resultant difficulties in the definition which is discussed in §11.4.

Later investigations have made it clear that Eskola's eight metamorphic facies do not completely cover the whole *P–T* range of metamorphism. This is particularly apparent in low-temperature metamorphism. Hence, Coombs et al. (1959) and Coombs (1960, 1961) proposed and established three additional metamorphic facies, which are **zeolite, prehnite–pumpellyite, and lawsonite–albite–chlorite facies**. These three facies are accepted in this book. Thus, we use 10 metamorphic facies in total, and we only use them as major facies in this book. These 10 metamorphic facies, together with the critical mineral assemblage of metabasites for each of them, are tabulated in Table 7.1.

7.2.2 Solid-solution minerals and metamorphic facies

Since almost all metamorphic minerals except quartz are solid solutions, the understanding of their stability in relation to the prevailing physical and chemical conditions is important for the study of metamorphism. Eskola had no idea of how to deal with solid-solution minerals. He thought, for example, that hornblende and biotite are stable in the amphibolite facies, but unstable in the granulite facies. Actually the compositions of hornblende and biotite change continuously in transitional *P–T* conditions from the amphibolite to the granulite facies, and these two minerals are stable in the latter facies in certain composition ranges. Ramberg (1944a, 1949, 1952) and Ramberg & DeVore (1951) were the first to introduce into metamorphic petrology an adequate thermodynamic treatment of solid solutions.

As an example to visualize how the temperature of a reaction changes with the composition of the solid solutions involved, we consider the reaction that produces garnet at the expense of biotite, simplifying both minerals as Fe–Mg solid solutions. In the high-temperature part of the amphibolite facies, relatively FeO-rich metapelites show the assemblage biotite + sillimanite + garnet (+ muscovite + quartz), whereas relatively MgO-rich metapelites show the assemblage biotite + sillimanite (+ muscovite + quartz) without garnet, as shown in Figure 5.12c. Let us consider a metamorphic region made up of these relatively MgO-rich metapelites. With increasing temperature, the breakdown of associated muscovite in the presence of quartz may begin at the entrance to the granulite facies, producing K-feldspar by a reaction such as (11.19). With a further increase of temperature, the decomposition of biotite to produce garnet begins by a reaction such as:

$$\overset{\text{biotite}}{K(Fe,Mg)_3AlSi_3O_{10}(OH)_2} + \overset{\text{sillimanite}}{Al_2SiO_5} + 2\ \overset{\text{quartz}}{SiO_2}$$

$$= \overset{\text{garnet}}{(Fe,Mg)_3Al_2Si_3O_{12}} + \overset{\text{K-feldspar}}{KAlSi_3O_8} + H_2O \qquad (7.1)$$

The MnO-free garnet + K-feldspar assemblage on the right-hand side is characteristic of the granulite facies. Since biotite has a higher MgO/FeO ratio than the associated garnet, the relationship between the two minerals could be schematically shown by a diagram like that in Figure 7.2. In an area where the common metapelites have MgO/FeO ratios plotting in the range between A and B in the diagram, garnet begins to coexist with biotite at temperature T_1, and

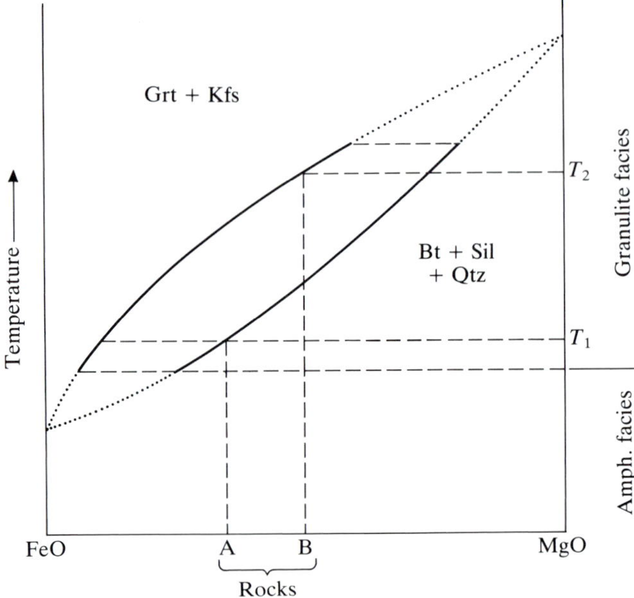

Figure 7.2 Schematic phase relation between the biotite + sillimanite + quartz assemblage and the garnet + K-feldspar assemblage in metapelites. Biotite and garnet are simplified as Fe–Mg solid solutions. Dotted lines represent metastable relations below and above the temperature range under consideration. In the amphibolite facies K-feldspar is unstable in the composition range of typical metapelites. Common metapelites are assumed to have compositions between A and B.

biotite disappears completely at T_2. The temperature range from T_1 to T_2 represents a transitional state, where the amphibolite-facies-like biotite-bearing assemblage gradually diminishes and is replaced by the biotite-free assemblage characteristic of the granulite facies with increasing temperature. The mineralogy and temperature ranges of such transitional states vary with the bulk chemical compositions of rocks.

In a transitional area between any two metamorphic facies, it is common that some metabasites show mineral assemblages characteristic of one facies, while other metabasites closely associated with them show those characteristic of the other facies. These differences could result from a difference in MgO/FeO ratio or some other parameters of the bulk-rock chemical compositions.

Therefore, the precise definitions of metamorphic facies boundaries must be based on an understanding of solid-solution effects. This was emphasized by Ramberg (1952: 136), who hence defined a metamorphic facies as follows: rocks formed within a certain *P–T* range, limited by the stability of certain critical minerals of defined composition, belong to the same metamorphic facies. However, boundaries between the individual metamorphic facies now in general use have not been precisely defined in this way, because many metabasites have high degrees of freedom, so that their mineral compositions cannot be uniquely correlated with external conditions (§5.2.6).

7.2.3 Metamorphic facies series in progressive metamorphism

Eskola's (1920) system of metamorphic facies was a static scheme of classification of metamorphic rocks, just as Grubenmann's depth-zone classification was. His claim for the existence of each facies was based on investigations of paragenetic relations carried out in different places in the world. For example, his specification of the amphibolite facies was based mainly on his own study in the Orijärvi region of Finland, and that of the pyroxene–hornfels facies on Goldschmidt's study in the Oslo area of Norway. Therefore, he did not, and could not, consider the problems of transitional states between different metamorphic facies.

Progressive metamorphism usually produces a sequence of zones belonging to two or more metamorphic facies. The whole sequence constitutes a **metamorphic facies series**. Tilley (1924), Vogt (1927) and Barth (1936) advanced the study of progressive metamorphism in this direction, and attempted to refine the definitions of individual metamorphic facies by defining facies boundaries by specific mineralogical changes. Miyashiro (1961) expanded the concept of metamorphic facies series so that it could be used in a classification of regional metamorphism on the basis of prevailing pressures (Ch. 8).

The compositions of solid-solution minerals change by continuous and discontinuous reactions with coexisting minerals with increase of temperature. The understanding of progressive mineralogical changes became possible only after the effects of solid solutions on the stability relations were realized, as illustrated above.

In progressive metamorphic regions, if we use some of the many mineralogical changes that were intentionally ignored by Eskola, we can subdivide each of Eskola's facies. In this way, Turner (1948) and Turner & Verhoogen (1951), in particular, proposed many subfacies, that is, subdivisions of Eskola's facies, and gave them many lengthy names which are difficult to remember. These authors based most of their subfacies on the progressive metamorphic sequence of zones defined for pelitic rocks in the well documented part of the Scottish Highlands, i.e. the Barrovian sequence (§1.6). However, progressive metamorphic sequences throughout the world are very diverse, and the Barrovian sequence is no more than a specific case. If we continue to propose a number of subfacies based on individual regions in this way, the names proliferate. Such subfacies names will simply be an increasing burden on our memory without providing any essential progress in our understanding. We should refrain from proposing subfacies for general use in this way. (This is not an objection to the use of special subfacies names for convenience of regional description or for the limited purpose of discussion.)

7.2.4 An attempt to define metamorphic facies in relation to external buffering

Eskola (1915, 1920, 1939) did not have a clear, internally consistent view of the behavior of H_2O in metamorphism. It appears that he imagined the ubiquitous presence of an aqueous fluid during metamorphism, just like virtually all geologists at that time, although he sometimes used relations that hold only in the presence of pure CO_2 fluid.

J. B. Thompson (1955) assumed the external buffering of the chemical potentials of H_2O and CO_2 in rocks undergoing metamorphism. Thus, rocks having the same chemical composition in terms of the fixed components (i.e. components other than the externally buffered) could crystallize into different mineralogical compositions under different external conditions (i.e. under different temperature, pressure, and chemical potentials of externally buffered components), and correspondingly come to have different amounts of externally buffered components. Under a definite set of external conditions, a definite relationship holds between the chemical composition of a rock in terms of its fixed components and its mineralogical composition, which can be represented by composition–paragenesis diagrams.

In this kind of case, if rocks undergoing metamorphism are externally buffered with respect to certain components, we come to the following definition of metamorphic facies, which essentially is that given by Korzhinskii (1959: 64): metamorphic rocks that formed under such similar external conditions that the relationship between their chemical composition, in terms of the fixed components and their mineralogical composition, is marked by the same regularities, belong to the same metamorphic facies.

This definition has real significance only when it is known which components are externally buffered, that is, show a nearly or completely uniform distribution of their chemical potentials in the pertinent small area during metamorphism. If only H_2O is externally buffered, a metamorphic facies should represent a volume in the *P–T*–(chemical potential of H_2O) space. If H_2O and CO_2 are externally buffered, and if their chemical potentials can vary independently, a facies should represent a "volume" in the *P–T*–(chemical potentials of H_2O and CO_2) space. However, it appears that H_2O and CO_2 are not always externally buffered, as seen in Chapter 6.

7.3 Difficulty in rigorously defining metamorphic facies

For clarity at this point in the discussion, we begin by reiterating Eskola's and Korzhinskii's ideas of metamorphic facies.

According to Eskola, we can take a small metamorphic area where all metamorphic rocks were crystallized under virtually the same *P–T* conditions. This conclusion is supported empirically by the construction of composition–paragenesis diagrams. Paragenetic relations as shown in these diagrams vary with *P–T* conditions. Each metamorphic facies is intended to represent a particular range of *P–T* conditions, and so is characterized by a particular set of one or more composition–paragenesis diagrams.

Subsequently it was understood that most metamorphic reactions are dehydration and decarbonation reactions, and so are controlled by not only *P–T* conditions but also the chemical potentials of H_2O and CO_2. According to Korzhinskii, the existence of particular paragenetic relations as expressed in composition–paragenesis diagrams requires the prevalence not only of a constant pressure and temperature but also of constant values of the chemical potentials of H_2O and CO_2 throughout the metamorphic area. Hence, Korzhinskii modified Eskola's definition, as stated in the preceding §7.2.4.

Korzhinskii's modification can save Eskola's intentions only when the chemical potentials of H_2O and CO_2 are genuinely uniform in an appropriately chosen small metamorphic area, say, 200 m across. Recent investigations, however, have made it clear that the chemical potentials of CO_2 and H_2O often vary within a metamorphic area, commonly over a short distance and even within a single outcrop, and that this causes great mineralogical variations, particularly in metabasites and calcareous rocks, even when metamorphosed under virtually the same *P-T* conditions (§6.6). This means that, even in a small area, metamorphic rocks cannot be assigned to a metamorphic facies as was defined by Korzhinskii. This has caused a fundamental difficulty in the definition of metamorphic facies.

If we hold rigorously to Korzhinskii's definition of metamorphic facies, two or more metamorphic facies occur mixed in a small area and even in a single outcrop. This eliminates the practical usefulness of the concept of metamorphic facies.

Alternatively, if we maintain the idea that a small area, if appropriately chosen, always belongs to a specific metamorphic facies, Korzhinskii's definition of metamorphic facies as a volume in the *P-T*-(chemical potential of H_2O) space or in the *P-T*-(chemical potentials of H_2O and CO_2) space does not hold in actual small metamorphic areas. On the other hand, if H_2O and CO_2 are always fixed components, a rigorous paragenetic analysis becomes possible by taking these components at apices of composition–paragenesis diagrams. However, this would not generally be the case.

In addition to this difficulty, we have already observed that the *ACF* and other existing composition–paragenesis diagrams are incomplete, and that the facies classification based mainly on the paragenetic relations in metabasites cannot be precise and rigorous (§5.7, §7.1.3).

Even if a rigorous definition of a metamorphic facies is impossible, we have to advocate the use of the system of metamorphic facies, because the system is very useful as a simple and convenient classification scheme for the great diversity of metamorphic rocks. This merit outweighs its defects. In the next section, a new practical definition of metamorphic facies is proposed to meet this situation.

7.4 Tentatively accepted definition of metamorphic facies

In spite of the theoretical and technical difficulties about the traditional concepts of metamorphic facies, a concept of metamorphic facies and a system of facies classification is still used in this book on the understanding that they give a simple and practical classification of metamorphic rocks roughly based on *P-T* conditions, which is particularly convenient for routine petrographic work. The system of facies classification shown in Table 7.1 is accepted. So, the corresponding tentative definition of metamorphic facies used in this book is discussed below.

7.4.1 Tentative definition of metamorphic facies based on two postulates

(a) Let us postulate that spatially closely associated metamorphic rocks belong to the same metamorphic facies. Temperature and pressure may then be considered to be virtually constant, or in small ranges, in these metamorphic rocks. This is the most basic postulate, because metamorphic rocks showing other than critical mineral assemblages can be assigned to a specific facies only under this postulate. For example, rocks such as biotite–quartz schist, pure quartzite and pure calcite marble are stable over a very wide range of *P–T* conditions and so belong to two or more metamorphic facies, whatever classificatory scheme of metamorphic facies is adopted. For such rocks, we could choose one specific metamorphic facies out of them, on the basis of more critical mineral assemblages that occur in other rocks closely associated with them spatially. There may be retrogradely metamorphosed rocks nearby that show clear evidence of their origin under distinctly different *P–T* conditions. These retrograde rocks should be excluded from the metamorphic facies under consideration.

(b) We further postulate that each metamorphic facies represents a volume in three-dimensional space *P–T*–(chemical potential of H_2O) that is characterized by the stability ranges of critical mineral assemblages. Thus, the effects of the chemical potentials of CO_2, O_2 and any other volatiles (if they vary independently of the chemical potential of H_2O) are ignored in the definition of metamorphic facies. Therefore, individual metamorphic facies are defined or characterized by a set of *ACF*, *A'KF*, *AFM* and other composition–paragenesis diagrams that hold under a specific value or specific range of the chemical potential of H_2O.

If the group of spatially closely associated metamorphic rocks under consideration is externally buffered with regard to the chemical potential of H_2O, there is no conflict between the two postulates. Although our petrographic experience shows that external buffering with regard to H_2O does not appear to hold true rigorously in most regional metamorphic complexes, the local variations in the chemical potential of H_2O are usually relatively small. So the above two postulates will be approximately true in usual cases of metamorphism.

In some cases, there may be so large a difference in the chemical potential of H_2O between spatially closely associated rocks as to cause a major difference in a critical mineral assemblage. In a case of this kind, only the first postulate is accepted, and the most common of the critical mineral assemblages is used to assign the closely associated metamorphic rocks to a specific metamorphic facies.

7.4.2 Disregarding the effects of CO_2 and O_2

There are many different possibilities with respect to the distribution of the chemical potential of CO_2. The chemical potential of CO_2 may be externally buffered in some cases. If graphite is present in contact with an intergranular C–H–O fluid, the variance of the system is 3, and so the chemical potential of CO_2 is a function of that of H_2O at a given temperature and pressure. In still other cases, the chemical potential of CO_2 may vary from rock to rock with variations in mineral assemblage. If the chemical potential of CO_2 varies locally independently of that of H_2O, this will cause variations of mineral assemblages even in the same metamorphic facies as defined above. It is postulated that we disregard all such differences in mineral assemblage caused by differences in chemical potential of CO_2, when we assign a group of rocks to a specific metamorphic facies. In the chlorite zone defined for metapelites, for example, the associated metabasites may show the following three types of mafic mineral assemblages: actinolite + epidote + chlorite, chlorite + epidote + calcite, and chlorite + ankerite. This difference is usually caused by local differences in the chemical potential of CO_2, which increases in the above-mentioned order of mineral assemblages (§11.3.3). These differences in mineral assemblage are excluded from our definition of metamorphic facies.

In the same metamorphic facies, some calcareous rocks may show the calcite + quartz assemblage, whereas others may have wollastonite, owing to a difference in the chemical potential of CO_2.

This definition of metamorphic facies will generally work well, with the possible exception of the zeolite facies. In the zeolite facies, the effect of CO_2 on mineralogical composition appears to be of vital importance, and so disregarding the effects of CO_2 may reduce the usefulness of the facies concept.

The fugacity of O_2 is usually internally buffered (§6.2), and so causes no difficulty in the definition of metamorphic facies. Although local variations in O_2 fugacity may cause marked mineralogical variations under a given temperature, pressure and chemical potential of H_2O (§3.6), these mineralogical variations will be ignored in the definition of metamorphic facies.

Approximate *P–T* ranges of some metamorphic facies are shown in Figure 7.3.

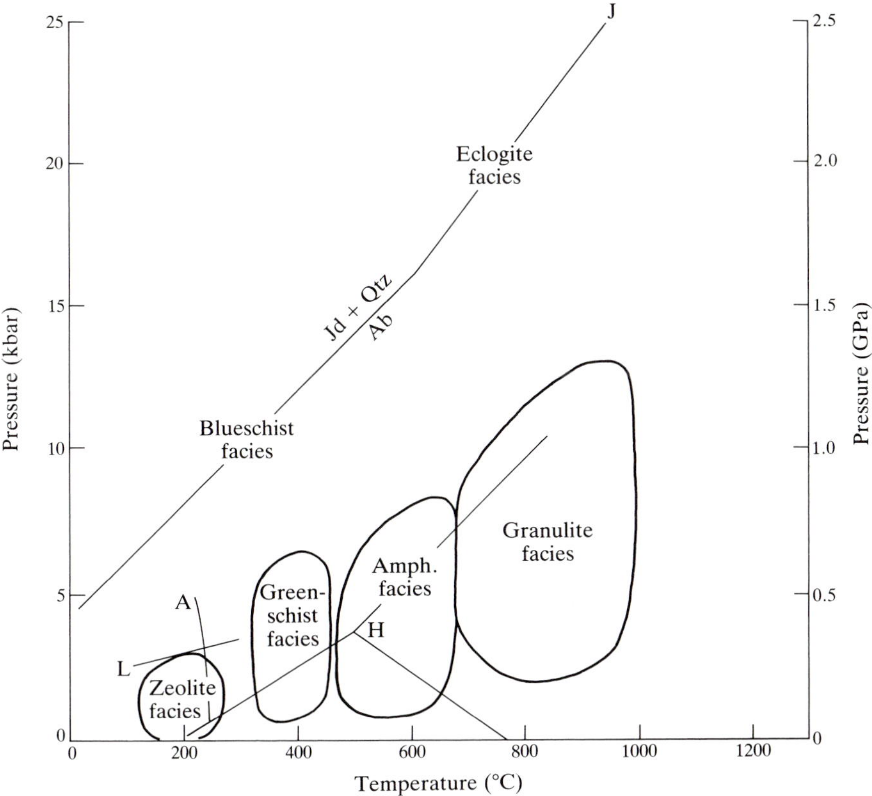

Figure 7.3 Approximate *P–T* fields of metamorphic facies. A: equilibrium curve for analcime + quartz = albite + H_2O (Liou 1971b). L: low-pressure limit for lawsonite +quartz in the presence of an H_2O fluid (Liou 1971a). H: triple point of Al_2SiO_5 minerals (Holdaway 1971). J: equilibrium curve for albite = jadeite + quartz (taken from Fig. 3.1).

8 Classification of regional metamorphism based on *P/T* ratio

8.1 The definition of three types of regional metamorphism based on *P/T* ratio

8.1.1 Introduction

Harker (1932) considered that all progressive regional metamorphic complexes in the world are more or less similar to one another in their mineralogical characteristics. He regarded the regional metamorphism of the Barrovian region of the Scottish Highlands (§1.6) as the most typical example, and called this type of metamorphism **normal regional metamorphism**. This view was once widely supported throughout the world. (Characteristics of the metamorphism in the Buchan region were not clearly known then.)

Mainly based on petrographic experiences in Japan, Miyashiro (1961) came to the different view that regional metamorphic complexes are actually very diverse in their mineralogical characteristics, owing to differences in pressure. Three polymorphs of Al_2SiO_5 (andalusite, kyanite and sillimanite) are widespread in pelitic rocks in middle- and high-grade zones of progressive metamorphic regions. The phase relations of the three polymorphs are a useful indicator of pressure in the temperature range 400–700°C (Fig. 8.1). Prior to any experimental determination of the phase relation of the three polymorphs, Miyashiro (1949, 1953) proposed such an inverted Y-shaped phase diagram, and used it to distinguish ranges of *P–T* conditions of metamorphism. He supported the idea that glaucophane is a high-pressure mineral, and therefore considered that andalusite-producing metamorphism, as well as glaucophane-producing metamorphism are significant kinds of regional metamorphism, and no less important than what was called normal regional metamorphism, which is characterized by the occurrence of kyanite. He therefore proposed a classification of progressive regional metamorphisms into three types (or three metamorphic facies series), and ascribed the differences between the types to differences in pressure.

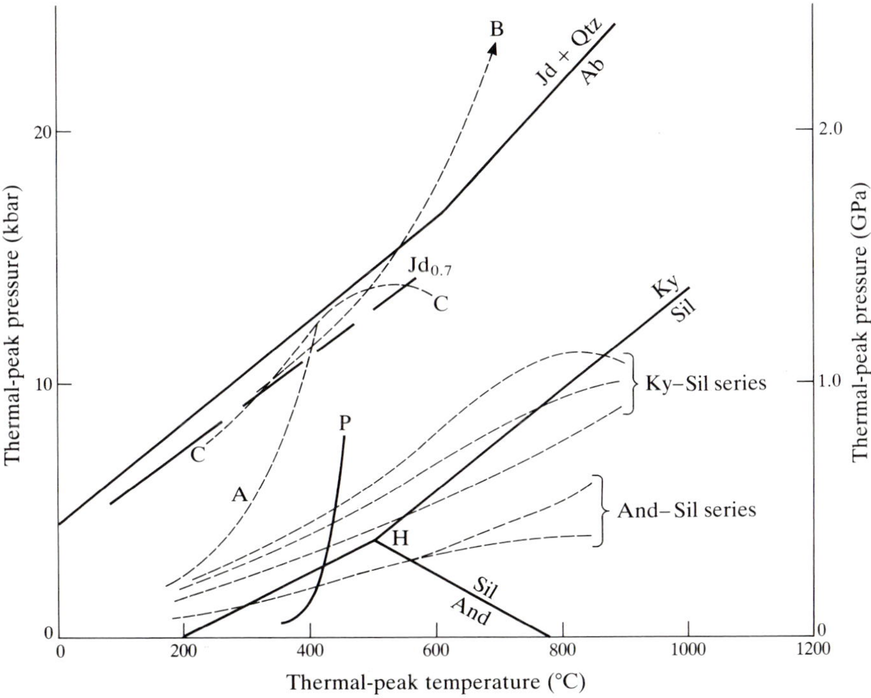

Figure 8.1 *P–T* diagram showing schematic field *P–T* curves (dashed lines) of the three main *P/T* ratio types of regional metamorphism. Curves A, B and C represent the high-*P/T* type: A (Franciscan), B (western Alps) and C (New Caledonia) as discussed in §8.2 and §8.3.3. The triple point (H) of andalusite, kyanite and sillimanite is after Holdaway (1971). The equilibrium curve for the reaction: jadeite + quartz = albite, has been taken from Figure 3.1. The low-pressure limit for the stability of jadeite solid solution with 70% jadeite component + quartz is also shown. Curve P represents the univariant curve for the reaction: pyrophyllite = andalusite (or kyanite) + 3 quartz + H_2O after Kerrick (1968).

8.1.2 Three-fold classification

The following three-fold classification is primarily a classification of regional metamorphic complexes according to the position of field *P–T* curves on the *P–T* diagram; it is modified only slightly from that of Miyashiro (1961). The field *P–T* curve shows the variation of thermal-peak *P–T* conditions (i.e. "*P–T* conditions of metamorphism") along a line in a direction perpendicular to the thermal-peak isotherm (or a dehydration-reaction isograd) at each point on it on the exposed surface of a metamorphic complex (§2.2.6). Field *P–T* curves of regional metamorphism usually show a positive slope, except for rather anomalous parts related to syn-metamorphic intrusions, faulting and folding. This classification deals also with metamorphic facies series in progressive regional metamorphic complexes.

The three categories are defined with reference to the stability relations of the Al_2SiO_5 polymorphs, glaucophanic amphibole and jadeite + quartz, as follows:

(a) *Low-P/T type (or andalusite–sillimanite series)*, in which andalusite is stable at lower temperatures, and sillimanite is stable at higher temperatures. Examples of possible field *P–T* curves are shown by dashed lines in Figure 8.1. The original type region of this series was the Ryoke metamorphic belt in Japan. The Buchan region of the Scottish Highlands (Fig. 1.4) also belongs to this series, and the well documented Acadian metamorphic region of this type in Maine is reviewed in §9.4.3.

(b) *Medium-P/T type (or kyanite–sillimanite series)*, in which kyanite is stable at lower temperatures, and sillimanite is stable at higher temperatures. Three possible field *P–T* curves are shown in Figure 8.1. They lie at higher pressures than the field *P–T* curves of the low-*P/T* type. The best documented example of this series is the Barrovian region of the Scottish Highlands (Fig. 1.4).

(c) *High-P/T type (or glaucophanic metamorphism)*, which is characterized by the occurrence of glaucophanic amphibole in some rocks in the progressive metamorphic sequence. Examples of observed field *P–T* curves of the high-*P/T* type are shown by dashed lines A, B and C in Figure 8.1. From observations of the modes of occurrence of glaucophanic amphibole and jadeitic pyroxene in many metamorphic regions, Miyashiro (1961) assumed that the formation of jadeitic pyroxene + quartz requires a pressure higher than that of glaucophanic amphibole. This assumption has been justified by later experiments (Fig. 12.1). Hence, this type of metamorphism may be subdivided into two classes.

   (i) *High-pressure transitional type*, which forms under pressure high enough to form a glaucophanic amphibole solid solution in some rocks (usually including some metabasites), but not high enough to produce jadeitic pyroxene (pyroxene high in jadeite content) in association with quartz. This type is intermediate between the kyanite–sillimanite series and the jadeite–glaucophane type.

   (ii) *Jadeite-glaucophane type*, which forms under pressures high enough to produce not only glaucophanic amphibole but also jadeitic pyroxene coexisting with quartz, at least in some part of the progressive sequence. Figure 8.1 shows the equilibrium curve for the reaction: pure jadeite + quartz = albite. Natural jadeitic pyroxene usually contains aegirine, diopside and hedenbergite components, and so is stable down to a considerably lower pressure, as shown in Figure 3.1. (A more detailed discussion on it is given in §12.3.2)

Thus, these *P/T* ratio types are believed to represent increasing thermal-peak *P/T* ratio values in the above-cited order.

Low-temperature limit for the formation of Al_2SiO_5 minerals and field P–T *curves.* In actual metamorphism, there is a low-temperature limit for the formation of andalusite, kyanite and sillimanite, on the low-temperature side of which hydrous Al-silicates form instead. In rocks unusually high in $Al_2O_3/(Na_2O + K_2O + CaO)$, the hydrous Al-silicate is pyrophyllite, and the low-temperature limit of Al_2SiO_5 is defined by the following reaction:

$$\underset{\text{pyrophyllite}}{Al_2Si_4O_{10}(OH)_2} = Al_2SiO_5 + 3\,\underset{\text{quartz}}{SiO_2} + H_2O \qquad (8.1)$$

In the presence of a pure H_2O fluid, the equilibrium temperature of this reaction, as shown by curve P in Figure 8.1, lies at temperatures somewhat higher than 400°C (Kerrick 1968). If the fluid is impure, or a fluid phase is not present, then andalusite or kyanite may be stable down to lower temperatures than curve P.

In pelitic rocks of more ordinary compositions, however, the first formation of andalusite or kyanite takes place by reactions involving pre-existing chloritoid or staurolite at a temperature considerably higher than curve P, say around 450°C. Thus, that part of the phase diagram of Al_2SiO_5 minerals which lies below about 400°C cannot be used for the definition of the three main types. The field *P–T* curve for the andalusite–sillimanite series probably passes through the kyanite field at these low temperatures.

In Figure 8.1, the low-temperature end of the field *P–T* curve of each metamorphic region is tentatively assumed to approach the *P–T* conditions on the Earth's surface (cf. §11.1.6). Although in some regions field *P–T* curves may show considerably high pressures at very low temperatures, such a part of the field *P–T* curve may have no real significance, because metamorphic crystallization may not occur at such very low temperatures.

Average geothermal gradients and actual depths. Figure 8.2 illustrates average geothermal gradients to give various sets of temperature and pressure within the crust. The average geothermal gradients that cause the low-, medium- and high-*P/T* types of metamorphism are, approximately, greater than $35°\,km^{-1}$, between 35° and $15°\,km^{-1}$, and smaller than $15°\,km^{-1}$, respectively. These values vary with the average density of the crustal rocks used in the calculation.

In addition, it is important that, because of the shape of its stability field, andalusite is stable only at depths of less than about 14 km.

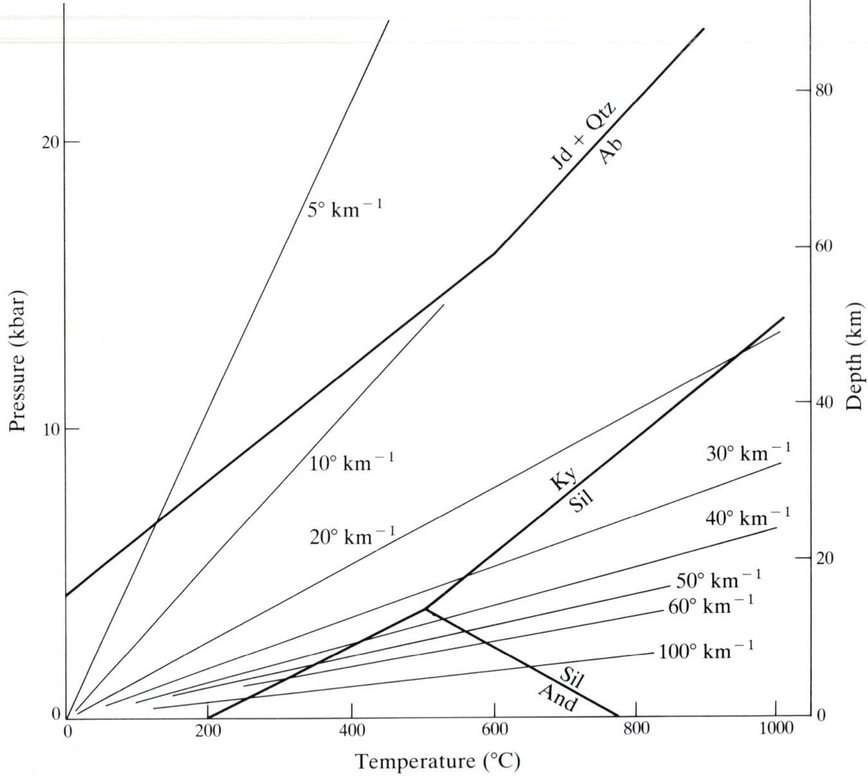

Figure 8.2 Average geothermal gradients and depths in relation to the stability fields of Al_2SiO_5 minerals and jadeite + quartz. Compare with Figure 8.1. (Such geothermal gradients vary with the assumed density of the crust.)

Relation of the three-fold classification to the classification of metamorphic facies. The three-fold classification of regional metamorphism based on P/T ratios is at the same time a classification of the sequences of metamorphic facies that occur in progressive metamorphic regions. Figure 8.3 summarizes the empirically observed relationship of the three-fold classification to the system of metamorphic facies. Detailed discussion of this relationship is given in Chapters 11 and 12.

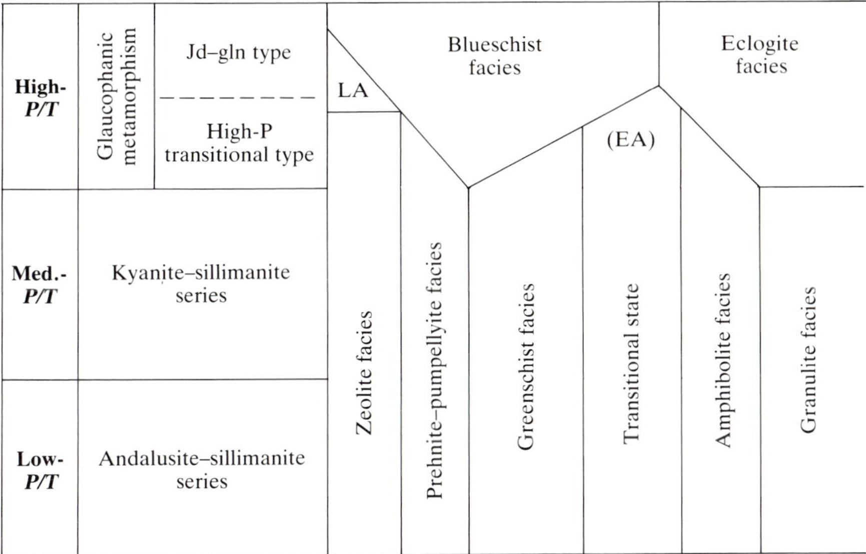

Figure 8.3 Schematic, empirical diagram showing the relationship between the *P/T* ratio types and metamorphic facies. Temperature increases to the right. LA represents the lawsonite– albite–chlorite facies. (EA) indicates the so-called epidote-amphibolite facies.

8.2 Diachronous progressiveness of metamorphism in accretionary complexes

Most high-*P/T* metamorphism and some medium- and low-*P/T* metamorphism take place in accretionary complexes that have been deposited in trench zones in arc–trench systems. Some metamorphic regions in accretionary complexes show a markedly distinct character with respect to the distribution of the age and grade of metamorphism, which is here called **diachronous progressiveness**. The most typical example is the Franciscan metamorphic complex in the Coast Ranges of California.

The Franciscan metamorphic complex is exposed in a zone, about 100 km in width, between the Pacific Coast and the Sierra Nevada Mountains (Fig. 9.7). It is a famous high-pressure metamorphic complex with glaucophane and jadeite, and is formed by metamorphism of accretionary complexes in a subduction zone along Mesozoic trench zones on the west side of the North American continent. The ages of deposition of the original sediments range from Jurassic to early Tertiary, and tend to become younger westwards. The ages of metamorphism range from Late Jurassic to Late Cretaceous, and tend to decrease westwards.

The temperature and pressure at thermal peak of metamorphism generally tend to decrease westwards across the region.

As the temperature of the Franciscan metamorphism tends to increase eastwards, the region is progressive-looking. However, the metamorphism was not a single event. The differences in the age of metamorphism between different parts of the region appear to be caused essentially by continued subduction and uplift of the metamorphic complex from Jurassic to Tertiary times. In the trench zone along the west coast of North America, accretionary complexes grew successively westwards through Paleozoic and Mesozoic times. Recently Maruyama et al. (1992) and Isozaki & Maruyama (1992) have shown, in northern California, that the Franciscan Complex is made up of the following four units (provinces) from east to west: 160–120 Ma blueschists; an Early Cretaceous accretionary complex that suffered high-P/T metamorphism 100 Ma ago, a Late Cretaceous accretionary complex; and a Tertiary accretionary complex. In other words, the complex may be divided into four distinct accretionary units (zones), showing progressive decrease in the age of sedimentation and metamorphism in response to the westward growth of the accretionary complexes. Metamorphism appears to have been episodic rather than continuous. The apparently systematic westward decrease of the grade of metamorphism seems to be related to the mechanism of uplift of the metamorphosed complexes.

It appears that some other high-P/T metamorphic regions also show a more or less similar oceanward decrease of the age and grade of metamorphism (Ernst 1975, 1977). Such metamorphism is not progressive in the proper sense of the term, because the metamorphism of the whole region was caused not by a single tectonothermal event but by a series of genetically related consecutive tectonothermal events that occurred over a long geologic time. Different parts of such an apparently progressive metamorphic sequence were metamorphosed at distinctly different times, not because of the slowness of thermal conduction, but because of the successive occurrence of a series of tectonothermal events. Since the whole series of tectonothermal events produces an apparently single progressive sequence of index-mineral zones, at least in crude petrographic work, this type of metamorphic region has usually been treated as if it represented a single progressive metamorphic sequence, as in Figures 8.1 and 8.7. In order to emphasize that it does not represent a single progressive metamorphic event in the proper sense, this type of metamorphism is called **diachronous-progressive metamorphism** in this book. The diachronous progressiveness appears to be a result of the progressive oceanward growth of the accretionary complex in the trench zone.

The oceanward growth of accretionary complexes is relatively common in Cordilleran-type orogenic belts such as in the Franciscan. In continental collision zones, high-pressure metamorphism occurs, not only in accretionary complexes, but also in subducting continental crust made up of older metamorphic and granitoid rocks. This may cause some irregularities in diachronous-progressive relations. An example of roughly diachronous-progressive subduction-zone metamorphism of the continental collision zone is observed in the western Alps, as is reviewed in Chapter 12 (Fig. 12.8).

Note that not all subduction-zone metamorphic regions show regular diachronous-progressive characters, but all known examples of diachronous-progressive metamorphism occur in trench-subduction zones. Most known diachronous-progressive metamorphism occurred under high-*P/T* conditions as in the Franciscan. However, some accretionary complexes show medium- and low-*P/T* diachronous-progressive metamorphism. An example of a low-*P/T* diachronous-progressive metamorphic complex was described from the Akaishi Mountains, central Japan, by Matsuda & Kuriyagawa (1965). For a review of the region, see Miyashiro (1973: 140–42).

In a diachronous-progressive metamorphic region, the field *P–T* curve is a compound curve connecting field *P–T* curves for all tectonic units (age provinces). It shows only a rough, general trend of variation of *P–T* conditions across the whole metamorphic region. Closer examination may reveal that each unit has its own trend of variation of *P–T* conditions across the unit, and this trend may not be harmonious with the general trend of the region as a whole.

8.3 Petrographic characteristics of progressive metamorphism of the three *P/T* ratio types

*8.3.1 Petrography of low-*P/T* progressive regional metamorphism*

The low-*P/T* type of regional metamorphism shows great diversity owing to differences in pressure as well as to variations in bulk-rock composition from area to area. For a more detailed review of descriptive data, see Miyashiro (1973: 167–80).

Some typical metamorphic regions of this type such as the Ryoke and Hidaka belts of Japan and the Cooma and other areas in eastern Australia, appear to have been metamorphosed at such low pressures that cordierite is common in metapelites at low as well as at high grades, and staurolite is absent or extremely rare. In ordinary metapelites, garnet is common only in the very high-grade part. In middle and low grades, garnet is confined to MnO-rich metapelites, and is high in the spessartine component. Figure 8.4 shows a progressive sequence of mineralogical changes of metapelites in such low-pressure regional metamorphic complexes (e.g. Harte & Hudson 1979, Morand 1990). The highest grade zone of this sequence is rare in Phanerozoic orogenic belts, but is widespread in Precambrian high-grade gneiss regions (e.g. Korsman 1977).

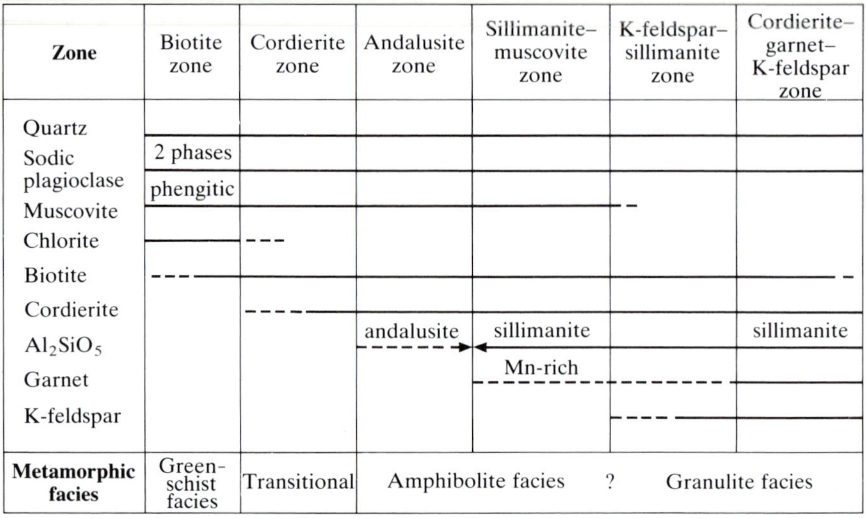

Zone	Biotite zone	Cordierite zone	Andalusite zone	Sillimanite–muscovite zone	K-feldspar–sillimanite zone	Cordierite–garnet–K-feldspar zone
Quartz						
Sodic plagioclase	2 phases					
Muscovite	phengitic					
Chlorite						
Biotite						
Cordierite						
Al_2SiO_5			andalusite	sillimanite		sillimanite
Garnet				Mn-rich		
K-feldspar						
Metamorphic facies	Green-schist facies	Transitional	Amphibolite facies		?	Granulite facies

Figure 8.4 Diagram showing progressive mineralogical changes of metapelites in regional metamorphism of the low-*P/T* type (andalusite–sillimanite series). This figure and Figures 8.5 and 8.6 are intended to show only the ranges of metamorphic grade where individual minerals occur, and not the mineral assemblages.

However, there are many other low-*P/T* metamorphic regions which were metamorphosed at pressures higher than in the above case, and which are transitional to the medium-*P/T* type to be described below. In these regions, staurolite occurs approximately in the andalusite zone and lower sillimanite-muscovite zone, garnet begins to occur at a lower grade, and cordierite begins to occur at a grade higher than in Figure 8.4. Some series show mineralogical characteristics very similar to the medium-*P/T* type, except for the occurrence of andalusite (e.g. Helms & Labotka 1991).

The metabasites in low-*P/T* regions are similar to those in medium-*P/T* regions, shown in Figure 8.6, except for the following differences: garnet does not occur or is very rare, and plagioclase becomes more calcic at relatively low grades (§11.4).

Low-*P/T* metamorphic regions show a sequence through the greenschist, amphibolite and granulite facies (Fig. 8.3). Zones of the zeolite and prehnite-pumpellyite facies may occur on the low-temperature side.

8.3.2 Petrography of medium-P/T progressive regional metamorphism

Figure 8.5 shows progressive mineralogical changes in some pelitic regions of this type. The Barrovian sequence in the Scottish Highlands (Fig. 1.4) coincides with the mineralogical relations in the range from the chlorite to the sillimanite-muscovite zone of this type. Chloritoid is virtually absent in Barrow's type area. (Progressive changes in paragenetic relations are discussed in §10.1.1.) However, there are many regional metamorphic complexes that have kyanite and sillimanite but show a considerable deviation in zone sequence from the typical Barrovian, mainly owing to differences in bulk-rock composition of the pelitic rocks exposed.

Figure 8.6 shows typical progressive mineralogical changes of metabasites in regional metamorphism of the medium-*P/T* type. The index-mineral zones of associated metapelites are shown for comparison at the foot of the diagram.

Medium-*P/T* metamorphic regions usually show a part of the sequence through the zeolite, prehnite–pumpellyite, greenschist, amphibolite and granulite facies in order of increasing temperature (Fig. 8.3).

Zone	Chlorite zone	Biotite zone	Garnet zone	Staurolite zone	Sillimanite–muscovite zone	K-feldspar–sillimanite zone	Cordierite–garnet–K-feldspar zone
Quartz							
Sodic plagioclase			2 phases				
Mucovite	phengitic						
Chlorite							
Chloritoid							
Biotite							
Garnet		Mn-rich					
Staurolite							
Al_2SiO_5				Ky	sillimanite		sillimanite
Cordierite							
K-feldspar						orthoclase	
Metamorphic facies	Greenschist facies		Transitional	Amphibolite facies		Granulite facies	

Figure 8.5 Diagram showing progressive mineralogical changes of metapelites in regional metamorphism of the medium-*P/T* type (kyanite–sillimanite series).

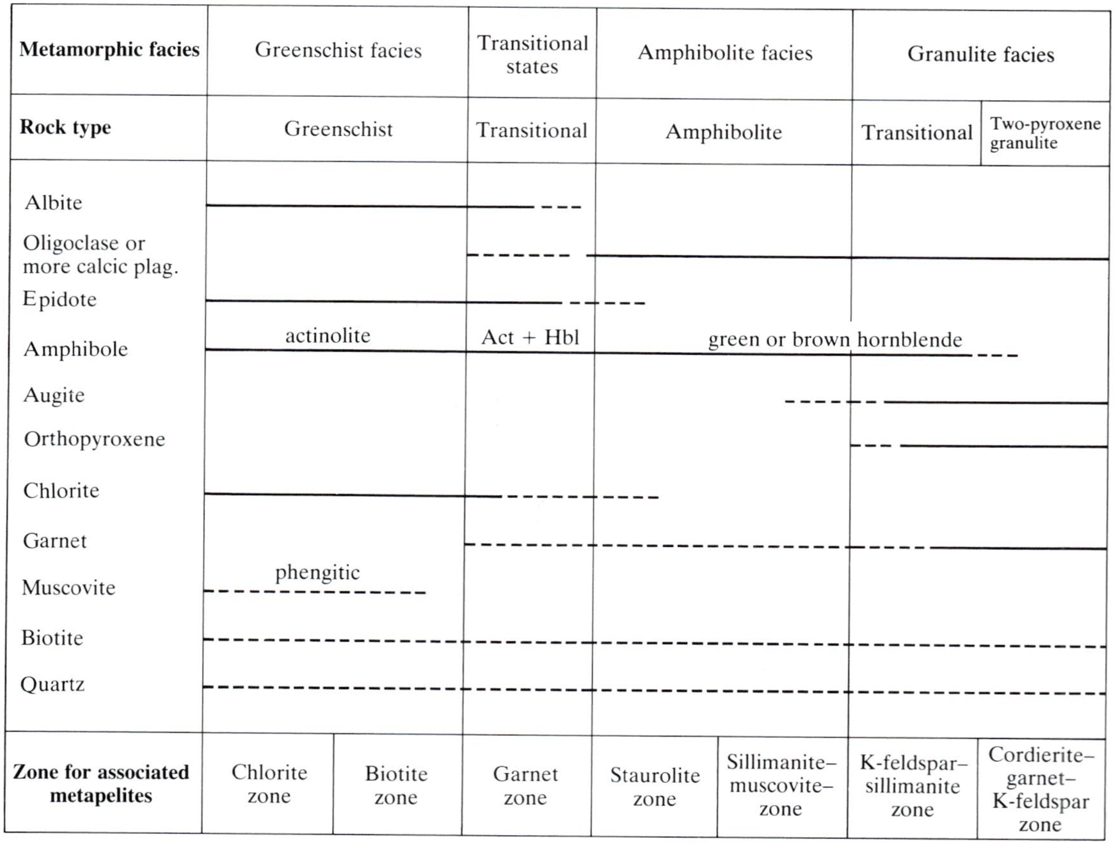

Metamorphic facies	Greenschist facies	Transitional states	Amphibolite facies	Granulite facies	
Rock type	Greenschist	Transitional	Amphibolite	Transitional	Two-pyroxene granulite
Albite					
Oligoclase or more calcic plag.					
Epidote					
Amphibole	actinolite	Act + Hbl	green or brown hornblende		
Augite					
Orthopyroxene					
Chlorite					
Garnet					
Muscovite	phengitic				
Biotite					
Quartz					

Zone for associated metapelites	Chlorite zone	Biotite zone	Garnet zone	Staurolite zone	Sillimanite–muscovite–zone	K-feldspar–sillimanite zone	Cordierite–garnet–K-feldspar zone

Figure 8.6 Generalized diagram showing progressive mineralogical changes of metabasites in medium-*P/T* regional metamorphism.

*8.3.3 Petrography of high-*P/T *progressive regional metamorphism*

The Franciscan as a whole is diachronous-progressive and it shows a sequence of zones belonging to the zeolite (laumontite), prehnite–pumpellyite, and blue-schist facies, which are exposed successively eastwards toward the Mesozoic continent, as shown in Figure 8.7 (Ernst 1975, 1977, Maruyama et al. 1992). Both pressure and temperature probably increase in this direction. The field *P–T*

	Zone	Zeolite-facies zone	Prehnite–pumpellyite-facies zone	Blueschist–greenschist-facies zone	Eclogite + albite-amphibolite blocks
Metaclastic rocks	albite				
	quartz				
	laumontite				
	pumpellyite				
	lawsonite				
	Na-amphibole			glaucophane	
	Na-pyroxene				jadeite
	white mica		phengite		
	chlorite				
	stilpnomelane				
	sphene				
	calcite				?
	aragonite				
Metabasaltic rocks	albite				?
	quartz				
	pumpellyite				
	lawsonite				
	prehnite				
	epidote				
	Na-amphibole			glaucophane–crossite	
	Ca-amphibole		actinolite		barroisitic hbl
	Na-pyroxene				omphacite
	garnet				
	white mica		phengite		
	chlorite				
	sphene				
	rutile				
	calcite				?
	aragonite				

Figure 8.7 Diachronous-progressive metamorphism of the Franciscan high-*P/T* metamorphic region in the Coast Ranges of California (Ernst 1977).

curve of this sequence is schematically illustrated as dashed line A in Figure 8.1. In addition, eclogite and albite amphibolite occur as blocks that were emplaced tectonically out of the regular sequence.

	Zone	Low-grade zone	Lws zone	Mn–Grt zone	LET	Epidote zone	Omphacite zone
Metasediments	aragonite						
	lawsonite						
	epidote						
	jadeite						
	omphacite						
	Na-amphibole						
	barroisite						
	spessartine						
	almandine						
	phengite						
	paragonite						
	chlorite						
	albite						
	quartz						
	graphite		disordered			ordered	
	sphene						
	rutile						
Metabasites	pumpellyite						
	lawsonite						
	epidote						
	omphacite						
	Na-amphibole						
	actinolite						
	barroisite						
	almandine						
	phengite						
	paragonite						
	chlorite						
	albite						
	quartz						
	sphene						
	rutile						
	Metamorphic facies		Blueschist facies				Eclogite facies

Figure 8.8 Progressive metamorphism in the Tertiary high-P/T metamorphic belt of northern New Caledonia (Yokoyama et al. 1986). LET: transitional zone where lawsonite coexists with epidote.

The metamorphism of the western Alps shows zones of the zeolite and prehnite–pumpellyite facies, blueschist facies and eclogite facies successively eastwards (Goffé & Chopin 1986). Although the sequence is cut by large faults and the eastward increase of metamorphic temperature is not continuous, it may be regarded as representing a diachronous-progressive change. The field *P–T* curve of the sequence is schematically shown by line B in Figure 8.1.

In northern New Caledonia, Brothers and co-workers discovered an apparently continuous progressive metamorphic sequence from the low blueschist facies to the low eclogite facies (Yokoyama et al. 1986; Fig. 8.8). The field *P–T* curve for this series is shown as line C in Figure 8.1.

8.4 Geologic characteristics of the three *P/T* ratio types

Very broadly speaking, the low- and medium-*P/T* types occur in ancient arcs, continental collision zones and continental extension zones, whereas the high-*P/T* type occurs in ancient subduction zones (§1.3). General discussions are given below.

8.4.1 Variation of P/T *ratio within a metamorphic belt*

It is common for an extensive regional metamorphic complex to show a considerable amount of variation in thermal-peak pressure. In some cases, this leads to the division of the region into two regions of different *P/T* ratio types. Figure 1.4 shows the Caledonian metamorphic region of the Scottish Highlands, which is divided into the Buchan and Barrovian regions, which represent low- and medium-*P/T* metamorphic tracts, respectively. Geologic formations and geologic structures are continuous between the two regions. The Acadian metamorphic region of the northern Appalachians is also divided into low- and medium-*P/T* regions (J. B. Thompson & Norton 1968, Chamberlain & Lyons 1983).

In such a thermal-peak temperature distribution as in Figure 2.9, the crustal layers above level M may undergo low-*P/T* metamorphism, whereas in the crustal layers below level M the average geothermal gradient decreases monotonically with increasing depth, and metamorphism at greater depths may become medium-*P/T* type (see also §9.4.3.4).

It appears that some subduction metamorphic complexes are made up of regions of the medium-*P/T* and high-pressure transitional types, whereas others have regions of the high-pressure transitional type and jadeite– glaucophane type.

*8.4.2 Geologic characteristics of the low-*P/T *type*

The low-*P/T* type (andalusite–sillimanite series) is produced in the crust where intense heating occurs at relatively shallow depths. In Phanerozoic time, this heating appears to have occurred in two major types of tectonic settings: (a) heating of shallow crust by igneous intrusion and associated fluid flows in arcs and continental collision zones, and (b) increase of heat flow from the mantle by crustal thinning and accompanying igneous intrusions in continental extension zones. Thermal models of low-*P/T* regional metamorphism are discussed particularly by De Yoreo et al. (1991).

Metamorphic regions of the low-*P/T* type are usually closely accompanied by abundant granitoid intrusions. In arcs, some of them are syn- or late-metamorphic intrusions within the pertinent metamorphic belts, whereas others are post-metamorphic intrusions into the inside as well as the outside of the belts. Although late-metamorphic intrusives continue their upward movement in the declining stage of metamorphism at the level of the present erosion surface, most of them may be syn-metamorphic intrusions at deeper crustal levels (e.g. Read 1957).

In continental extension regions, thinning of the crust or lithosphere will increase the upward heat flow from the mantle or asthenosphere, and will further lead to the formation and rise of basaltic and granitoid magmas. These may jointly cause high-temperature metamorphism of the low-*P/T* type.

Low-*P/T* metamorphism is particularly widespread in Archean and Early Proterozoic regions. These regions may have formed in tectonic settings entirely different from the Phanerozoic. Refer to Miyashiro et al. (1982: 156–66).

*8.4.3 Geologic characteristics of the medium-*P/T *type*

Metamorphic regions of the medium-*P/T* type may be classified into three categories from the viewpoint of their geologic association, as follows.

First, in the Scottish Highlands as well as in the northern Appalachians, for example, a medium-*P/T* region passes continuously into an adjacent low-*P/T* region. The age of metamorphism is the same between the two regions, and geologic formations and structures are continuous through the boundary. These associated regions differ only in their *P/T* ratios. Medium-*P/T* regions of this class are accompanied by abundant granitoid intrusions, and appear to have formed by the same mechanism as the associated low-*P/T* regions at depth in either arcs or continental collision zones, but with lower geothermal gradients and at greater depths. Lower geothermal gradients could be realized either at a margin of an intrusion region or in metamorphic regions where major intrusion occurred in deeper levels in the crust (Fig. 9.3). In some regions, a considerable volume of gabbroic rocks, derived from the mantle, also were intruded approximately during metamorphism.

Secondly, some medium-*P/T* regions are associated with high-pressure transitional type regions, e.g. the Haast Schists of New Zealand. It appears that regional-scale variations in *P–T* conditions produce medium-*P/T* series in some parts and high-*P/T* series in other parts of the same subduction-zone metamorphic belt. Granitoid intrusions are lacking in such regions.

Thirdly, there appear to exist medium-*P/T* metamorphic regions that show no connection to the low- or high-*P/T* regions. An example is the Eocene metamorphism in the Lepontine Alps and adjacent regions in the Alps. Bickle et al. (1975) claimed that the heating occurred by thickening of the continental crust by overthrusting during continental collision (§9.4).

*8.4.4 Geologic characteristics of the high-*P/T *type*

Most of the high-*P/T* metamorphic regions appear to have formed by subduction of a cold oceanic plate beneath a continental lithosphere, as in the case of the Franciscan in California (Ernst 1970, 1975). Subduction of a cold oceanic plate will produce an inverted temperature distribution within, and immediately above, the subducting slab. If subduction occurs at a constant rate for a long period, the inverted temperature distribution becomes stationary. Whether such primary thermal structures can be preserved in the distribution of metamorphic mineral assemblages depends on the mechanism of uplift of the metamorphosed complex, which is not well understood. These metamorphic regions usually show diachronous-progressive characteristics (§8.2). The field *P–T* curve for the Franciscan has a positive slope as shown by curve A in Figure 8.1. This feature appears to have resulted from the mechanism of uplift.

If an ocean basin between two continental masses has been completely consumed by subduction beneath one of the masses, the other continental mass, which is connected to the subducting ocean floor, may also be subducted to a limited extent in the earliest phase of continental collision. In this type of case, high-*P/T* metamorphism takes place in the subducting continental plate. The high-*P/T* metamorphic series in the western Alps (curve B in Fig. 8.1) is believed to have formed by subduction of a continental crust in this way (Ernst 1975, Goffé & Chopin 1986).

High-*P/T* metamorphic rocks of a continental origin now exposed in the western Alps were metamorphosed at much higher pressures, and so probably at greater depths, than high-*P/T* metamorphic belts in the circum-Pacific regions. This suggests that the amount of post-metamorphic uplift of metamorphosed complexes was greater in the western Alps. This difference might well be ascribed to a difference in buoyancy between the subducted continental mass of the western Alps and the subducted trench sedimentary piles lying on an oceanic crust in the circum-Pacific regions.

High-P/T metamorphic regions are not accompanied by granitoid intrusions, but usually contain a great abundance of metabasites and ultramafic rocks instead. Although a small amount of metabasites is commonly present in the other two types of regions, the great abundance of such rocks, particularly of ultramafic compositions, is characteristic of this type.

8.4.5 Geologic ages of glaucophane schists in relation to the changes in the types of global tectonic processes through the ages

Among the metamorphic facies of the high-P/T type, the blueschist facies covers the largest area. So the problem of the geologic age of high-P/T metamorphism has usually been discussed in the literature as that of the age of glaucophane schists or the blueschist facies.

Petrographic descriptions of glaucophane schists began in the Alps, Greece and other European localities and then extended to California, Japan and other places in the late 19th century. Glaucophane schists in Indonesia were described by W. P. de Roever in the late 1940s. The ages of glaucophanic metamorphism in most of these places appeared to be Mesozoic or Cenozoic. So, W. P. de Roever (1956) made the bold claim that glaucophane and lawsonite are minerals characteristic of the Mesozoic and Cenozoic orogenies. This claim meant that the P-T conditions of regional metamorphism had changed secularly with geologic time.

In the 1960s and 1970s, investigation of glaucophane schists in the circum-Pacific regions made great progress, and it was found that Paleozoic glaucophane schists are as common as Mesozoic glaucophane schists in these regions. So Ernst (1972b) modified W. P. de Roever's claim and concluded that glaucophane schists are characteristic of the Phanerozoic age, even though some glaucophane schists formed in the Late Proterozoic. Shortly afterwards, Dobretsov & Sobolev (1975) and others reported Early and Middle Proterozoic radiometric ages for glaucophane schists in some areas in the interior of the Asiatic continent.

Maruyama et al. (1989) have made a new worldwide survey on the occurrence of glaucophane. They doubt the reliability of the determinations of ages older than 1000 Ma, and have proposed the existence of three main periods of glaucophanic metamorphism: 80–130 Ma (Cretaceous), 400–500 Ma (Ordovician–Silurian), and 800–850 Ma (near the boundary between the Middle and Late

Proterozoic). The last-mentioned age group occurs much less frequently than the other two. Mesozoic and Cenozoic glaucophane schists occur along the margins of the present-day continents, whereas older ones occur commonly in ancient continental collision zones within the present-day Eurasian and African continents.

Even if the existence of Early and Middle Proterozoic glaucophane schists turns out to be valid, it is still true that the dominant majority of glaucophane schists in the world are Phanerozoic, and the formation of glaucophane schists is characteristic of Phanerozoic orogenies. The first appearance of glaucophane schists in the Proterozoic and the marked increase in the Phanerozoic must be the result of some secular change in the condition and style of tectonic movement on the Earth.

A possibility is that the first appearance and the marked increase of glaucophane schists may indicate the beginning of subduction and the dominance of the plate tectonics regime, respectively. Another possibility is that in the Archean and Early and Middle Proterozoic, subduction of plates occurred, but produced not high-*P/T*, but middle- and low-*P/T* metamorphic belts because of different tectonic conditions that prevailed at those times.

Recently a few authors have begun to claim an accretionary complex origin for supracrustal rocks of Archean cratons (Card 1990, Maruyama & Isozaki 1992). This view supports the second possibility above, implying that oceanic trenches and subduction were widespread in the Archean. Hence, the absence of glaucophane schists and other high-*P/T* metamorphic rocks in the Archean is ascribed to some special conditions accompanying Archean subduction, e.g. hotter Archean plates and mantle.

9 Tectonothermal evolution of metamorphic belts

The tectonic settings of Phanerozoic regional metamorphic belts are briefly summarized in §1.3. For the history and circumstance of the formulation of plate-tectonic models of regional metamorphism, the reader is referred to Miyashiro (1991).

This chapter reviews only a few exceptionally informative examples of well investigated metamorphic regions in various tectonic environments, so that the reader will understand basic features of tectonic–metamorphic processes.

9.1 Paired metamorphic belts

Miyashiro (1961, 1967a,b, 1972) pointed out the occurrence of paired metamorphic belts in Japan, California and other circum-Pacific regions. A pair is composed of a **high-P/T** and a **lower P/T ratio** metamorphic belt, which run parallel to each other generally along a continental margin, with the high-*P/T* belt lying on the oceanic side. The lower *P/T* ratio belt belongs to the low- or medium-*P/T* type or a mixture of them. Miyashiro claimed that the high-*P/T* metamorphic belt represents an ancient trench zone, whereas the lower *P/T* ratio metamorphic belt represents a belt of granitoid plutons and arc volcanism at an active continental margin or island arc. Metamorphism in the two belts of a pair was presumed to be broadly coeval. Paired metamorphic belts have been widely accepted as evidence for arc–trench systems in the geologic past.

Figure 9.1 and Table 9.1 show the distribution of some paired metamorphic belts in the circum-Pacific regions on the basis of recent re-examinations (e.g. Ernst 1975, 1977). Paired metamorphic belts also occur in some regions outside the circum-Pacific, e.g. Hispaniola (Nagle 1974), Jamaica (Draper et al. 1976), and the eastern Alps (Hawkesworth et al. 1975, Bickle et al. 1975). The reader is referred to Ernst (1977) and Yardley (1989) for more detailed discussions of geologic and geophysical characteristics of paired metamorphic belts.

Table 9.1 Paired metamorphic belts in the circum-Pacific and other regions.

Region	High-P–T belt	Lower P/T ratio belt	Age of metamorphism	References
1. New Zealand	Wakatipu	Tasman	Jurassic–Cretaceous	Landis & Coombs (1967)
2. Sulawesi (Celebes)	—	—	Tertiary	Miyashiro (1961), Lee (1984)
3. Western Japan	Sanbagawa	Ryoke	Cretaceous	Miyashiro (1961), Banno (1986), Isozaki & Maruyama (1991)
4. Washington State	Shuksan	Skagit	Triassic	Misch (1966)
5. California	Franciscan	Sierra Nevada	Jurassic–Cretaceous	Hamilton (1969), Ernst (1975,1983)
6. Jamaica	Mt Hibernia	Westphalia	Cretaceous	Draper et al. (1976)
7. Hispaniola	—	—	Cretaceous or older	Nagle (1974)
8. Chile	Pichilemn	Curepto	Late Paleozoic	Gonzales-Bonorino (1971), Ernst (1975)

Not all known high-P/T metamorphic belts are accompanied by a belt of lower P/T ratio metamorphism and granitoid intrusion. The absence of the latter in some cases may be ascribed to the amount of subduction being too small to cause large-scale magma generation and the resultant transfer of heat above the subducting plate (Ernst 1975).

There is a *large fault separating the two metamorphic belts* in many pairs. Example are the Median Tectonic Line in western Japan (Fig. 9.2) and the Coast Range Thrust in California (Fig. 9.7). The highest-grade edge of the high-P/T metamorphic belt is directly cut by a fault of this kind in all paired belts, whereas the lower P/T ratio metamorphic belt is at a distance from the fault in some pairs, and is in direct contact with the fault in others.

Initially the fault was probably the boundary between the subducting plate and the overlying plate. Afterwards, it may have experienced a series of different fault movements, possibly including a large lateral movement.

It should be noted here that all metamorphic events in the trench zone are not of the high-P/T type. Rocks brought to great depths by subduction of the underlying oceanic plate undergo high-P/T metamorphism, whereas many trench-zone sedimentary piles remain at shallow depths, and are unmetamorphosed, or undergo low- or medium-P/T type metamorphism.

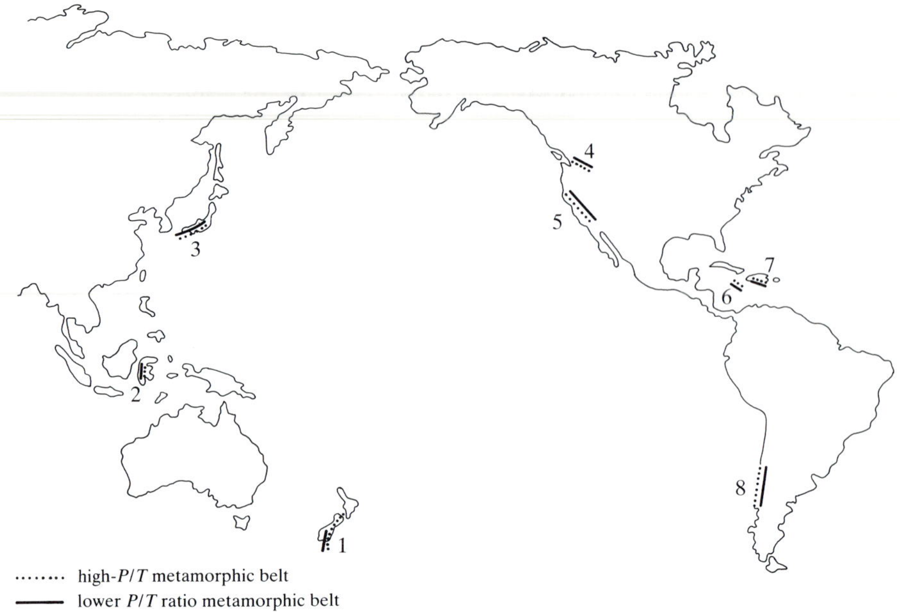

Figure 9.1 Some of the paired metamorphic belts in the circum-Pacific regions. Refer to Table 9.1 for the numbers of pairs.

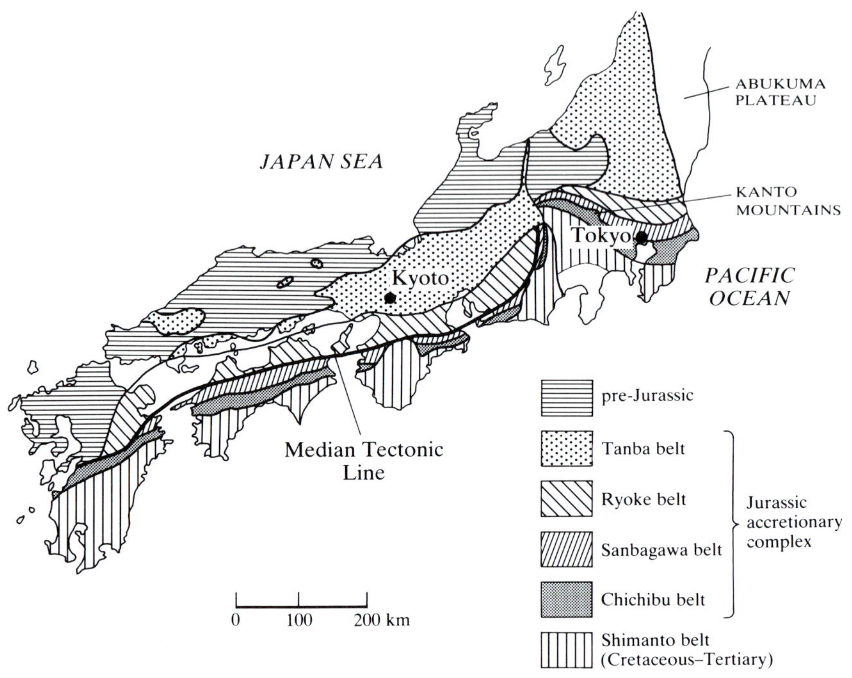

Figure 9.2 Tectonic sketch map of Japan. Japan is composed of accretionary complexes of geologic ages as shown. The Jurassic accretionary complex is made up of unmetamorphosed or very weakly metamorphosed Tanba and Chichibu belts as well as the Ryke low-P/T metamorphic belt and the Sanbagawa high-P/T metamorphic belt. Modified from Banno & Sakai (1989) and many others.

9.2 Lower *P/T* ratio metamorphic belts in arc zones

9.2.1 General characteristics

This section discusses the metamorphic belts on the ancient continental side in paired metamorphic belts, which belong entirely to the low-*P/T* type in some cases, as in the Ryoke belt of Japan (Fig. 9.2), and are a mixture of low- and medium-*P/T* areas in other cases, as in the Sierra Nevada of California (Fig. 9.7). These belts formed in the depths of ancient arc zones (island arcs and active continental margins).

In some pairs, the lower *P/T* ratio metamorphic belt is situated at a distance from the above-mentioned large fault separating the two belts. In these cases, the belts may show a broadly symmetrical thermal structure, with metamorphic grade decreasing from a median axial zone toward both margins of the belt. In other pairs (e.g. the Ryoke belt) the ancient ocean-side half of the belt has been lost by movement along the large fault, so that metamorphic grade increases monotonically toward the fault.

The lower *P/T* ratio metamorphic belt is usually accompanied by a great abundance of granitoid intrusions, and often also by coeval extensive felsic volcanic rocks. The occurrence of these volcanic rocks suggests that the total amount of uplift and erosion was small. In many parts of the belt, exposures of granitoid rocks are more extensive than those of metamorphic rocks, and many relatively small isolated metamorphic areas are surrounded by an ocean of granitoid rocks. This suggests that the heat that caused the metamorphism was largely transported by granitoid magmas (Barton & Hanson 1989, Hanson & Barton 1989, De Yoreo et al. 1991).

Andalusite is stable only at depths of less than about 14 km (Fig. 8.2). Hence, the andalusite-bearing zone, at least, of low-*P/T* metamorphic regions was not more than 14 km deep during the metamorphic crystallization (e.g. Dodge 1971). More generally the present surface of low-*P/T* metamorphic regions was probably at depths not much greater, and commonly considerably smaller, than about 14 km during metamorphism, as schematically shown in Figure 9.3a and b. With increasing depth, the metamorphic region passes into a medium-*P/T* metamorphic complex (§8.4.1).

In some low-*P/T* metamorphic regions, a group of major plutons was intruded into relatively shallow depths in the crust, causing a very high geothermal gradient (Hamilton & Myers 1967; Fig. 9.3a). In these regions, thermal-peak temperature increases with pressure above the intrusion level, and shows a thermal maximum at the intrusion level, as schematically shown by the solid line in Figure 2.9. In the *P–T* diagram of Figure 9.4, field *P–T* curves of this type of progressive metamorphism show a temperature maximum in a position corresponding to the depth of the intrusions, like curves (1) and (2). Except in the vicinity of the maximum, the field *P–T* curve as a whole still shows a positive slope with d*P*/d*T* increasing with increasing temperature.

In other low-*P/T* metamorphic regions, major plutons were intruded into both middle and deep levels of the crust, as schematically shown in Figure 9.3b. These intrusions generally fit into Daly's (1933) concept of batholiths. In this case, thermal-peak temperature tended to increase monotonically with increasing depth, as shown by curve (3) in Figure 9.4.

Possible stability fields of andalusite, kyanite and sillimanite in the crust are schematically shown in Figure 9.3a and b. Within the crust above the intrusions, regional metamorphism of the low-*P/T* type may occur, and later erosion may expose it on the surface, for example along line A. However, immediately

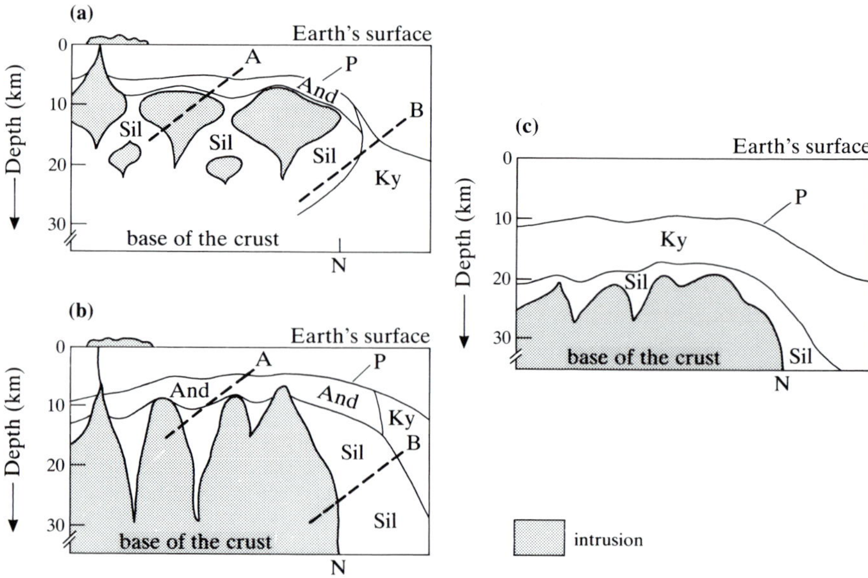

Figure 9.3 Schematic cross sections showing possible relationships between intrusions (stippled) and regional metamorphism in island arcs and active continental margins. (a) Where the major intrusions occur at the middle depths of the crust and reach a depth as shallow as about 8 km. (b) Where the major intrusions occur in all depths greater than about 8 km. (c) Where the major intrusions are more than 20 km deep. Line P indicates the outer limit for the possible formation of Al_2SiO_5 minerals (corresponding to curve P in Figure 8.1). And, Ky and Sil indicate the stability fields of andalusite, kyanite and sillimanite, respectively, within the crust. N indicates the periphery of the intrusion-rich region. A and B represent possible original positions of present-day erosion surfaces of the low- and medium-*P/T* types, respectively.

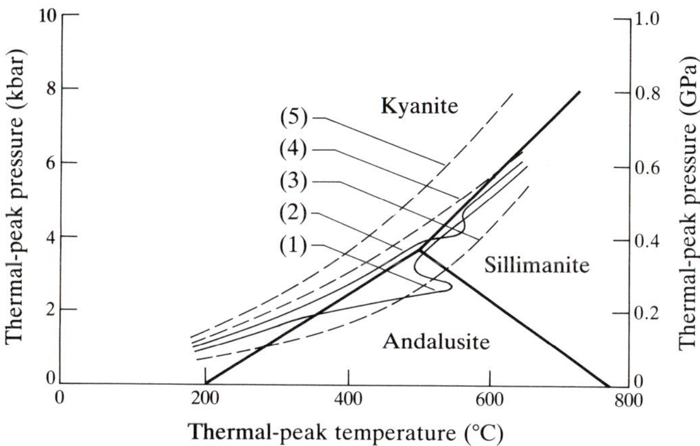

Figure 9.4 Possible examples of field P–T curves in progressive regional-metamorphic complexes. Curves (1), (2) and (3) are discussed in Section 9.2.1, and curves (3), (4) and (5) in Section 9.4.2. As a reference framework of P–T conditions, the stability field boundaries between andalusite, kyanite and sillimanite are shown by solid straight lines.

outside the intrusive region, progressive metamorphism of the medium-P/T type may take place along an erosion surface like line B.

In still other regions, however, intrusions may stop rising at a much greater depth within the crust, as schematically illustrated in Figure 9.3c. In this case the geothermal gradient is much smaller than in the above two cases, and so metamorphism may be of the medium-P/T type only.

9.2.2 Thermal-peak isobaric and isothermal surfaces in the Caledonian metamorphic region of the southeastern Scottish Highlands

9.2.2.1 Introduction. The Caledonian metamorphic belt in the Scottish Highlands is the best investigated example of a lower P/T ratio metamorphic belt from an ancient continental margin (Figs 1.4, 1.5). The metamorphic region is divided into low- and medium-P/T regions (called the Buchan and Barrovian regions, respectively). It is paired with a high-P/T metamorphic belt lying to the south.

Chinner (1966, 1980), Harte & Hudson (1979) and Harte & Dempster (1987) made an elaborate semi-quantitative analysis of thermal-peak isobaric and isothermal surfaces for a wide Dalradian region in the southeastern Scottish Highlands. Figure 9.5 shows Harte & Hudson's map, which was derived mainly from their two petrogenetic grids for metapelites: one based on Schreinemakers' principle and the other on the Mg/(Fe + Mg) ratios of biotite in two mineral assemblages representing continuous reactions.

The differences in pressure between different parts of this region are as great as several kilobars, enough to be measured by ordinary geothermobarometers. Thus, Baker (1985, 1987) published a map showing thermal-peak isotherms and isobars mainly based on geothermobarometric measurements, as already shown in Figure 2.4. Figures 9.5 and 2.4 show generally similar configurations of isothermal and isobaric surfaces, except in the northwestern margin of the region.

It appears that different parts of the metamorphic region had different histories of heating, burial, uplift and erosion (Harte & Dempster 1987). The genetic

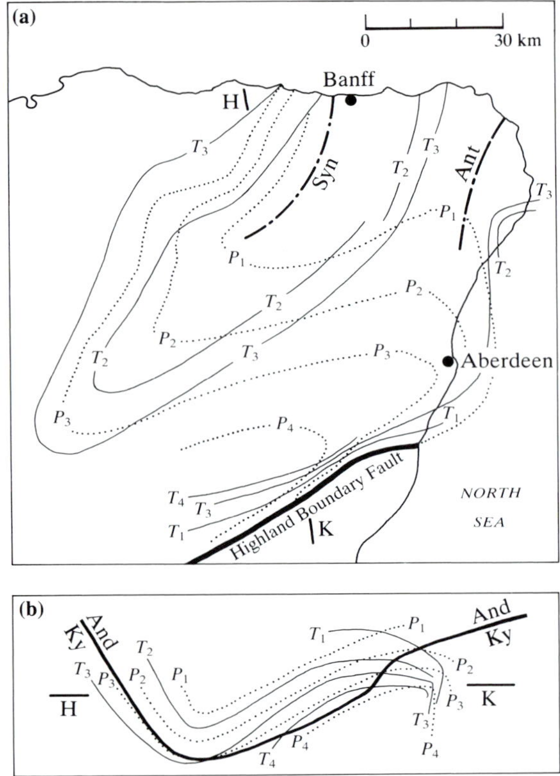

Figure 9.5 (a) Semi-quantitative thermal-peak isobars and isotherms in the Dalradian metamorphic region. Temperature increases in the order of T_1, T_2, T_3 and T_4. Pressure increases in the order of P_1, P_2, P_3 and P_4. Chain lines "Syn" and "Ant" indicate the Boyndie syncline and the Buchan anticline respectively. (b) The vertical cross section along line H–K in (a). Line "And/Ky" indicates the andalusite–kyanite isogradic surface. After Harte & Hudson (1979).

interpretation of the distribution of thermal-peak temperature and pressure in such a large metamorphic region is difficult and equivocal. In the following section some of the main features are summarized without entering into the intricacies of interpretation.

9.2.2.2 Isobaric and isothermal surfaces in the central part of the region. A wide central part of the Dalradian region between Banff and Glen Muick (Fig. 2.4) shows thermal-peak isothermal and isobaric surfaces with a relatively gentle slope. At the present-day erosion surface thermal-peak pressure tends to increase southwestwards. The recorded *P–T* conditions are about 2 kbar and below 500 °C in the Buchan region around Banff and reach about 8 kbar and 750 °C in a wide Barrovian region to the southwest.

The isothermal surfaces are either subparallel to the isobaric surfaces or they intersect them at a small angle. The isobaric surfaces make a synform with an NE-trending axis near Banff. They were probably flat during metamorphism, and their present shapes are mainly a result of post-metamorphic deformations. Before these deformations, the thermal-peak isothermal surfaces were also probably flat with temperature increasing downward.

Hence, the field *P–T* curves in directions perpendicular to isotherms show a positive slope.

9.2.2.3 A zone adjacent to the Highland Boundary Fault. A zone near the Highland Boundary Fault, including Barrow's classic area in Glen Clova and Glen Esk, shows unusually steep dips in bedding and isothermal surfaces, which are both partly inverted. As a consequence, metamorphic grade on the Earth's surface increases rapidly away from the fault. By contrast, the isobaric surfaces appear to be relatively flat.

Harte & Dempster (1987) ascribe the steep dips of isothermal surfaces to the cooling effects of large, continuing syn-metamorphic vertical movement on a nearly vertical fault zone along the present Highland Boundary Fault.

9.3 High-*P/T* metamorphic complexes in subduction zones

*9.3.1 Strong deformations and small-scale thermal structures in high-*P/T* metamorphic complexes*

As discussed above, it appears that the low- and medium-*P/T* regional metamorphic complexes in the depths of ancient arc zones preserve their original structures fairly well. This is probably because these complexes generally suffered only relatively weak syn- and post-metamorphic deformations and relatively small amounts of uplift leading to exposure on the surface of the Earth. In contrast, there are reasons for thinking that high-*P/T* metamorphic complexes, which form in subduction zones along trench zones, usually suffered very strong syn- and post-metamorphic deformations so that most of their structures now observed on the surface may be the results of these deformations.

9.3.1.1 Shuffled-cards structures of accretionary complexes. In Cordilleran-type orogenic belts, subduction-zone metamorphism occurs mostly or entirely in accretionary complexes in trench zones. Some accretionary complexes are unmetamorphosed. In the Jurassic accretionary complex of western Japan (Fig. 9.2), recent structural–paleontological investigations have produced most instructive evidence for the fact that even unmetamorphosed, apparently undisturbed-looking sedimentary masses actually suffered a great amount of deformation. Apparently conformably stratified sequences composed of alternations of chert layers and clastic (sandstone–mudstone) layers are widespread. However, radiolarian and conodont fossils in them indicate that the cherts are much older than the associated clastic rocks. Thus, the alternations were formed not by true stratigraphic accumulation but by some mechanical process. It appears that cherts and clastic rocks of different geologic ages were sliced into layers, which were thrust between one another, just as playing cards are shuffled (Isozaki & Maruyama 1991, Maruyama & Isozaki 1992).

Hashimoto et al. (1992) have made a detailed investigation of a Sanbagawa high-*P/T* metamorphic region in the Kanto Mountains (Fig. 9.2), Japan, and have discovered that the metamorphic complex shows a shuffled-cards structure more or less similar to that described above in unmetamorphosed sediments. They have found that in metapelites the Fe/Mg distribution coefficients between chlorite and garnet show a good correlation with the degree of graphitization of carbonaceous substances, when the degree is expressed numerically by Tagiri's (1981) method. Hence the degree of graphitization is a good indicator of metamorphic temperature. Graphitic substances are present in virtually all metapelites, and the degree of graphitization is more sensitive to metamorphic temperature than distribution coefficients. Therefore, they used as quantitative measures of the temperature of metamorphism not only mineral assemblages and distribution coefficients but also the degree of graphitization.

In this way, Hashimoto et al. have divided the metamorphic region into three progressive metamorphic zones: I, II and III, in order of increasing temperature (Fig. 9.6a). Although the grade of metamorphism tends to increase generally to the northwest in the region, this increase is not regular. Each zone does not make a single area clearly separated from the areas of the other zones, but occurs as a stripe inserted between the other zones. For example, a stripe of

(a) Progressive metamorphism

	Zone I	Zone II	Zone III
Biotite		······	——
Garnet	·····	——	——
Chlorite	——	——	——
Muscovite	——	——	——
Albite porphyroblast	······——	——	——
Degree of graphitization	—— 27	28 —— 35	36 ——

(b)

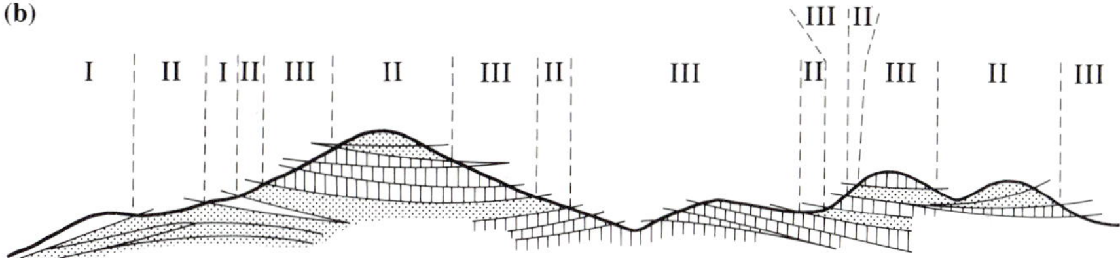

Figure 9.6 (a) Progressive metamorphism of pelitic rocks in the Sanbagawa schists in the Kanto Mountains. (b) Idealized cross section of the metamorphic region in the Kanto Mountains. Thin solid lines represent faults. Schistosity is parallel to the faults. After Hashimoto et al. (1992).

zone III is commonly intercalated between two stripes of zone II, and metamorphic grade changes abruptly at the boundary between the stripes. This indicates that the region is made up of gently inclined, stacked sheets of metamorphic rocks, each sheet belonging to one of the three zones (Fig. 9.6b). The stacking appears to have formed by mechanical shuffling of sheets of metamorphic rocks. If all rocks in this region are considered, virtually all values of the degree of graphitization have been observed without any clear gap. Hence, we may deduce that originally there was a metamorphic complex showing a continuous gradation of temperature of metamorphism, and the complex was later mechanically sliced and shuffled.

Continued subduction of a large oceanic plate will cause a very great amount of syn- or post-metamorphic deformation in rocks along the subduction zone. Moreover, for a metamorphic complex in subduction zones to be exhumed and exposed, the amount of necessary uplift will be of the order of 20–50 km, an amount much greater than that necessary for metamorphism of the low- and medium-*P/T* types in arc zones. The structural disturbance accompanying this uplift may be so great that original thermal structures will be totally destroyed.

9.3.1.2 Small-scale thermal structures. A regular increase of metamorphic (thermal-peak) temperature in a definite direction on a small scale (usually over a distance of a few kilometers in a direction perpendicular to the schistosity) has been reported from many places in high-P/T metamorphic complexes. Some of them may have formed in subduction zones, or at least may reflect some primary features of temperature distribution in subduction zones. However, it is usually not clear how far they were deformed at some later time.

In the Sanbagawa high-P/T metamorphic regions of Japan, small-scale progressive metamorphic zones of this kind have been mapped in many places. In their present structural relations, the temperature of metamorphism increases downward in some cases and upward in others (that is, there are inverted metamorphic zones). Apparent isograds are parallel to the schistosity. However, in some areas where a detailed structural analysis has been made the apparent isograds have been found to be fault lines between two rock sheets showing different metamorphic grades, e.g. in the Kanto Mountains (Hashimoto et al. 1992) as described above, and in central Shikoku (Higashino 1990, Hara et al. 1990).

Beautiful, apparently progressive metamorphic zones on a small scale have been reported to occur in some areas of the Franciscan and other parts of the western United States. For example, Maruyama et al. (1985) described a progressive metamorphic sequence from a chlorite–albite zone through jadeite–lawsonite zone to a glaucophane–jadeite–lawsonite zone, showing downward increasing temperatures in the Pacheco Pass areas of California. The lowest- and the highest-grade part are only a few kilometers apart. For other examples, see Maruyama & Liou (1988) and Peacock (1987b). Many of them show inverted metamorphic zones.

Inverted metamorphic zones of this kind may preserve the inverted temperature distribution that prevailed in the upper layer of the subducting plate and its overlying hanging wall. The estimated temperature gradients in these inverted metamorphic zones are extremely high: commonly one to a few hundred degrees per kilometer. Such high gradients are not impossible if we assume appropriate models of subduction. However, it is equally possible that they are secondary structures formed by subsequent deformations.

9.3.2 *Large-scale distribution patterns of grade and age of metamorphism*

9.3.2.1 Diachronous-progressive metamorphism. The Franciscan and some other metamorphic complexes show a diachronous-progressive distribution pattern of the age and grade of metamorphism (§8.2). It is particularly clear in the Franciscan of the Coast Ranges of California (Fig. 9.7). There, the age of sedimentation of the original rocks as well as the age and grade of metamorphism tends to increase eastward, that is, toward the ancient continent (Ernst 1975, 1977).

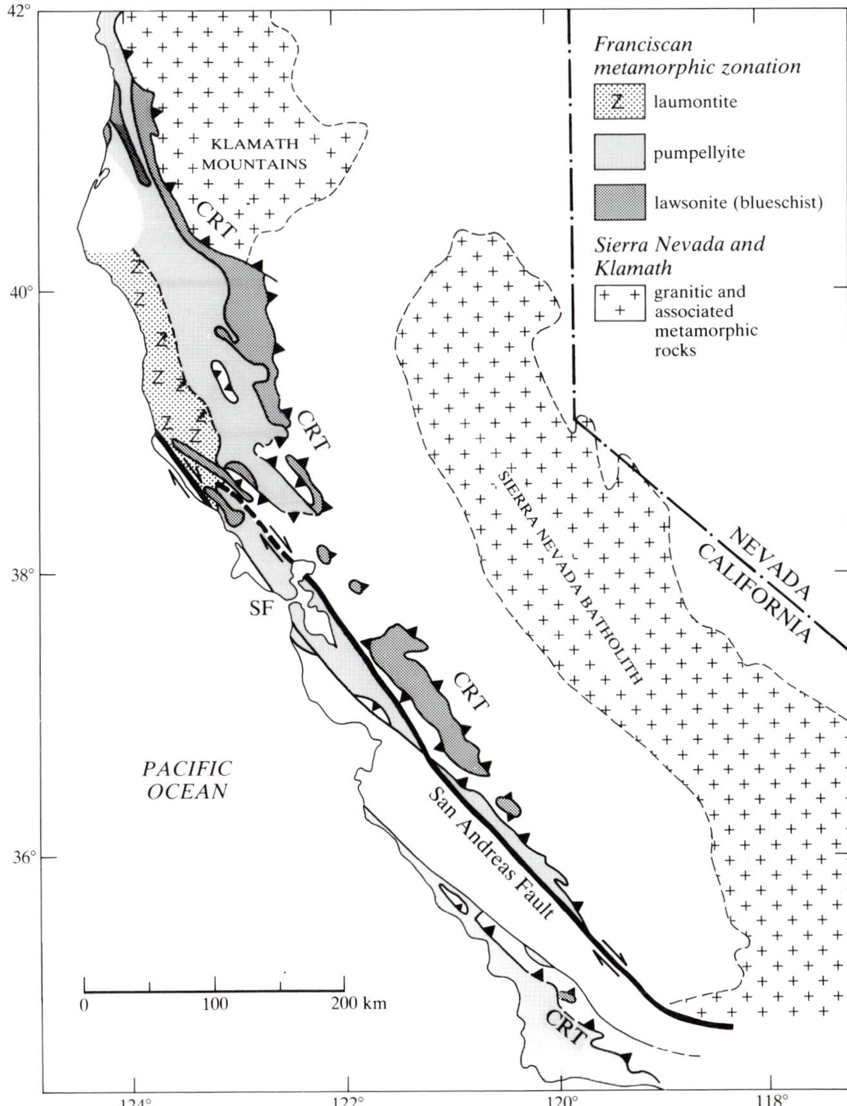

Figure 9.7 The Franciscan Complex in the Coast Ranges of California and the granitic-metamorphic belt of the Sierra Nevada and Klamath Mountains. CRT: Coast Range Thrust that marks the eastern limit of the Franciscan region. SF: San Francisco. Based on Ernst (1975) and others. For the progressive mineral changes in the Franciscan, see Figure 8.7. The Franciscan and the Sierra Nevada belt form an example of paired metamorphic belts of the circum-Pacific regions (Fig. 9.1, Table 9.1).

The increase of the age of metamorphism across a diachronous-progressive metamorphic region seems to be discontinuous (Isozaki & Maruyama 1992, Maruyama et al. 1992). In other words, the region is composed of a number of distinct age provinces (§8.2). Each province appears to show within it its own small-scale distribution pattern of metamorphic grades, as described above. Hence, the increase of metamorphic grade across the whole of a diachronous-progressive metamorphic region appears actually to be no more than a statistical tendency. The field *P–T* curve across a diachronous-progressive metamorphic region is a compound curve showing a statistical trend of change in *P–T* conditions.

These large-scale patterns of distribution of geologic ages and grades of metamorphism are probably related not only to their primary characteristics but also to the mechanism of uplift. It is likely that they were mainly controlled by the progressive oceanward growth of the accretionary complex in the trench zone (§8.2).

Many hypotheses have been proposed for the mechanism of uplift (e.g. see Platt 1987). In the following, only Ernst's tectonic model for the mechanism of uplift is reviewed, because it is particularly intended to explain diachronous–progressive relations.

9.3.2.2 Ernst's (1975) model. Ernst (1975, 1977) has proposed a tectonic model on the basis of his studies in the Franciscan, as follows (Fig. 9.8). In the Mesozoic trench along the west coast of North America, new sediments were accumulated on the oceanic side of the existing sedimentary pile, resulting in oceanward growth of the accretionary complex. The subducting cold oceanic plate dragged these younger sediments deeper, beneath the existing sedimentary pile, causing underplating of the pile by younger sediments. The temperature rose and metamorphism took place in the deep part of the accretionary complex. Successive underplating caused the rise of older, more strongly metamorphosed parts of the complex to the Earth's surface, particularly on the continental side. On the surface, it resulted in a general tendency for the age of deposition and of metamorphism to increase progressively toward the continent. The subducting cold slab maintained the overlying accretionary complex at low temperatures, thus producing high-*P/T* metamorphic conditions. The tectonic–metamorphic process continued for as long as the subduction of the oceanic plate continued.

Ernst considered that most other large high-*P/T* metamorphic complexes had formed in a similar way. This model can explain the regular relations among large-scale spatial distributions of ages of deposition and metamorphism, as well as of metamorphic grade in relation to subduction and uplift.

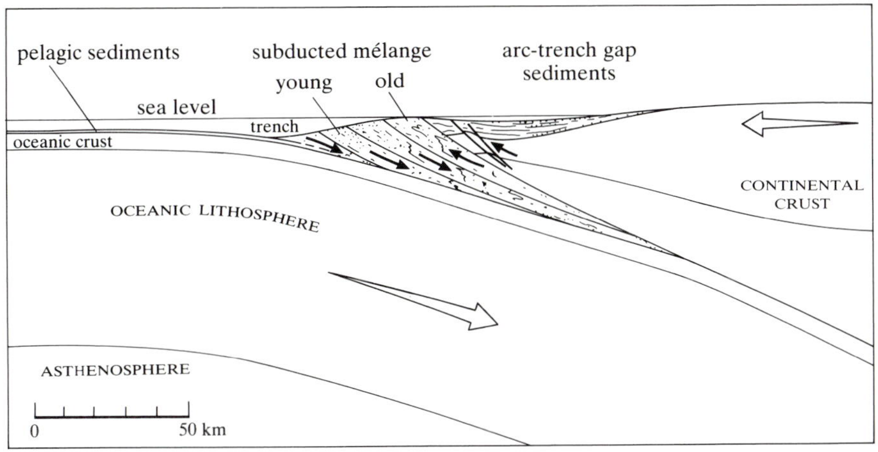

Figure 9.8 Ernst's (1975, 1977) model for subduction-zone metamorphism.

9.4 Regional metamorphism in continental collision zones

9.4.1 Introduction

Continental collision zones are usually accompanied by granitoid intrusions and regional metamorphic belts. Since a continental collision is preceded by a stage of subduction of the ocean basin lying between the two continents, a high-*P/T* metamorphic belt may form in the subduction stage and be paired with an associated coeval medium- or low-*P/T* metamorphic belt. Other medium- and low-*P/T* belts may form in the succeeding stages of collision.

In collision zones, however, detailed interpretation of the tectonic history is usually highly ambiguous. This is particularly true for Paleozoic and older collision zones. Two examples of well investigated continental collision zones are reviewed below. One is the Cretaceous–Tertiary collision zone of the eastern Alps, which is exceptional in having no granitoid intrusions. The other is the Acadian (Devonian) collision zone in the northern Appalachians, which is accompanied by abundant granitoid intrusions.

It appears that, in the absence of granitoid intrusions, the thickened crust of collision zones usually undergoes medium-*P/T* regional metamorphism, whereas if there are associated granitoid intrusions, the shallow crust may undergo low-*P/T* regional metamorphism.

9.4.2 The continental collision zone of the eastern Alps

9.4.2.1 Four major tectonic units of the Alps. The Alps are made up of four main paleogeographical–tectonic units (called *zones* or *domains*; Fig. 9.9).

The southern Alps are mainly located in northern Italy. The pre-Alpine crystalline basement, metamorphosed in the amphibolite and greenschist facies, is covered by folded but unmetamorphosed Permian rocks and Mesozoic sedimentary formations. On the north side, this domain is in contact with the Austro-Alpine domain on the peri-Adriatic tectonic line.

The Austro-Alpine domain is mainly exposed in the eastern Alps in Austria and southern Germany. The southern part of this domain is a pre-Alpine crystalline basement, called the *Altkristallin*, mainly in the amphibolite facies and accompanied by granitoid rocks. The Altkristallin itself is made up of nappes which were thrust northward onto, and hence now cover, the metamorphic region of the Penninic domain. In the eastern Alps, the metamorphic region of the Penninic domain is exposed only in windows such as the Tauern.

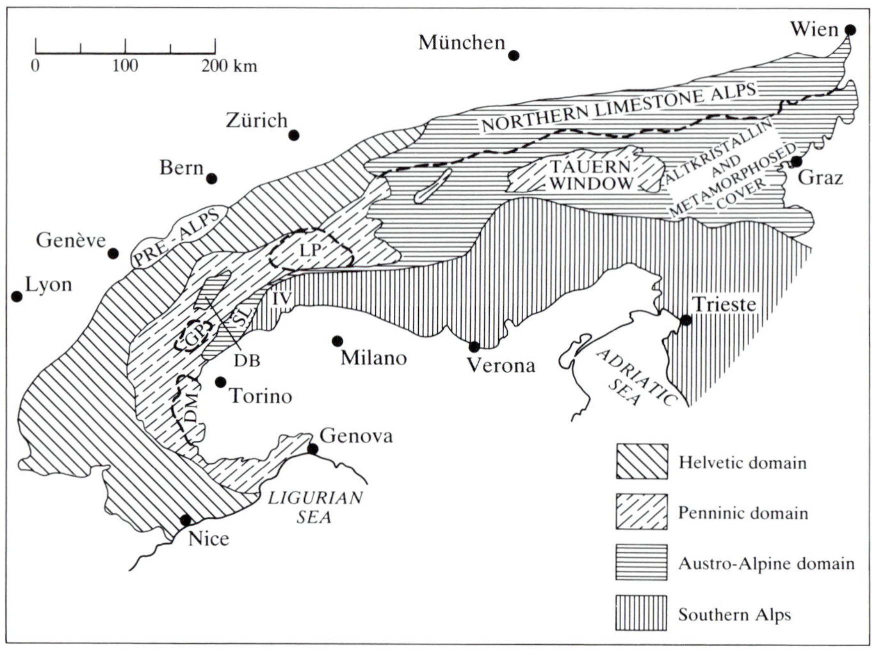

Figure 9.9 Tectonic map of the Alps. LP: Lepontine thermal dome; GP: Gran Paradiso massif; DM: Dora Maria massif; DB: Dent Blanche nappe; SL: Sesia–Lanzo zone; IV: Ivrea zone. Simplified from Frey et al. (1974). A metamorphic map of the western Alps is given in Figure 12.8.

On the other hand, the northern part of the Austro-Alpine domain is composed of the *northern limestone Alps* and others, which are Paleozoic and Mesozoic sedimentary covers that are folded, and partly unmetamorphosed and partly metamorphosed in the greenschist facies. This northern zone was originally deposited on the south side of the Altkristallin, but was afterwards transported in Tertiary time as nappes northward to the present position.

The Penninic domain is widely exposed in the western Alps in Switzerland, southeastern France and the westernmost part of northern Italy, because the overlying Austro-Alpine nappes were mostly removed by erosion there. This domain consists of nappes made up of a pre-Mesozoic granitoid and metamorphic basement and a Permian–Mesozoic metasedimentary cover. The major part of the cover is Jurassic–Cretaceous "*Bündnerschiefer*" (*schistes lustrés*) that were metamorphosed in the blueschist facies in Cretaceous time and accompanied by ophiolitic rocks. In large parts of the Penninic domain (especially in the Lepontine thermal dome), the Penninic nappes show strong overprinting of Eocene medium-P/T metamorphism up to the amphibolite facies.

The Helvetic domain lies on the northern side of the Penninic zone in the western Alps, mainly in Switzerland and southeastern France. This domain consists of a pre-Alpine crystalline basement, and a Permian–Mesozoic–Tertiary sedimentary cover. The cover is partly unmetamorphosed and partly metamorphosed in the greenschist and lower-temperature facies.

9.4.2.2 Sequence of tectonic and metamorphic events in the eastern Alps. In Jurassic time, the pre-Mesozoic crystalline basement of the Helvetic and Penninic domains was part of a northern continental plate (Eurasia), whereas the Austro-Alpine domain was part of a southern continental plate (Apulian continent). Between the two continents, there was a small ocean basin, where Mesozoic sediments of the Penninic domain were deposited (e.g. Hawkesworth et al. 1975, Compagnoni et al. 1977, Gillet et al. 1986).

Subduction stage. Presumably in Early Cretaceous time, the northern and southern continental plates began to approach each other by southward subduction of the small ocean floor in between. In the subduction zone, Mesozoic sediments and volcanic rocks on the oceanic plate were subjected to Cretaceous high-P/T metamorphism, which is now seen in the Bündnerschiefer. Hawkesworth et al. (1975) found that roughly coeval (i.e. Middle and Late Cretaceous) low-P/T metamorphism took place in the southern continental plate (Austro-Alpine unit). This formed andalusite, staurolite and sillimanite, as now observed along the southern margin of the Tauern window. This zone in combination with the zone of high-P/T metamorphism on the north side formed Cretaceous paired metamorphic belts.

Continental collision stage. The upper layer of the southern continental plate, perhaps about 20 km thick, was separated from the underlying layer along a nearly horizontal surface, presumably during the Cretaceous low-*P/T* metamorphism (Armstrong & Dick 1974). The upper layer was thrust northward onto the pre-existing high-*P/T* metamorphic belt of the Penninic zone to form the Austro-Alpine sheet. Hawkesworth et al. (1975) considered that this emplacement of the Austro-Alpine thrust sheet occurred near the Cretaceous/Tertiary boundary (about 90–65 Ma).

Post-collision heating of the thickened crust. Subsequently, a new metamorphic event of the medium-*P/T* type occurred at about 40–35 Ma ago (Eocene) in large parts of the Penninic domain, and produced greenschist- and amphibolite-facies metamorphic rocks, which are now exposed in the Lepontine Alps, Hohe Tauern and other places.

Some authors considered that this metamorphic event was caused by the heat supplied from unexposed plutonic masses at depth. On the other hand, Bickle et al. (1975) claimed that the temperature, pressure and thermal gradient of this metamorphism can be explained simply as a result of temperature rise within the crust thickened by the overthrusting of the Austro-Alpine sheet. This rise of temperature was thought to have occurred during the interval of about 30 million years following the tectonic burial under the Austro-Alpine thrust sheet (cf. Droop 1985).

The overthrusting of the Austro-Alpine sheet thickened the continental crust, and appears to have caused an isostatic rise of the surface level by a few kilometers. This probably increased the rate of erosion of the surface and decreased the thickness of the continental crust, with consequent cooling of the crust including the metamorphic complex. The observed heat-flow values at the present surface give a constraint for calculation of the thermal history of the region (England 1978).

9.4.2.3 P–T–t *paths of the Eocene metamorphic rocks in the eastern Alps.* The thermal-model studies of the Eocene metamorphic rocks in the eastern Alps began with Oxburgh & Turcotte (1974), and opened a new vista in metamorphic studies in general. The collision zone of the eastern Alps had only one major thrusting event, and no coeval granitoid intrusions. In these respects, it represents an extreme, simple collision zone case rather than a typical case. For this reason, it is amenable to a simple theoretical analysis.

Bickle et al. (1975) examined the effects of variations of parameters included in the calculation of the thermal history of such an overthrust region, and concluded that the Eocene regional metamorphism that reached thermal peak about 30 million years after the overthrusting can be explained by radioactive heating in crust thickened by the overthrusting under normal heat flow from the underlying mantle. This suggests that the Eocene regional metamorphism occurred with no magmatism and no abnormally high heat-flow from the mantle. The *P–T–t* paths of individual Eocene metamorphic rocks in the eastern Alps and other similar collision zones have been discussed by England & Richardson (1977), Oxburgh & England (1980), England & Thompson (1984) and A. B. Thompson & England (1984), as mentioned in various parts of this book.

A thick thrust sheet (the Austro-Alpine sheet) was probably emplaced at a rate of centimeters per year (a velocity characteristic of plate motion). This means that a thrust movement of 100 km occurred in 2 million years, a period negligibly short compared with the time required for thermal conduction and radiogenic heating. So, to a first approximation, the thrusting may be regarded as having taken place instantaneously. Any rock which was buried beneath the sheet experienced a very rapid increase of pressure with little increase of temperature, because there was little time for conductive heating. After having reached its burial depth, the rock underwent conductive heating. If uplift and erosion had not taken place, this process of heating would have occurred at a constant depth and pressure. Since uplift actually occurred in the eastern Alps, erosion began, and so pressure decreased, as shown by curves such as (b) in Figure 2.2. Temperature increase slowed and then terminated. Then, temperature began to fall. The *P–T–t* paths after the pressure maximum vary with the rate of erosion. The temperature value of the thermal peak varies accordingly as shown schematically by curves B_1 and B_2 in Figure 2.2.

Field *P–T* curves of rocks in such continental collision zones are schematically shown by curves (3), (4) and (5) in Figure 9.4. The curves show a positive slope, having dP/dT increasing with increasing T. The slope of the curve varies with the heat-source distribution and conductivity of rocks. Such metamorphism is either of the medium- or low-*P/T* type.

9.4.3 The Acadian metamorphic belt in the northern Appalachians

9.4.3.1 Introduction. The Acadian metamorphic belt, about 100 km wide, occurs along the central axis of the northern Appalachians, and trends NE–SW in central Massachusetts, New Hampshire and western Maine, generally along the Merrimack synclinorium. The Acadian (Devonian) orogeny is said to have taken place by the collision of the North American continent, with the Avalonian microcontinent on its east side. The Acadian metamorphic region is accompanied by many granitoid intrusions, which appear to have played an essential rôle in the thermal budget of regional metamorphism.

The metamorphism belongs to the medium-*P/T* type in the southwestern half, and to the low-*P/T* type in the northeastern half of the Acadian metamorphic region (J. B. Thompson & Norton 1968, Carmichael 1978, Chamberlain &

Lyons 1983). In the following section, attention is focused on the low-*P/T* regional metamorphism in the northeastern half. It is probably the best investigated low-*P/T* metamorphic region in the world. The petrology of the region has been studied over many years by Guidotti, Holdaway and their co-workers (e.g. Evans & Guidotti 1966, Guidotti 1970, 1974, Holdaway et al. 1982, 1988) as well as by Ferry (§6.7). Lux et al. (1986) and De Yoreo et al. (1989) have published tectonic syntheses with special emphasis on the thermal history of the metamorphic region, on which the following review is mainly based.

9.4.3.2 Geologic relations. Geologic, petrologic and geothermobarometric data indicate that the present-day erosion surface of the Acadian metamorphic region was situated at depths within the crust which increased in the present southwestward direction during the Devonian metamorphism, as schematically shown in Figure 9.10a. Near the northern end, the present-day and Middle Devonian surfaces of erosion nearly coincide. The present-day crust has an almost uniform thickness of about 35 km in this and adjacent regions. So the crust produced by the Acadian deformations was initially much thicker in the southwestern part than in the northeastern part.

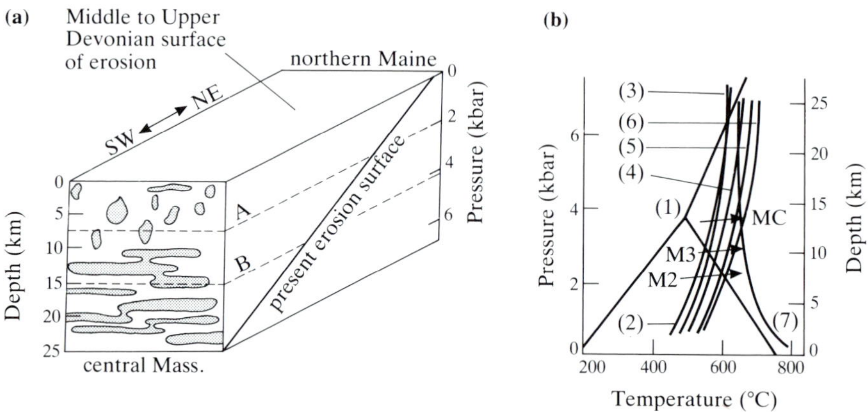

Figure 9.10 (a) Schematic diagram showing the crustal structure of the Acadian metamorphic region. Granitoid plutons are shown stippled. The vertical scale is exaggerated. The actual angle between the Devonian and the present-day surface is 3.4°. (b) Field *P–T* curves (heavy arrows) for low *P/T* metamorphic regions of M2, M3 and MC in the Acadian metamorphic regions of western Maine. (1) triple point for the Al_2SiO_5 minerals; (2) paragonite breakdown curve; (3) curve for the reaction: St + Chl + Ms + Qtz = Als + Bt; (4) curve for the reaction: St + Ms + Qtz = Als + Grt+ Bt; (5) curve for the reaction: Ab + Ms + Qtz = Kfs + Als; (6) muscovite breakdown curve; (7) minimum melting curve of granite. Presence of virtually pure H_2O fluid is assumed. After De Yoreo et al. (1989).

On the present-day erosion surface, the density of plutons increases southwestward: about 13% of the surface area near the northeastern end, about 28% in western Maine, and about 35% in the southwestern part of the region. Field relations and gravity data show that the plutons near the northeastern end are roughly equidimensional, and are surrounded by narrow contact metamorphic aureoles, whereas those in the central and southwestern parts are laterally extensive, sheet-like bodies, less than 2 km in thickness (Fig. 9.10a), accompanied by regional-scale metamorphism. Most of the plutons are granitoid and peraluminous, being characterized by the presence of muscovite and/or garnet. Chemical and isotopic studies suggest a lower crust or a mixed crust–mantle source for these granites.

Thermal-model calculations have shown that, if the crust was thickened to 55 km by the Acadian deformation, the temperatures in lower layers of the crust exceeded the minimum melting point of granite within a few tens of millions of years of thickening. Thus, granitoid magmas formed in the lower crust and intruded the middle and upper crust.

*9.4.3.3 Low-*P/T *metamorphic events in western Maine.* Guidotti, Holdaway and their co-workers have shown the existence of at least four distinct metamorphic events: M1, M2, M3 and MC in chronological order, in the Acadian region of western Maine. These events were separated by periods of cooling.

M1 (slightly older than 400 Ma, Devonian). Synchronous with major deformations, and produced widespread chlorite-zone regional metamorphism.

M2 (400–395 Ma, Devonian). Post-tectonic metamorphic crystallization, which produced widespread staurolite and andalusite in pelitic rocks.

M3 (395–370 Ma, Devonian). Post-tectonic metamorphic crystallization, which produced a progressive sequence of metamorphic zones, characterized by the index minerals: biotite→almandine→staurolite→sillimanite, all accompanied by quartz and muscovite. M3 did not produce andalusite. In many areas, M3 metamorphism overprinted M1 and M2, producing polymetamorphic rocks.

MC (332 Ma, Carboniferous). The MC metamorphism has been distinguished from other metamorphic events, only where the metamorphic grades are in the sillimanite–muscovite or K-feldspar–sillimanite zones.

M3 and MC metamorphism produced regional-scale metamorphic regions with high-grade zones often extending tens of kilometers from the nearest outcrop of plutonic bodies. However, the shapes of the isograds still generally show correspondence to outcrop patterns of associated granitoid plutons. It appears that intrusion of granitoid plutons occurred episodically and caused the temporally separate thermal events M2, M3 and MC. From K/Ar dates of muscovite and biotite, the rate of regional exhumation has been estimated at 0.05 km Ma^{-1} in the period from 400–300 Ma.

M2, M3 and MC metamorphisms occurred at pressures near, or lower than, that of the triple point of Al_2SiO_5 minerals (Fig. 9.10b). The field *P–T* curves of these metamorphic regions were found to be nearly horizontal (with d*P*/d*T* around 0.3 kbar per 100°C or less).

*9.4.3.4 Thermal models for the low-*P/T *regional metamorphism.* Figure 9.10a shows a schematic diagram indicating the crustal structure during the metamorphism. The crust above dashed line A is characterized by contact metamorphism around equidimensional granitoid plutons, whereas that between lines A and B is characterized by low-*P/T* regional metamorphism accompanying the flat-lying granitoid sheets, as now exposed in western Maine. The crust below line B is a regime of medium-*P/T* regional metamorphism now exposed in the south-western part of the region.

De Yoreo et al. (1989) studied one-dimensional thermal models for the low-*P/T* Acadian metamorphism of western Maine. Since the nature of the basement is uncertain, they estimated lower and upper limits for the crustal heat production. First, they calculated the change of temperature in the crust in the absence of igneous intrusions. Then, they calculated the change of temperature caused by intrusion of a horizontal magmatic sheet. The metamorphic regions now under consideration were at depths shallower than about 15 km (Fig. 9.10b), that is, between lines A and B in Figure 9.10a.

If granitoid intrusions had not occurred, the thermal-peak conditions at depths less than 15 km would have been less than 500°C, and within the stability field of kyanite (in the greenschist facies or in the transitional state between the greenschist facies and the amphibolite facies). Thus, the low-*P/T* condition observed in this region cannot be explained in the absence of granitoid intrusions.

Next, they studied thermal models in which a horizontal granitoid sheet, 2 km thick, was intruded into various depths (8–16 km) tens of millions of years after the crustal thickening by the Acadian deformation. In these cases the thermal-peak conditions at depths shallower than the intrusive level have been found to lie in the stability fields of kyanite, andalusite and then sillimanite, with increasing temperature and increasing depth. The highest temperature could reach that of the sillimanite–muscovite or K-feldspar–sillimanite zones. Thus, these models with a granitoid intrusion explain the formation of the low-*P/T* metamorphic rocks.

9.5 Regional metamorphism in continental extension regions

9.5.1 Possible examples of extension-region metamorphism

Heat flow through continental crust is much greater in present-day extension regions (e.g. the Basin & Range province of the western United States) than in typical stable continental regions. This suggests that the lower crust in continental extension regions is experiencing metamorphism at temperatures as high as 700–1000 °C. In the past two decades, some of the granulite-facies complexes of the low-*P/T* type, in particular, have been ascribed to metamorphism in extensional tectonic environments.

The Late Paleozoic (Hercynian) orogenic belts in western Europe have many granulite-facies metamorphic complexes. The radiometric ages of at least some of the complexes are not Late but Early Paleozoic (Caledonian). In the belts, the Late Paleozoic was a time of crustal compression and resultant deformations, and therefore a time of orogeny in the ordinary meaning of the term, whereas the Early Paleozoic was a time of granulite-facies metamorphism accompanied by granitoid intrusions. Weber (1984) developed the hypothesis that, in Early Paleozoic time, continental rifting occurred above an anomalous mantle, and the heat transferred from the mantle to the lower crust caused granulite-facies metamorphism and formation of associated granitoid plutons.

A possible sequence of events that causes formation and exposure of granulite-facies metamorphic complexes of this kind might be as follows (Weber 1984). Extension of continental crust causes crustal thinning by brittle rifting and graben formation in the upper crust, and ductile stretching in the lower crust. The crustal rifting is accompanied by rifting in the underlying lithospheric mantle, which promotes the ascent of hot lithospheric mantle material to the mantle/crust boundary. Basaltic magmas formed by partial melting of the asthenospheric material may intrude into the lower crust. Intense heating by the asthenospheric material and magmas causes granulite-facies metamorphism in the lower crust at 7–11 kbar and 700–850 °C at the base of a crust, approximately 30–40 km thick. The heating of the crustal rocks results in the formation of granitoid magmas.

During these processes at depth, sedimentation of a Paleozoic sequence continued on the surface. This means that tectonic uplift did not take place, presumably because the processes were extensional and did not cause crustal thickening. In Late Paleozoic time, crustal compression occurred, producing nappe and fold structures that involved already-formed granulite-facies rocks and their associated plutons. This compressional phase thickened the continental crust, and caused new metamorphic and plutonic events, followed by uplift and deep-reaching erosion, resulting in the exposure of Early Paleozoic granulite-facies metamorphic complexes on the surface.

Wickham & Oxburgh (1985, 1987) published a more or less similar version of origin for the Late Paleozoic low-*P/T* metamorphic region of the eastern Pyrenees, which was subsequently uplifted and exposed by the Tertiary orogeny. A sequence of Lower and Upper Paleozoic pelitic rocks occurs there, and the lower part of it has been metamorphosed. The high-grade metamorphic sequence passes downward through a migmatite zone into anatectic granites. The *P-T* condition of metamorphism reached 700°C at 3.0–3.5 kbar (about 11–13 km in depth). These high-temperature conditions at shallow depths were ascribed to heat transfer by magmas or other fluids rising from the lower crust or the mantle. Wickham & Oxburgh have speculated that the Late Paleozoic Pyrenees were an extension region. Extension caused thinning of the lithosphere and hence the rise of asthenospheric substance and the generation and rise of basaltic magmas. The crustal thinning caused subsidence of the crust, resulting in continuous sedimentary deposition on the Earth's surface.

9.5.2 P–T–t paths in extension-region metamorphism

Sandiford & Powell (1986) discussed continental extension and consequent *P–T–t* paths of metamorphic rocks. Extension produces relatively narrow linear rift zones in some cases and wide extension regions in others. The areal extent of metamorphism differs accordingly. Increase of temperature in the crust tends to be high in areas where mantle lithosphere has become thin, convective heat transfer in the mantle is active, and magmas rise into the crust. In areas where crustal thickness remains constant, prograde metamorphism may occur at constant pressure, whereas in areas where crustal thinning takes place, prograde metamorphism may be accompanied by decreasing pressure.

Generally the region will not be greatly uplifted, and so erosional denudation will be negligible. Hence, syn-metamorphic volcanic rocks will be well preserved, and the cooling of metamorphic complexes will take place under a virtually constant pressure. From petrologic studies of retrograde mineral changes, Sandiford (1985, 1989) and Sandiford et al. (1987) claimed that some granulite-facies metamorphic regions experienced post-peak cooling under a nearly constant pressure. They regarded this as evidence for the continental extension origin of these metamorphic complexes.

10 Changes of paragenetic relations and of chemical compositions of solid-solution minerals in metapelites with temperature and pressure

10.1 Progressive changes of paragenetic relations in metapelites expressed in *AFM* diagrams

The mineralogical composition of metamorphic rocks varies not only with *P–T* conditions but also with the chemical composition of rocks exposed in metamorphic regions. Even among pelitic rocks, there are considerable variations in bulk chemical composition, which may lead to major variations in mineralogical composition under the same *P–T* conditions. The effects of bulk chemical composition on mineralogical composition can be clarified by constructing various types of composition–paragenesis diagrams for successive index-mineral zones. This section deals with this problem with respect to metapelites.

10.1.1 The Barrovian sequence of the Scottish Highlands

10.1.1.1 The range of bulk-rock compositions. We begin with the Barrovian sequence of progressive metamorphic zones in the Dalradian of the Scottish Highlands (§1.6). The petrology of this sequence has been exceptionally well documented (e.g. Barrow 1893, 1912, Tilley 1925, Chinner 1960, 1961, 1965, 1966, 1967, 1980, Atherton 1968, 1977, Harte & Johnson 1969, Harte & Hudson 1979, Baker & Droop 1983, Baker 1985, Dempster 1985, McLellan 1985).

In order of increasing temperature, the Barrovian sequence is composed of the zones of chlorite, biotite, garnet (almandine), staurolite, kyanite and sillimanite defined for pelitic rocks, as shown in Figures 1.4 and 1.5. It belongs to the kyanite–sillimanite series (medium-P/T type). It should be noted, however, that all kyanite–sillimanite series regions do not develop a Barrovian sequence, because the formation of this sequence is the result of a combination of P–T conditions of this series with a certain limited range of bulk-rock chemical compositions, as outlined below.

Atherton (1977) made a comprehensive study of the bulk-rock chemical composition of Dalradian metapelites. In the Thompson *AFM* diagram, the dominant majority of the Dalradian metapelites plot in the field between the almandine–chlorite line and the biotite line. The molar MgO/(FeO + MgO) ratios are in the range 0.35–0.55. This very limited range of bulk-rock chemical composition is essential for the formation of the regular mineralogical characteristics of the Barrovian sequence.

10.1.1.2 Progressive changes of paragenetic relations in the Barrovian sequence. Figure 10.1 shows a series of *AFM* diagrams representing individual zones of the Barrovian sequence. In the Barrovian sequence, muscovite is stable in the presence of quartz up to the highest grade, i.e. the sillimanite zone. Therefore, we can use ordinary *AFM* diagrams in which mineral assemblages are plotted from the muscovite point. In the following discussion, all ferromagnesian minerals are simplified as binary Fe–Mg solid solutions, unless otherwise stated.

In pelitic rocks free of carbonate, the formation of biotite in the transition from the chlorite to the *biotite zone* is essentially related to a change in the extent of Tschermak substitution particularly in muscovite, and so cannot be analyzed by the *AFM* diagram. This reaction is given as reaction (2.1) of §2.3.3 in a simplified form and as reaction (5.4) of §5.4.3. It may be visualized by use of Eskola's *A'KF* diagram, which can display the extent of Tschermak substitution in muscovite as well as in chlorite and biotite (Fig. 11.3). For a more general discussion, the reader is referred to Ernst (1963b), Mather (1970) and Miyashiro & Shido (1985).

In some parts of the Dalradian, many metapelites in the chlorite zone contain calcite and dolomite (Mather 1970: 272). Therefore, biotite-producing reactions involving carbonate, such as reactions (2.2)–(2.5), may have occurred. Some segments of the observed biotite isograd may have been controlled by reactions of this kind. (The effects of carbonate minerals on the course of progressive metamorphism of pelitic rocks is discussed in §10.2.)

In the biotite zone, the typical metapelites show the biotite + chlorite assemblage accompanied by muscovite, quartz and albite. In the *garnet zone*, the typical metapelites show the almandine + biotite + chlorite assemblage. In this grade, chlorite has a little lower $A/(F + M)$ ratio than the associated garnet, and

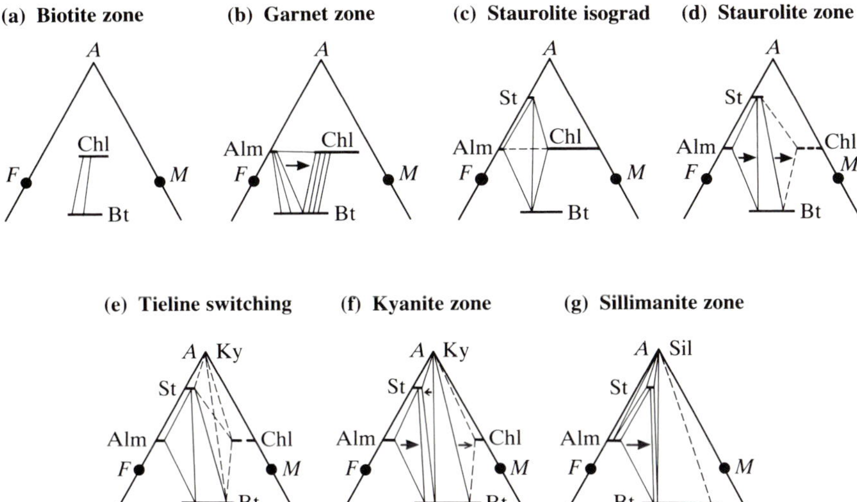

Figure 10.1 A series of *AFM* diagrams showing the progressive changes of paragenetic relations in pelitic metamorphic rocks in the Barrovian sequence in Barrow's classic area (Glens Clova and Esk) in the southeastern part of the Scottish Highlands. Arrows indicate the direction of movement of paragenesis triangles with increasing temperature. It is essential that metapelites contain a "normal" amount of MnO. Based on Chinner (1965), Harte & Hudson (1979) and others.

so the reaction is of the type:

chlorite + muscovite +quartz = garnet + biotite + H_2O (Fig. 10.1b).

The *staurolite isograd* is characterized by a tieline-switching reaction from almandine–chlorite to staurolite–biotite, leading to the formation of the staurolite +biotite assemblage (Fig. 10.1c). With increasing temperature within the *staurolite zone*, both staurolite–almandine–biotite and staurolite–biotite–chlorite triangles move to the right (Fig. 10.1d).

Next, the tieline-switching reaction from staurolite–chlorite to kyanite–biotite occurs (Fig. 10.1e). In some areas (e.g. Whetstone Lake area of Ontario; Carmichael 1970) this reaction is said to mark the *kyanite isograd*.

In Barrow's classic area of the Scottish Highlands, however, Chinner (1965) pointed out that the metapelites exposed there have compositions plotting on the left side of the staurolite–biotite side of the staurolite–biotite–kyanite triangle, and so metapelites do not show the effect of the above tieline-switching reaction. With further increase of temperature, the staurolite–biotite side of the staurolite–biotite–kyanite triangle moves to the left, and so some rocks in Barrow's classic area begin to form kyanite. This produces the kyanite isograd in this area. On this kyanite isograd, the molecular MgO/(FeO + MgO) ratio of biotite in the staurolite–biotite–kyanite triangle was found to be 0.50 (§5.4.6).

The Barrovian staurolite and kyanite zones are characterized not by the mere occurrence of staurolite and kyanite, respectively, but by the formation of the staurolite–biotite and kyanite–biotite tielines, respectively. Because of the appearance of these tielines, metapelites on the less aluminous side of the almandine–chlorite line come to contain the highly aluminous minerals staurolite and kyanite. Chlorite disappears at the staurolite isograd in Barrow's classic area, because the MgO/(FeO + MgO) ratios of the exposed metapelites are relatively low, whereas chlorite persists until the tieline switching from staurolite–chlorite to kyanite–biotite occurs in the Whetstone Lake area and many other places, where some metapelites have higher MgO/(FeO + MgO) ratios.

The *sillimanite isograd* in the Barrovian sequence is the line on which sillimanite becomes stable in place of kyanite. In high-grade parts of the sillimanite zone, staurolite may disappear by the terminal reaction (5.15).

10.1.1.3 Additional comments. There are many metapelites that contain four- or even five *AFM* phases in violation of the basic rule of the *AFM* diagram. These assemblages usually contain garnet with a considerable amount of MnO (and CaO). So they have been ascribed to the presence of these components. Symmes & Ferry (1992) have shown that the widespread occurrence of garnet in metapelites is itself a result of the presence of a small amount of MnO (§5.5.3). As a consequence, the *AFM* diagrams in Figure 10.1 are for metapelites containing a "normal" amount of MnO and not for the six-component system SiO_2–Al_2O_3–FeO–MgO–K_2O–H_2O.

In Figure 10.1, most of the paragenesis triangles move to the right with increasing temperature. This is due to two factors related to chlorite and garnet. Low-grade metapelites usually contain chlorite, which has a large content of H_2O and a higher MgO/(FeO + MgO) ratio than coexisting ferromagnesian minerals. Progressive dehydration reactions produce ferromagnesian minerals at the expense of chlorite, and so increase the MgO/(FeO + MgO) ratio of all the coexisting ferromagnesian minerals. Garnet is an anhydrous mineral and has a MgO/(FeO + MgO) ratio lower than any coexisting ferromagnesian minerals. So progressive dehydration reactions tend to produce garnet at the expense of other ferromagnesian minerals, resulting in an increase of the MgO/(FeO + MgO) ratios of all the coexisting ferromagnesian minerals (§5.4.4.3).

10.1.2 Unusual pelitic rocks plotting on the more aluminous side of the almandine–chlorite line

Virtually all pelitic rocks of the Scottish Dalradian plot on the lower side of the almandine–chlorite line in the *AFM* diagram. This is a significant reason for the origin of the Barrovian sequence. In other regions, such as parts of the northern Appalachians, there are pelitic rocks plotting on either side of the almandine–chlorite line, and this causes some deviations in the sequence of appearance of index minerals with increasing metamorphic grade (e.g. Albee 1968, 1972, Zen 1981).

In *AFM* diagrams for low- and middle-grade metapelites, the mineral assemblages on the lower and upper sides of the almandine–chlorite line are usually separated by the garnet–chlorite join. The formation of the garnet–chlorite paragenesis is an important factor controlling mineral assemblages of metapelites. It should be noted that the garnet–chlorite paragenesis forms because of the presence of a "normal" amount of MnO in pelitic rocks (Symmes & Ferry 1992), as discussed in §5.5.3.

Figure 10.2 shows a simplified series of schematic *AFM* diagrams for pelitic rocks in some areas with metapelites plotting on both sides of the almandine–chlorite line. At low temperatures, the garnet–chlorite join divides the diagram into two parts: a composition field of ordinary metapelites on the lower side, and one of unusually aluminous metapelites on the upper side. The progressive sequence of the appearance of index minerals in ordinary metapelites plotting below the garnet–chlorite join is: biotite, garnet and staurolite, just as in the Barrovian zones. For convenience of comparison with the Dalradian, the biotite, garnet and staurolite zones are defined here on the basis of metapelites plotting below the garnet–chlorite line.

However, in pelitic rocks plotting above the garnet–chlorite join, chloritoid occurs in the biotite zone, and kyanite (or andalusite) and staurolite begin to occur in the garnet zone. At higher temperatures, the garnet–chlorite join is replaced by the staurolite–biotite join. So staurolite occurs in metapelites on either side of the almandine–chlorite line, as shown in Figure 10.2d, whereas kyanite (or andalusite) is still confined to metapelites above the almandine–chlorite line. At still higher temperatures, kyanite (or andalusite) occurs in metapelites that plot on either side of the almandine–chlorite line, as shown by Figure 10.2e.

In some metamorphic areas, the four-phase assemblage garnet (manganiferous) + chlorite + biotite + chloritoid that straddles the almandine–chlorite line occurs commonly in metapelites at temperatures lower than that of formation of the staurolite–biotite join. This obscures the distinction between metapelites on the lower and upper sides of the almandine–chlorite line. Actually observed paragenetic relations are more diverse than those shown in Figure 10.2 (e.g.

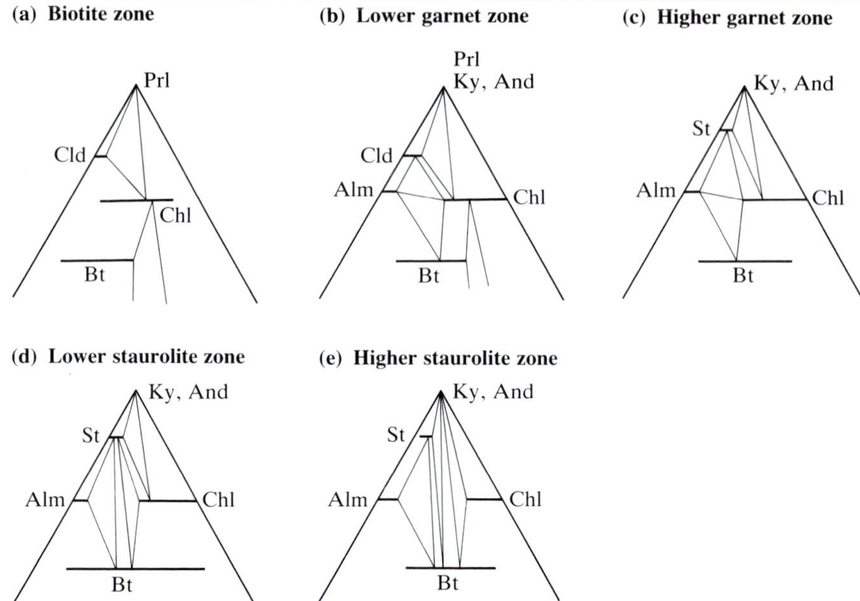

Figure 10.2 Simplified, schematic *AFM* diagrams showing progressive changes in paragenetic relations in pelitic rocks plotting on the lower and upper sides of the almandine–chlorite line. The zone name for each diagram is based on an index mineral in ordinary metapelites plotting on the lower side of the line. Metapelites are assumed to contain a "normal" amount of MnO.

Albee 1972), and are controlled largely at least by the MnO content (Spear & Cheney 1989, Symmes & Ferry 1992).

10.1.3 Progressive changes of metapelites at temperatures higher than in the Barrovian sequence

In the Barrovian sequence, metapelites even in the highest-temperature part contain muscovite, and belong to the sillimanite–muscovite zone in Figure 8.5. Temperatures higher than those in the Barrovian sequence are known to occur in some limited regions within Phanerozoic orogenic belts, but occur very widely in high-grade gneiss regions of the Precambrian.

The progressive changes of metapelites in this temperature range is schematically shown in Figure 10.3. Here, the ordinary *AFM* diagram is used for metapelites below the breakdown temperature of muscovite, whereas the modified *AFM* diagram projected from the K-feldspar point is used at higher temperatures.

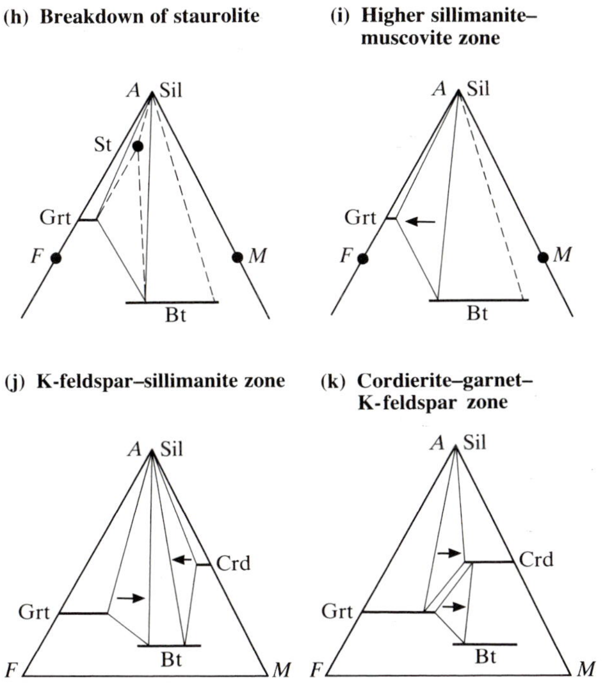

(h) Breakdown of staurolite

(i) Higher sillimanite–muscovite zone

(j) K-feldspar–sillimanite zone

(k) Cordierite–garnet–K-feldspar zone

Figure 10.3 Schematic *AFM* diagrams showing progressive changes of paragenetic relations in metapelites at temperatures higher than those of the Barrovian sequence. Diagrams (h) and (i) are ordinary *AFM* diagrams projected from the muscovite point, whereas diagrams (j) and (k) are modified *AFM* diagrams projected from the K-feldspar point (§5.4.1). Diagram (h) succeeds diagram (g) of Figure 10.1.

The following changes occur with increasing temperature:

(a) Breakdown of staurolite occurs in the middle of the sillimanite–muscovite zone by a terminal reaction (5.15) shown in Figures 5.12 and 10.3h. (This reaction may occur in the highest temperature part of the sillimanite zone in the Barrovian sequence.)

(b) Breakdown of muscovite occurs with the resultant formation of K-feldspar by reactions such as: muscovite + quartz = K-feldspar + sillimanite + H_2O (Fig. 10.5e). This begins near the amphibolite–granulite facies boundary, and in actual cases continues over a zone with a considerable width, presumably because of ionic substitutions ignored in the *AFM* diagram. Finally muscovite disappears in the K-feldspar–sillimanite zone (Fig. 10.3j).

(c) Tieline-switching reaction from biotite–sillimanite to garnet–cordierite occurs by a reaction of the type: biotite + sillimanite + quartz = garnet + cordierite + K-feldspar + H_2O, as shown by the change from (j) to (k).

The topologies of Figure 10.3j and k occur in both low- and medium-*P/T* metamorphic regions (see also Figs 8.4, 8.5). In the granulite facies, metapelites relatively high in FeO/MgO ratio may contain orthopyroxene, as shown in Figure 11.7 (Froese 1978, Lonker 1980).

10.2 Effects of the presence of carbonates and the infiltration of externally derived aqueous fluid on low- and medium-*P/T* metamorphism of pelitic rocks

*10.2.1 Carbonates in pelitic rocks in low- and medium-*P/T* metamorphic regions*

10.2.1 Introduction. In the preceding section the discussion of metamorphism of pelitic rocks with increasing temperature ignored the effects of the presence of carbonate minerals in pelitic sedimentary and metamorphic rocks. Actually average pelitic sedimentary rocks contain several per cent of carbonate minerals. In chlorite and biotite zones of low- and medium-*P/T* metamorphic regions, many or most pelitic metamorphic rocks contain a small amount of carbonates: calcite, dolomite, ankerite (FeO-rich dolomite) and siderite.

These carbonates may participate in biotite-forming reactions, as already discussed for reactions (2.2)–(2.5) in §2.3.3. In carbonate-bearing rocks, decarbonation reactions produce CO_2 and cause an increase of the chemical potential of CO_2 in the intergranular fluid. This may change the type and order of reactions in prograde metamorphism.

It is well known that small amounts of carbonates are widespread in low-grade pelitic metamorphic rocks. When special attention was not paid to the carbonates, they may have beeen mistakenly enumerated as calcite. In the literature, there are a considerable number of relatively reliable, but fragmental, descriptions of the occurrence of dolomite, ankerite and siderite in metapelites. For example, the occurrence of these carbonates in the Dalradian is referred to in §10.1.1.2.

10.2.1.2 Carbonates in metapelites and metamorphic grades. Our knowledge is still very incomplete about the range of metamorphic grades over which these carbonates occur in metapelites and about the frequencies of their occurrence in each grade. Metapelites of chlorite zones appear commonly to contain both siderite and ankerite, whereas metapelites in the lower grade part of biotite zones would contain ankerite. Calcite is probably common in metapelites of both chlorite and biotite zones, but not in higher grade zones. These relations suggest that some reactions consuming siderite and ankerite usually occur on biotite isograds and in lower biotite zones, whereas reactions consuming calcite usually occur in biotite zones (and possibly also in higher grade zones if calcite is present there). Since epidote is common in chlorite zones, however, some carbonates may be consumed by chlorite-zone reactions like the following: dolomite + amesite component (chlorite) + quartz + H_2O = clinozoisite + antigorite component (chlorite) + CO_2; and calcite + amesite component (chlorite) + quartz = clinozoisite + antigorite component (chlorite) + H_2O + CO_2 (with coefficients ignored).

Carbonates are consumed by various reactions with associated minerals. Each reaction consumes a carbonate (or carbonates) and some other minerals, and produces other minerals in definite proportions. If a rock contains a relatively large amount of a carbonate to be consumed as compared with other minerals participating in the reaction, part of the carbonate may survive the reaction, and persist to a higher grade. The grades to which carbonates persist vary with the bulk-rock chemical composition, and, in particular, with the proportions of carbonates and other minerals in the initial rock.

A most systematic investigation of carbonate minerals in metapelites in a chlorite zone and a biotite zone has been made by Ferry (1984) in Maine. Metapelites of the chlorite zone there contain siderite, ankerite and calcite. Production of biotite on the biotite isograd was due to a number of decarbonation reactions consuming siderite and ankerite. Calcite is the only carbonate in the biotite zone.

10.2.2 Symmes & Ferry's (1991) calculation of the effects of carbonates on the course of prograde metamorphism of average pelite

Symmes & Ferry (1991) made a thermodynamic calculation of the effects of the presence of carbonates on the course of prograde metamorphism of pelites. For the original rock, they assumed an "average pelite", with a chemical composition very close to, but slightly modified from, the average low-grade pelite of Shaw (1956), and with a mineralogical composition corresponding to a chlorite-zone condition. All fluids under consideration are assumed to be mixtures of H_2O and CO_2. They calculated the successive changes of mineralogical composition of the rock with increasing temperature for three models of the mode of fluid–rock interaction. The three models are described below in the order of increasing flux of an externally derived aqueous fluid.

The no-infiltration case. This model represents the course of metamorphism with increasing temperature under the condition of no infiltration of an externally derived fluid into the rock. If the rock-fluid system is closed, dehydration and decarbonation reactions will greatly increase the amount of fluid. This is unlikely in natural metamorphic rocks. So, the model assumes that the volume of fluid in equilibrium with the rock is kept at a constant value (1%) by allowing fluid to leak out of the system.

The intermediate-flux case. This model represents the course of metamorphism with increasing temperature with infiltration of a time-integrated volume flux of 10^4 cm^3 cm^{-2} of fluid (corresponding to 487 mol cm^{-2} at the biotite isograd). Fluid with a composition of 95 mole % H_2O and 5 mole % CO_2 is assumed to infiltrate at the low-temperature end of the rock column. The volume of fluid in the system is always kept at 1%. Hence, the infiltrating fluid displaces each successive fluid through the rock column. Because of reactions with rocks, the composition of fluid changes as it passes through the column.

The large-flux case. This model represents the course of metamorphism with increasing temperature with infiltration of sufficiently large quantities of externally derived fluid of a specified composition (95% H_2O and 5% CO_2). The time-integrated fluid flux is greater than 5×10^4 cm^3 cm^{-2}. Since the number of moles of externally derived fluid is much greater than the number of moles of minerals in the system, equilibrium fluid composition remains virtually constant (5% CO_2) at all times during metamorphism.

Results of calculation for 5 kbar and the initial fluid composition with 5% CO_2. The stable mineral assemblages, mineral modes and the compositions of minerals and fluid were calculated for the temperature range 350–650°C (roughly corresponding to grades from the chlorite to the sillimanite zone). All reactions encountered are thermodynamically continuous; that is, they occur over a range of temperatures. Figure 10.4 illustrates the calculated changes of fluid compositions. Table 10.1 summarizes the calculated order of appearance and disappearance of minerals with increasing temperature for each of the three models of fluid–rock interaction.

In the no-infiltration case, the first reaction involves the decarbonation of siderite with the formation of biotite. It is followed by another reaction involving the decarbonation of ankerite with the formation of biotite, plagioclase and siderite. Further increases of temperature cause reactions involving the decarbonation of the siderite and dehydration of silicates, producing both CO_2 and H_2O. The CO_2 content of the fluid increases with temperature, reaching 85% at 531°C (Fig. 10.4). Since these reactions eliminate all carbonates, a further temperature increase causes only a dehydration reaction, resulting in a decrease of the CO_2 content to 33%. In this case biotite coexists with carbonate(s) over a wide range of temperature.

In the intermediate-flux case, the first reaction involves the decarbonation of siderite with the formation of biotite. It is followed by another reaction involving the decarbonation of ankerite with the formation of biotite and pla-gioclase. These two reactions increase the CO_2 content of fluid up to 27% at 484°C. Since these reactions eliminate all carbonates, further temperature increases cause reactions involving H_2O only with the fluid composition kept nearly constant. The order of appearance and disappearance of minerals in this case is the same as in the large-flux case (Table 10.1). The order of reactions that cause these mineralogical changes is shown in Table 10.2.

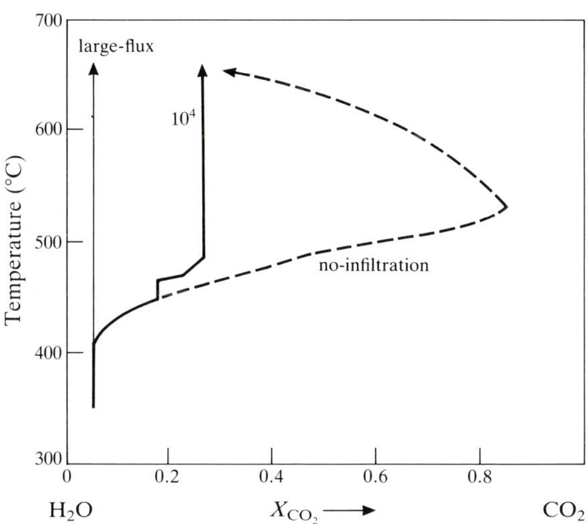

Figure 10.4 Calculated compositional changes of intergranular fluid with increasing temperature at 5 kbar with an initial fluid composition of 5% CO_2. In the intermediate-flux case, the time-integrated fluid flux is of 10^4 cm^3 cm^{-2} (labeled "10^4"). Kinks in the paths of the no-infiltration and intermediate-flux cases result from changes in mineral assemblages (after Symmes & Ferry 1991).

Table 10.1 Sequence of mineralogical changes in average metapelite with increasing temperature at 5 kbar, with an initial fluid composition of 5% CO_2 (Symmes & Ferry 1991).

No-infiltration		Intermediate flux		Large flux	
°C		°C		°C	
405	Biotite appears	405	Biotite appears	405	Biotite appears
475	Plagioclase appears	447	Siderite disappears	406	Siderite disappears
489	Chlorite disappears	464	Plagioclase appears	440	Plagioclase appears
515	K-feldspar appears	484	Ankerite disappears	451	Ankerite disappears
516	Ankerite disappears	492	Garnet appears	507	Garnet appears
530	Kyanite appears	516	Staurolite appears	532	Staurolite appears
531	Garnet appears	517	Chlorite disappears	532	Chlorite disappears
531	Siderite disappears	598	Sillimanite appears	608	Sillimanite appears
		598	Staurolite disappears	609	Staurolite disappears

Note: The average metapelite has the following chlorite-zone mineralogical composition: 5.60 quartz, 0.75 muscovite, 0.56 albite, 0.16 chlorite, 0.39 ankerite, 0.40 siderite (in mol kg^{-1} rock). All rocks at all temperatures contain quartz and muscovite.

Table 10.2 Sequence of net-transfer reactions in average metapelite with increasing temperature in the large-flux case at 5 kbar, with an initial fluid composition of 5% CO_2 (Symmes & Ferry 1991).

Net-transfer reaction	Temperature range (°C)
Qtz + Ms + Sd + H_2O = Bt + Chl + CO_2	405–406
Qtz + Ms + Chl + Ank = Pl + Bt + CO_2 + H_2O	440–51
Qtz + Pl + Chl + Bt = Ms + Grt + H_2O	507–32
Ms + Chl + Grt = Qtz + Pl + Bt +St + H_2O	532
Qtz + Bt + St = Ms + Pl + Grt + H_2O	532–608
Qtz + Ms + Pl + St = Bt + Grt + Sil + H_2O	608–609
Ms + Grt + H_2O = Qtz + Pl + Bt + Sil	609–50

Note: Ank = ankerite, Sd = siderite; for other symbols, see Appendix I.

Since the CO_2 content of fluid reaches a very high value in the no-infiltration case as compared with the other two cases, the sequence of reactions that take place and the resultant order of the appearance and disappearance of minerals with increasing temperature differ greatly between the no-infiltration case and the intermediate- and large-flux cases. The latter two cases show the same sequence of reactions, and hence the same sequence of appearance and disappearance of minerals.

In the no-infiltration case, the order of appearance of minerals is: biotite, plagioclase (25% An), K-feldspar, kyanite, garnet. In this case, K-feldspar appears at a lower temperature than sillimanite, and carbonates persist to a relatively high temperature (Table 10.1). In the intermediate- and large-flux cases, on the other hand, the order of appearance of minerals is: biotite, plagioclase (25% An), garnet, staurolite, sillimanite. In this case, carbonates disappear at a low temperature before the disappearance of chlorite, and sillimanite appears at a lower temperature than K-feldspar (Table 10.1).

The observed orders of appearance and disappearance of minerals in ordinary regional metamorphism of pelitic rocks with increasing temperature are very different from that in the no-infiltration case, but show a considerable resemblance to those in the intermediate- and large-flux cases. Therefore, Symmes & Ferry have concluded that ordinary regional metamorphism of pelitic rocks occurs in the presence of fluid infiltration, and that fluid infiltration is a general driving force of regional metamorphism of pelitic rocks.

This view is tentatively accepted here, although the validity of this conclusion depends on the distribution and amount of carbonates in pelitic metamorphic rocks, and in this respect our data are still very incomplete.

10.3 Progressive changes of paragenetic relations expressed in Al_2O_3–$NaAlO_2$–$KAlO_2$ diagrams

J. B. Thompson & A. B. Thompson (1976) gave a detailed analysis of progressive changes of paragenetic relations in pelitic rocks by use of their Al_2O_3–$NaAlO_2$–$KAlO_2$ diagram (§5.6). A brief review is given below.

*10.3.1 In low- or medium-*P/T *metamorphism*

Figure 10.5a–f shows progressive changes of paragenetic relations with increasing temperature under a constant pressure and chemical potential of H_2O. Increasing temperature causes continuous changes of composition of minerals in each paragenesis triangle. When a specific temperature is passed, a discontinuous change occurs, which is the appearance of a new phase, or the disappearance of an old phase, or a tieline switching. Each of diagrams (a) to (f) shows paragenetic relations in a temperature range limited by such discontinuous changes.

Thus, a discontinuous change from pyrophyllite to Al-silicate occurs between (a) and (b), the disappearance of the Na end-member of paragonite occurs between (b) and (c), and the terminal reaction for the disappearance of paragonite occurs between (c) and (d). The change from (d) to (e) is caused by a tieline-switching reaction, while that from (e) to (f) is caused by the disappearance of the K end-member of muscovite. J. B. Thompson & A. B. Thompson (1976) actually observed assemblages (a), (b), (d), (e) and (f), while they inferred the existence of (c) as the intermediate step between (b) and (d).

It should be noted that the progressive dehydration changes from (c) to (f) occur only at pressures below about 4 kbar (Fig. 11.5). At higher pressures, muscovite shows dehydration-melting in the presence or absence of an aqueous vapor (§11.6.3).

At very high temperature, the miscibility gap in alkali feldspar should close. At this temperature, if muscovite is still stable, the paragenetic relations should be as shown in diagram (g), while if muscovite is no longer stable, they should be as shown in diagram (g′). The latter case was thought to be more likely.

The stability boundary between diagrams (a) and (b) is the breakdown curve of pyrophyllite (curve P in Figure 8.1). Diagram (b) is valid at temperatures around the triple point of Al_2SiO_5. Diagrams (c)–(g′) hold at higher temperatures. The change from (d) to (e) is the tieline-switching reaction by which sillimanite (or andalusite) begins to coexist with K-feldspar (Fig. 5.16). This marks the entrance to the K-feldspar–sillimanite zone of Figures 8.4 and 8.5.

The tieline-switching reaction between (d) and (e) generally occurs over a

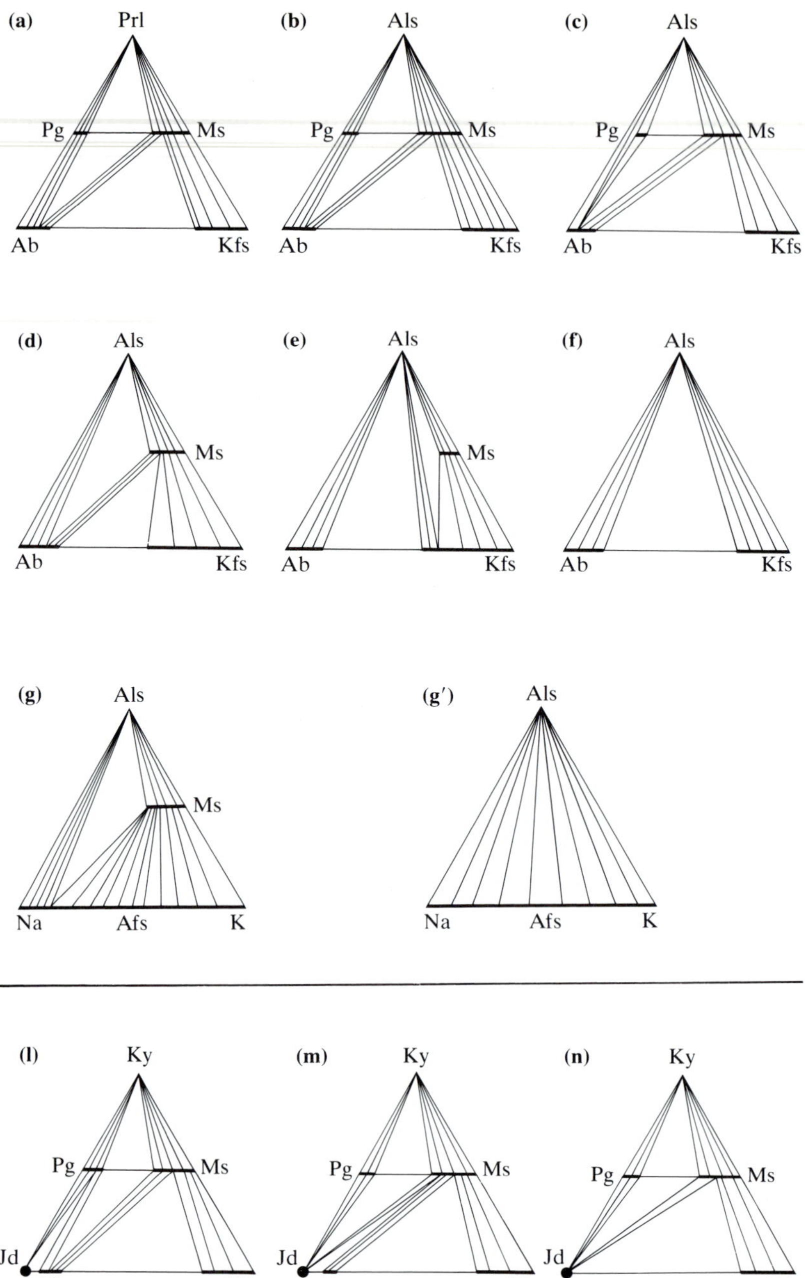

Figure 10.5 Progressive changes of paragenetic relations in Al_2O_3–$NaAlO_2$–$KAlO_2$ diagrams after J. B. Thompson & A. B. Thompson (1976). (a–f) Progressive changes of paragenetic relations in quartz-bearing rocks with increasing temperature in this order from the greenschist to the granulite facies. (g, g′) Paragenetic relations with complete solid solution in the alkali feldspars. (l)–(n) Paragenetic relations with jadeite, which are derived by increase of pressure from (b) above. As pressure increases in the order (l→m→n), jadeite forms in rocks with an increasingly wider range of composition.

fairly wide range of temperatures probably because of the CaO content of plagioclase (e.g. Tracy 1978). In Maine, Evans & Guidotti (1966) described a zone, about 10 km wide, where the four phases muscovite + K-feldspar + sillimanite + plagioclase coexist. They visualized the equilibrium relations of minerals by use of a sillimanite–Or–Ab–An tetrahedron.

At temperatures lower than that of diagram (a), mineral assemblages should involve kaolinite, montmorillonite and zeolite. Interested readers are referred to J. B. Thompson & A. B. Thompson (1976).

*10.3.2 In high-*P/T *metamorphism*

At very high pressures, jadeite may form in addition to, or instead of, albite.

We start from diagram (b) of Figure 10.5. With increases of pressure, the following three reactions occur successively, resulting in the formation of paragenetic relations given in (l), (m) and (n) in Figure 10.5.

$$\text{Na-albite} = \text{jadeite} + \text{quartz} \tag{10.1}$$

$$\text{paragonite} + \text{albite} = \text{jadeite} + \text{muscovite} + \text{quartz} \tag{10.2}$$

$$\text{albite} = \text{jadeite} + \text{K-feldspar} + \text{quartz} \tag{10.3}$$

Starting from diagrams other than (b) in Figure 10.5, we can derive various high-pressure paragenetic relations with jadeite in analogous ways. However, the existence of mineral assemblages derived in these ways in natural rocks has not been well ascertained.

10.4 Petrogenetic grids

10.4.1 Petrogenetic grids in general

10.4.1.1 Introduction. When a group of rocks falling within a limited range of chemical compositions are metamorphosed in a range of P–T conditions, the rocks usually undergo a number of reactions with increasing temperature. Univariant curves for these reactions intersect with one another in the P–T diagram, as illustrated, for example, in Figure 2.2. A network of intersecting univariant curves in P–T space is called a **petrogenetic grid** (Bowen 1940). Bowen imagined that the univariant curves of solid–solid reactions generally will have a much greater slope than, and so intersect with, those of dehydration and decarbonation reactions. On the other hand, J. B. Thompson (1955) pointed out that the former usually have a much smaller slope than, and so intersect with, the latter.

In the presence of pure H_2O fluid or under a constant chemical potential of H_2O, a discontinuous solid–solid or dehydration reaction (§5.4) takes place on a univariant curve in the P–T diagram, and causes a change in the mineral paragenesis of the rock. The paragenesis change may be visualized by use of an appropriate composition–paragenesis diagram. Hence, all variations of mineral parageneses caused by changes in P–T conditions and in bulk-rock composition may be visualized by constructing a petrogenetic grid combined with composition–paragenesis diagrams for the individual P–T fields surrounded by univariant curves.

10.4.1.2 Schreinemakers' rule. The phase rule tells us that if a system composed of c components is made up of $c + 2$ phases, the variance (number of the degrees of freedom), F, is zero (§3.1.1). This type of equilibrium holds only at some specific combination of temperature and pressure characteristic of the system. In the P–T diagram of Figure 10.6, an assemblage of $c + 2$ phases is equilibrated at a specific point (**invariant point**), e.g. I_1 and I_2. If any one of the $c + 2$ phases is removed from the assemblage, the system comes to have $c + 1$ phases, and hence $F = 1$, and should be stable on a univariant curve in the diagram. There are $c + 2$ different ways to remove a phase. Thus, in general, $c + 2$ univariant curves exist, which radiate from the invariant point, as illustrated in Figure 10.6. Such a group of univariant curves around an invariant point has been called a **Schreinemakers bundle**. Each of the univariant curves is divided into two parts by the invariant point. One part represents a stable equilibrium and the other part a metastable equilibrium, between $c + 1$ phases.

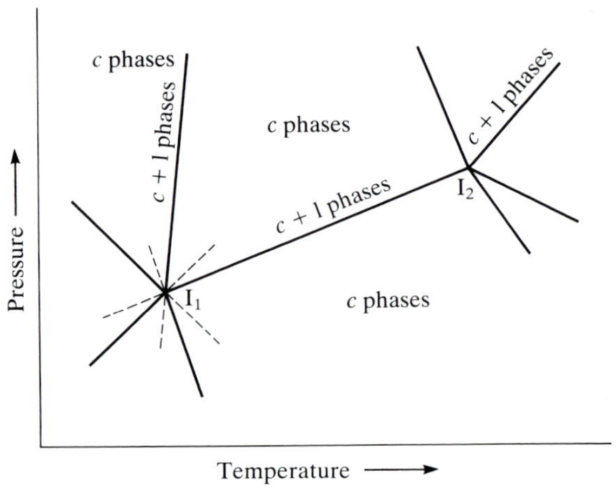

Figure 10.6 Schematic diagram showing Schreinemakers bundles. See text.

We can always write a chemical equation relating the compositions of the $c + 1$ phases coexisting on a univariant curve. Thus, each curve corresponds to the reaction represented by the equation. The slope of the curve is given by the Clausius–Clapeyron equation (eq. 3.8). There is a general rule governing the sequence of univariant curves around an invariant point. It was discovered by the Dutch chemist F. A. H. Schreinemakers in the 1910s, and is well described by Zen (1966). The sequence of curves depends only on the chemical compositions of the phases coexisting at the invariant point.

Univariant curves of this kind divide the *P–T* diagram around the invariant point into $c + 2$ fields (divariant fields), in each of which c or less phases can coexist in equilibrium. Under the *P–T* conditions within each field, mineral assemblages of rocks vary with their bulk-rock chemical composition, and this variation could be shown by a composition–paragenesis diagram.

As a simple example, we may consider the one-component system Al_2SiO_5. As shown in Figure 3.1 (or Fig. 8.1), the three phases (kyanite, sillimanite and andalusite) coexist at the triple point (invariant point), from which three univariant curves representing two-phase equilibria radiate. Only one phase is stable in each of the fields bounded by these curves.

If $c + 3$ minerals, for example, are known to form in the range of chemical compositions and *P–T* conditions under consideration, the number of possible selections of $c + 2$ phases out of the $c + 3$ minerals is indicated by the combination ${}_{(c+3)}C_{(c+2)} = c + 3$. This means that $c + 3$ invariant points exist, although some of them may be unstable.

If a univariant curve radiating from an invariant point represents the same reaction (same mineral assemblage) as a univariant curve radiating from another invariant point, these two univariant curves are actually the same curve, which connects the two invariant points under consideration (Fig. 10.6). The intersection of two univariant curves becomes an invariant point, if the mineral assemblages of the two univariant curves can form in the same rock.

10.4.2 Petrogenetic grids for pelitic metamorphic rocks

10.4.2.1 Construction of petrogenetic grids for metapelites. Construction of a petrogenetic grid for metapelites began with Albee (1965b), and was technically developed particularly by A. B. Thompson (1976) and Guiraud et al. (1990).

In the early attempts to construct petrogenetic grids, the stability of possible invariant points and univariant curves as well as the relative positions of invariant points were inferred from petrological data. The slopes of univariant curves were estimated from the Clausius–Clapeyron equation. Ferromagnesian minerals were assumed to have a fixed MgO/(FeO + MgO) ratio (e.g. Albee 1965b, Hess 1969). With increasing data based on synthetic experiments, experimentally determined or constrained univariant curves and invariant points came to be incorporated into grids. Ferromagnesian minerals with variable MgO/(FeO + MgO) ratios were considered (e.g. A. B. Thompson 1976).

Some grids constructed in this way can account fairly well for observations in natural metapelites. In particular, the grid constructed by Harte & Hudson (1979) has been widely used. Koons & Thompson (1985) have constructed a petrogenetic grid for metapelites with special application to high-P/T metamorphism.

A recent trend of progress lies in the use of univariant curves and invariant points calculated from internally consistent thermodynamic data sets, as seen in the grids of Spear & Cheney (1989), Powell & Holland (1990) and Guiraud et al. (1990).

10.4.2.2 Spear & Cheney's (1989) petrogenetic grids for metapelites. As an example of recently developed petrogenetic grids for metapelites, Figure 10.7 shows one of Spear & Cheney's (1989) grids, which is based mainly on thermodynamic calculations, although they have used some synthetic data and inferences from natural parageneses. Pelitic rocks are regarded as belonging to the six-component system Al_2O_3–FeO–MgO–K_2O–SiO_2–H_2O. All the assemblages considered contain quartz, muscovite (or K-feldspar instead at high temperature and low pressure) and an aqueous fluid.

At an invariant point, five minerals coexist in addition to quartz, muscovite and an aqueous fluid, and all the coexisting minerals have specific compositions. Generally, five univariant curves radiate from each invariant point, although only some of these curves are shown in Figure 10.7.

Along a univariant curve, the compositions of coexisting solid-solution minerals are functions of temperature (or pressure) alone. A topological change in the AFM diagram occurs at a univariant curve.

In a divariant field between univariant curves, the compositions of solid-solution minerals are functions of both temperature and pressure. In each field of this kind, dashed lines show a set of isopleths indicating the FeO/(FeO + MgO) ratio of garnet in a three-phase assemblage containing garnet and biotite.

It is known that the garnet + chlorite assemblage is widespread in low-grade metamorphic rocks. Naturally it had generally been considered that this assemblage is stable in low-grade metamorphic conditions, and that the tieline-switching reaction: garnet + chlorite → chloritoid + biotite occurs with increasing temperature under some limited conditions (§10.1). On the contrary, Spear & Cheney claimed that in the six-component system under consideration the chloritoid + biotite assemblage is stable over the whole pressure range of regional metamorphism at low temperatures, and that the tieline-switching reaction: chloritoid + biotite → garnet + chlorite occurs with increasing temperature over a limited range of pressure. They ascribed the discrepancy of their claim from the observed natural relations to the effects of MnO and CaO. These minor components are highly concentrated in garnet, resulting in an expansion of the stability field of the garnet + chlorite assemblage to lower and higher temperatures. This view has been confirmed by Symmes & Ferry (1992; §5.5.3).

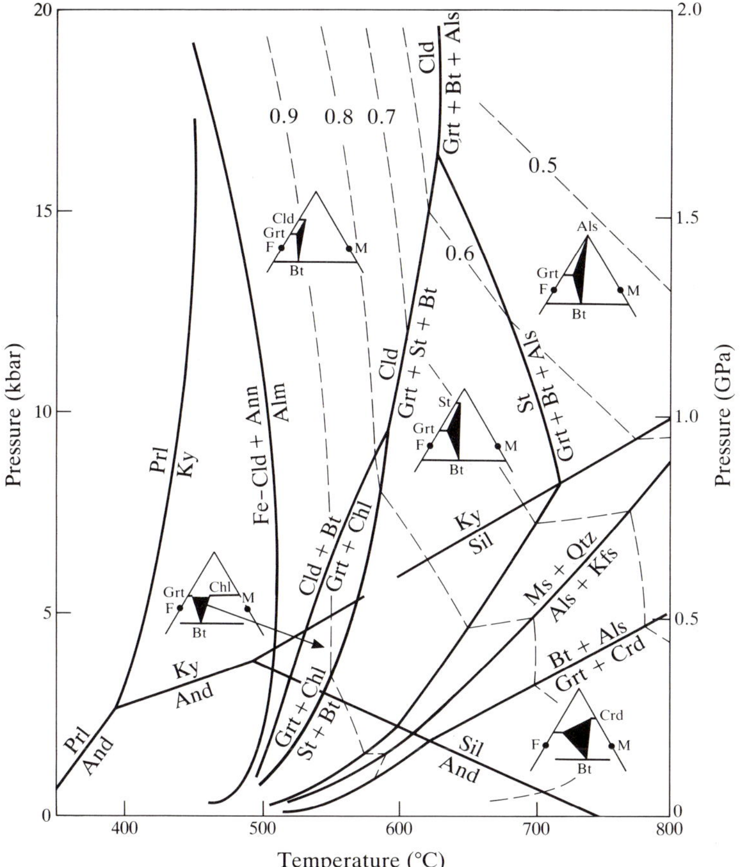

Figure 10.7 Spear & Cheney's (1989) petrogenetic grid for metapelites of the six-component system Al_2O_3–FeO–MgO–K_2O–SiO_2–H_2O in the presence of quartz, muscovite (or K-feldspar) and aqueous fluid. The minerals participating in univariant reactions are shown on the low- and high-temperature side of the curves. The ferromagnesian minerals are Fe–Mg solid solutions, except for the curve of Fe–Cld + Ann→Alm. Dashed lines are isopleths for the FeO/(FeO + MgO) ratio of garnet in the three-phase assemblages with garnet and biotite that are indicated on the *AFM* diagram in each divariant field.

10.5 Prograde and progressive compositional changes of solid-solution minerals

10.5.1 Variance as a constraint to the chemical compositions of solid-solution minerals

If the temperature, pressure and the chemical potential of H_2O in a group of metamorphic rocks are externally controlled, the variance of stable mineral assemblages undergoing dehydration reactions must be 3 or greater (§5.1). In a rock with a variance of 3, the chemical composition of each solid-solution mineral is a function of temperature, pressure, and the chemical potential of H_2O. If we assume that pressure and chemical potential of H_2O are kept constant, the composition of a solid-solution mineral changes with temperature. Such compositional changes take place by reactions (net-transfer and exchange reactions) of the mineral with coexisting minerals.

If metamorphic rocks contain an intergranular fluid phase with a virtually pure H_2O composition, the chemical potential of H_2O is a function of temperature and pressure. In this case, the variance of a stable mineral assemblage must be 2 or greater. In a rock with a variance of 2, the chemical composition of each solid-solution mineral is a function of temperature and pressure. If pressure is kept constant, the chemical composition of each mineral is a function of temperature alone, and may be used as a measure of the temperature of metamorphism.

These two cases, with and without an aqueous fluid phase, can be discussed in analogous ways. So, in all following discussions in this section, we consider only the case with an aqueous fluid phase.

There are many metamorphic rocks which have a variance greater than 2 in the presence of an aqueous fluid phase. In rocks where this is the case, the chemical compositions of solid-solution minerals vary not only with temperature and pressure but also with bulk-rock chemical composition. So the chemical compositions of these minerals cannot be a reliable measure of metamorphic grade.

Theoretical treatment of prograde and progressive chemical changes of minerals is easier for metapelites than for metabasites, because metapelites tend to show a lower variance (§5.2.6).

10.5.2 Solid-solution minerals in metapelites of the six-component system Al_2O_3–FeO–MgO–K_2O–SiO_2–H_2O

Metapelites are usually simplified as being composed of six components: Al_2O_3,

FeO, MgO, K_2O, SiO_2 and H_2O, and the mineral assemblages of the metapelites may be expressed by Thompson's (1957) AFM diagram (§5.4).

10.5.2.1 Three-phase assemblages. Metapelites containing three AFM phases (in addition to muscovite, quartz and an aqueous fluid) are represented by a paragenesis triangle in the AFM diagram. The variance of such a system is 2. The state of the system including the composition of each solid-solution mineral is a function of temperature and pressure alone. Variations in bulk-rock chemical composition cause no change in the compositions of solid-solution minerals. In other words, if we take a particular three-phase paragenesis, the composition of any solid-solution mineral therein varies with temperature and pressure alone, irrespective of bulk-rock composition. The same kind of mineral (say, biotite) shows different compositions in different three-phase parageneses at the same temperature and pressure.

As an example, Figure 10.7 shows how the FeO/(FeO + MgO) ratio of garnet in relatively Al-poor metapelitic rocks with various three AFM phases is controlled by parageneses and temperature and pressure. Stable mineral assemblages change by discontinuous reactions, and hence the direction of isopleths showing FeO/(FeO + MgO) ratio changes at the univariant curves for such reactions. Although the FeO/(FeO + MgO) ratios of garnet shown in Figure 10.7 generally tend to decrease with increasing temperature and pressure, it should be noted that a reverse trend occurs in certain mineral assemblages.

In a metapelite showing a three-phase assemblage, there is one net-transfer reaction among the three AFM phases, quartz, muscovite and H_2O, if solid solutions are regarded as being based on Fe–Mg substitution (§5.4.4). In addition, there is an exchange reaction between any two minerals showing Fe–Mg substitution (§3.7). The equilibrium conditions for these reactions control the chemical compositions, and hence the prograde and progressive chemical changes, of individual minerals.

10.5.2.2 Two- and one-phase assemblages. In metapelites showing two- or one-phase assemblages in the AFM diagram, the variance is 3 or 4, respectively. At a given temperature and pressure, the state of the system still changes with one or two independent compositional variables of minerals, respectively. In other words, the compositions of solid-solution minerals vary not only with temperature and pressure but also with bulk-rock composition.

In a metapelite showing a two- or one-phase assemblage, there is no net-transfer reaction, so far as solid solutions are regarded as being based on Fe–Mg substitution (§5.4.3). So, the mole proportions of constituent minerals remain constant, and the compositions of solid-solution minerals change only by an exchange reaction, if it occurs.

If both Fe–Mg and Tschermak substitution occur, there is one net-transfer reaction in two-phase assemblages (§5.4.3; Miyashiro & Shido 1985).

10.5.3 Solid-solution minerals in metapelites with more than six components

Garnet commonly contains a considerable amount of MnO and CaO. Biotite and chlorite may contain a large amount of TiO_2 and of Fe_2O_3, respectively. These minor components have considerable effects on the stability and the paragenetic relations. So some metapelites must be treated as belonging to a seven- or eight-component system. Among these additional components, the rôle of MnO in relation to garnet is probably the most important, and has been discussed by many authors, including Spear (1988a,b) and Spear & Cheney (1989), as reviewed below.

10.5.3.1 Garnet-bearing metapelites and MnO. If we regard a garnet-bearing metapelite as belonging to the seven-component system composed of MnO in addition to the six components considered above, the variance (in the presence of an aqueous fluid) of a three-phase assemblage in the *AFM* diagram is 3. At equilibrium, the composition of garnet should vary not only with temperature and pressure but also with the bulk-rock chemical composition. Two- and one-phase assemblages containing garnet have a still greater variance.

If the MnO content of a metapelite showing a three-phase assemblage increases under a constant *P–T* condition, the MnO content of each Mn-bearing solid-solution mineral will increase, while the Mn/Fe and Mn/Mg distribution coefficients between garnet and associated Mn-bearing solid-solution minerals will remain nearly unchanged. Net-transfer reactions change the proportion of constituent minerals, including garnet. The MnO/(FeO + MgO) ratio is much higher in garnet than in any associated ferromagnesian minerals. An increase of the bulk-rock MnO content widens the *P–T* range where garnet-bearing assemblages are stable (Spear & Cheney 1989: Fig. 4).

If solid solutions are based on the Fe–Mg–Mn substitution alone, a three-phase assemblage has one net-transfer reaction, while a two- or one-phase assemblage does not (§3.1.2). If both Fe–Mg–Mn and Tschermak substitution are considered, a two-phase assemblage has one net-transfer reaction. In either case, exchange reactions exist between minerals that show the same type of substitution. It is empirically known that four-phase metapelitic assemblages containing garnet are not rare. This may be due to the presence of MnO, or MnO and CaO.

10.5.3.2 Progressive chemical changes of garnet in equilibrium crystallization. In garnet-bearing metamorphic rocks most, or a large part, of the MnO in the rock is concentrated in garnet. If the amount of garnet increases or decreases by a net-transfer reaction, the MnO content of garnet is diluted or augmented, respectively. Thus, the progress of net-transfer reactions is a main factor controlling the progressive chemical change of garnet.

Metapelites of the garnet zone of the Barrovian sequence commonly show the three-phase assemblage: chlorite + garnet + biotite. The net-transfer reaction in this assemblage is of the type: chlorite + muscovite + quartz = garnet + biotite + H_2O. This causes a progressive increase of garnet, and hence a progressive decrease of the MnO content in garnet. Although the actual MnO content of garnet at a specific temperature varies with the bulk-rock composition, garnet in every such rock shows a trend of decreasing MnO content with increasing temperature. The MgO/FeO ratio of garnet increases with the same ratio of the host rock (Spear 1988b: Fig. 1).

In the staurolite and kyanite zones of the Barrovian sequence, on the other hand, garnet-producing reactions occur in some pelitic rocks, while garnet-consuming reactions occur in others. This causes a decrease or increase of the MnO content of garnet, respectively (Spear 1988b).

All the discussions given above assume complete equilibrium in the metamorphic rocks. This is far from true in metapelites in low- and middle-temperature zones. Deviations from equilibria are considered in the next section.

10.5.4 Fractional crystallization caused by growth zoning of garnet

Most minerals in regional-metamorphic rocks, such as chlorite and biotite, are relatively uniform in composition throughout their grains, and appear to have crystallized in near equilibrium with associated minerals of the same rocks. There are a few exceptional minerals that commonly show strong zoning. The most remarkable example is garnet in low- and middle-grade metamorphic rocks. This mineral generally shows a strong zonal structure with an MnO-rich core surrounded by a rim of MnO-poor almandine, as illustrated in Figure 10.8. In this case, fractional crystallization occurs during the growth, as was first pointed out by Hollister (1966).

Strong zoning in garnet indicates that once garnet is crystallized, it is preserved and no longer reacts with the surrounding minerals, because the rate of atomic diffusion is negligibly small within garnet. Garnet has an MnO content tens of times greater than those of coexisting chlorite and biotite. So the first formed garnet has a very high MnO content as compared with the bulk-

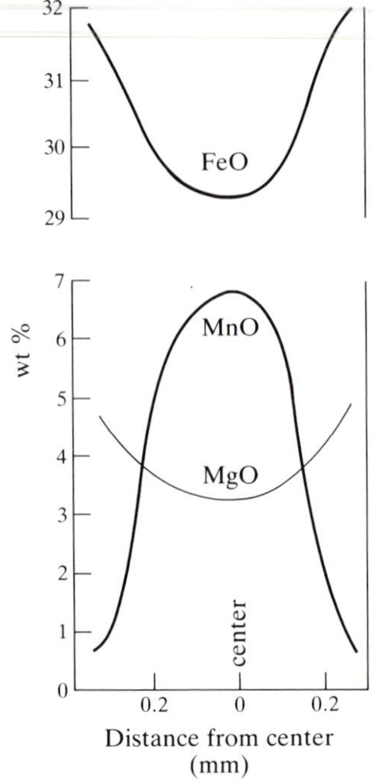

Figure 10.8 A typical zonal structure of a garnet grain in a middle-grade metapelite.

rock MnO content. The MnO of the rock is highly concentrated in the early-formed central core of garnet. Once the central core is formed, it is removed from the reacting system in the rock so that the reacting system is depleted in MnO. This decrease of MnO in the reacting system causes a decrease of the MnO content of the next layer of garnet to form, because the variance of metapelites is usually 3 or larger, as discussed in the preceding section. As the growth of garnet continues, producing successive layers around the central core, the MnO content of the reacting system decreases progressively, resulting in the formation of a zoned crystal showing a progressive outward decrease in MnO content. Because of this effect, the actual progressive decrease of the MnO content of garnet is more rapid in this case (fractional crystallization) than in the case of complete equilibrium.

Spear (1988b) has performed thermodynamic calculations taking all these factors into consideration for garnet-bearing mineral assemblages of the Barrovian sequence.

PART III

INDIVIDUAL METAMORPHIC FACIES: THEIR PETROLOGICAL AND MINERALOGICAL CHARACTERISTICS

11 Metamorphic facies of low- and medium-*P/T* regional metamorphism

As shown in Figure 8.3 and Table 7.1, low- and medium-*P/T* regional metamorphism usually produces a metamorphic facies series composed of the zeolite, prehnite–pumpellyite, greenschist, amphibolite and granulite facies in order of increasing temperature. These facies, however, are not confined to low- and medium-*P/T* regional metamorphism. The first three facies occur in some glaucophanic metamorphic regions as well, and the greenschist and amphibolite facies occur in many contact metamorphic aureoles also. These metamorphic facies in all these modes of occurrence are described in this chapter in order of increasing temperature.

11.1 Zeolite facies

11.1.1 General characteristics

11.1.1.1 Early studies. Eskola (1939) considered that the greenschist facies represented the lowest temperature in ordinary metamorphism. Figure 7.1 reproduces his synoptical table of metamorphic facies, where he wrote in the words "formation of zeolites" in a position indicating lower pressure at low temperature. Nevertheless, he did not set up a new metamorphic facies for the formation of zeolites, because he thought that zeolite-bearing rocks do not represent chemical equilibria, and that zeolites crystallize successively out of flowing hot solution, the compositional changes of which cause variations in the kinds of zeolites formed.

In the 1950s, Coombs (1954) studied a thick Triassic sedimentary pile in the Southland Syncline of New Zealand, and discovered that zeolites are formed there on a regional scale, and the kinds of zeolites vary regularly with increasing depth, and so probably with increasing pressure and temperature. Although the rocks of the pile still contain original minerals that did not undergo metamorphic crystallization, zeolites occur in mineral assemblages that formed by metamorphic crystallization, especially in the lower half of the sedimentary pile, and so the zeolite-bearing parts of rocks may well be regarded as representing a metamorphic facies for P–T conditions intermediate between the diagenesis of sedimentary rocks and traditional metamorphism. Thus, the name **zeolite facies** (Coombs et al. 1959) was proposed for the parts of rocks that underwent metamorphic crystallization and are characterized by the zeolite + quartz assemblage. Subsequently, Coombs (1960, 1961) proceeded to define the **prehnite–pumpellyite facies**, representing the condition intermediate between the zeolite and the greenschist facies.

Since then, many zeolite-bearing metamorphic areas have been reported from many parts of the world (e.g. Boles 1977). Most of them are Cenozoic or Mesozoic formations in the circum-Pacific regions such as New Zealand, Japan and western North America, although some Paleozoic zeolite-bearing regions are also known. It has become clear that some progressive metamorphic regions have zones of the zeolite and prehnite–pumpellyite facies, or only of the latter facies (or, rarely, only of the former facies) between an uncrystallized area and a greenschist-facies zone, whereas others have no such low-grade zone.

11.1.1.2 More recent studies. Early studies generally supported the above idea that zeolite-bearing assemblages are controlled by P–T conditions. From this viewpoint, different varieties of zeolites were assigned to various positions in the progressive sequence of mineral changes in the zeolite facies (e.g. Seki et al. 1969).

However, detailed studies have gradually made it clear that the conditions controlling zeolite formation in thick sedimentary piles are more complicated than was originally thought. Formation of zeolites is controlled not only by depth or P–T conditions, but also by the composition and pressure of intergranular fluids and the original rock materials.

Zeolites are mostly high in Ca, Na, or both. Hence, zeolites form in metamorphosed volcanic and volcaniclastic rocks such as those in the Southland Syncline, New Zealand, but cannot form in typical metapelites very low in Ca and Na. Well developed zeolite-facies areas are usually abundant in volcanic materials. The most common original materials from which zeolites easily form are volcanic glass and plagioclase.

In areas where volcanic materials are metamorphosed in the zeolite and/or prehnite–pumpellyite facies, associated pelitic sedimentary rocks may be too poorly crystallized or weakly metamorphosed with quartz, albite, and clay minerals, such as kaolinite, illite–smectite mixed-layer minerals and chlorite. Detailed mineralogical investigations, however, would clarify subtle progressive mineral changes (Zen & Thompson 1974).

It was Eskola's intention that the system of metamorphic facies constitute a scheme of classification of metamorphic rocks showing equilibrium mineral assemblages. On the other hand, sedimentary piles that underwent regional-scale zeolitization contain a large amount of original materials that did not suffer metamorphic crystallization, and must be excluded from zeolite-facies assemblages.

Even newly formed minerals may not be in equilibrium, because zeolites commonly form in sequences over time, and so coexistence in the same rock does not necessarily mean their equilibration. Some zeolites (for example, heulandite) show compositional variations not only between different rocks but also between different parts of the same rock, and this compositional variation appears to reflect variation in composition of the original glass from which the zeolites formed (Boles & Coombs 1975). These observations indicate that equilibrium was not approached even on a thin-section scale.

11.1.1.3 Burial metamorphism. Zeolite-facies rocks are usually non-schistose, and they preserve the texture of original rocks to a large extent. This has usually been interpreted as meaning that zeolite-facies metamorphism involves little or no penetrative deformational movements. Coombs (1961) coined the term **burial metamorphism** for this type of case. Burial metamorphism in this sense includes not only the zeolite facies but also parts of the prehnite–pumpellyite and blueschist facies. In this book, regional metamorphism is defined simply as regional-scale metamorphism, regardless of whether it produces schistosity or not (§1.3.1). So burial metamorphism is treated in this book as a subdivision of regional metamorphism, when it occurs on a regional scale.

It has been discovered that some such zeolite-bearing progressive metamorphic regions are actually diachronous-progressive (§8.2). In other words, the metamorphism of the zeolite-facies zone is younger than that of the associated greenschist facies or higher-temperature zones.

11.1.2 Examples of zeolite-facies areas

11.1.2.1 Southland Syncline, New Zealand. The *Taringatura Hills*, where the existence of regional-scale zeolitization was for the first time discovered by Coombs (1954), are in the northwestern part of the Southland Syncline near the southern end of South Island, New Zealand. The thick Triassic and Jurassic sedimentary pile of the syncline is made up almost entirely of volcanic materials (volcanic greywackes, siltstones and tuffs) of basalt–andesite–dacite–rhyolite affinities (i.e. a volcanic arc type), and has been subjected to regional-scale zeolite-facies metamorphism. Deformation is mild, and there is no igneous intrusion. So the temperature and pressure during metamorphism are thought to have been controlled generally by initial depth of burial.

In the Taringatura Hills, the Triassic section is about 10 km thick. In the upper part of the section, volcanic glass in tuff beds has mainly been replaced by heulandite or less commonly by analcime. Both zeolites coexist with newly crystallized quartz and fine-grained smectite (Coombs 1954, Coombs et al. 1959). Detrital Ca-bearing plagioclases are mostly fresh in the upper part, but are increasingly albitized in successively lower horizons. In the lower half of the sequence, analcime and heulandite almost disappear. It seems that analcime is mainly replaced by albite, whereas heulandite is replaced by laumontite plus quartz. Smectite, chlorite, sericite and mixed-layer minerals occur. Prehnite and pumpellyite appear as accessory minerals. These minerals occur as cements, fine-grained matrix and replacements of detrital minerals and glass shards. Adularia occurs in some rocks.

This observation appeared to suggest the existence of two depth stages (or subfacies) in the zeolite facies: a heulandite–analcime stage in the poorly crystallized upper half of the section, and a laumontite stage characterized by the laumontite + albite + quartz assemblage in the lower half. The stages may be connected by such reactions as:

$$\underset{\text{heulandite}}{CaAl_2Si_7O_{18}.6\,H_2O} = \underset{\text{laumontite}}{CaAl_2Si_4O_{12}.4\,H_2O} + \underset{\text{quartz}}{3\,SiO_2} + 2\,H_2O \qquad (11.1)$$

$$\underset{\text{analcime}}{NaAlSi_2O_6.H_2O} + \underset{\text{quartz}}{SiO_2} = \underset{\text{albite}}{NaAlSi_3O_8} + H_2O \qquad (11.2)$$

$$\text{laumontite} + \text{calcite} = \text{prehnite} + \text{quartz} + 3\,H_2O + CO_2 \qquad (11.3)$$

Since H_2O and CO_2 are on the right-hand sides, it is natural to regard these as representing a higher temperature and so a greater depth than the left. Coombs (1971) and Miyashiro & Shido (1970) attempted to understand zeolite and other mineral assemblages in the zeolite facies by regarding them as representing progressive sequences of dehydration and decarbonation.

Later, however, Boles & Coombs (1975, 1977) extended the work to more easterly parts of the Southland Syncline including the *Hokonui Hills*, where not only a Triassic but also an overlying Jurassic sedimentary sequence, mainly composed of mafic and intermediate volcanic materials, is exposed with a total thickness of 9 km. In this area, analcime occurs in relatively higher horizons, pumpellyite is confined to the lower horizons and albitization of plagioclase tends to increase with increasing depth just as in the Taringatura Hills, whereas heulandite, laumontite, prehnite and calcite occur in virtually all horizons. In other words, the distinction of two depth stages (or subfacies) as suggested in Taringatura does not hold in the Hokonui Hills. This complication probably means that zeolite-facies metamorphism is controlled not only by temperature and depth but also by many other factors, as discussed later.

11.1.2.2 Diachronous–progressive metamorphism in the Tanzawa Mountains, Japan. Some metamorphic areas show much more diverse zeolite minerals than the Southland Syncline. Moreover, a zeolite-facies zone commonly forms the lowest-grade part of a diachronous-progressive metamorphic sequence, the high-grade part of which reaches the greenschist or amphibolite facies. Detailed petrographical data are available from a diachronous-progressive metamorphic area of this kind in the Tanzawa Mountains, about 70 km southwest of Tokyo (Seki et al. 1969).

In this area, the metamorphic grade ranges from the zeolite to the amphibolite facies. The higher-grade parts of the metamorphic complex are strongly deformed and schistose, and even partly overturned, whereas the low-grade rocks are non-schistose. The original rocks were mostly pyroclastics and lavas of basaltic and andesitic compositions that formed in early Miocene to early Pliocene time. Seki et al. have divided the region into the following five zones in order of increasing temperature of metamorphism.

Zone I (low zeolite facies). Both mafic igneous and sedimentary rocks preserve their original textures and show no schistosity. Only the fine-grained groundmass and matrix suffered metamorphic crystallization. The parts subjected to metamorphic crystallization are characterized by the formation of clinoptilolite, heulandite, stilbite and mordenite, as well as by mixed-layer smectite–vermiculite and vermiculite–chlorite. Chlorite proper does not occur.

Zone II (high zeolite facies). The rocks are still usually non-schistose and incompletely crystallized. This zone is characterized by the occurrence of the laumontite + quartz assemblage. A mixed-layer clay mineral is common in the low-temperature half of this zone, whereas chlorite is common instead in the high-temperature half. In the higher half, wairakite and yugawaralite (zeolites) were occasionally found to occur in association with quartz.

As shown by the occurrence of opal, some metamorphic fluids were oversaturated with SiO_2, and this probably had an influence on the modes of occurrence of zeolites. On the other hand, some zeolites observed may have been stable only in the absence of quartz. Wairakite occurs only in the highest-temperature part of the zeolite-facies zone.

Zone III (prehnite–pumpellyite facies).

Zone IV (greenschist facies).

Zone V (amphibolite facies). The first appearance of plagioclase with 20% An has been regarded as the lower limit of this zone. The maximum An content observed is 73%.

The original rocks of Zones V, IV and III and a part of Zone II (called the Tanzawa Group) were deposited and then metamorphosed in the early and middle Miocene. Later uplift and erosion had exposed the metamorphic and plutonic complex by late Miocene time. A new depositional basin was formed to the south of the uplifted mountains, and fragments of the exposed rocks were transported southward and deposited. A new metamorphic event occurred in the newly deposited sediments, forming Zone I and a part of Zone II there in late Miocene and Pliocene time. Decisive evidence of this time relation is that metamorphic rock fragments derived from Zones III, IV and V (as well as quartz diorite) occur as pebbles in conglomerates of Zone I and the newly added area of Zone II. A mass of quartz diorite was intruded into Zone V slightly after the metamorphism, giving some contact metamorphic effect on the adjacent part in Zone V.

11.1.3 Stability relations of zeolites with respect to temperature and pressure

If there is an intergranular fluid composed of virtually pure H_2O at the same temperature and pressure as the solid phases, the stability relations of zeolites are controlled by temperature and pressure alone. Some experimental data collected under such conditions are reviewed below.

Figure 11.1 shows that the high-temperature limit of the stability field of the analcime + quartz assemblage, as given by reaction (11.2), lies at about 200°C under an H_2O pressure equal to a solid-phase pressure of lower than 5 kbar. In the absence of quartz, however, analcime is stable up to about 600°C at fluid pressures of less than a few kilobar. Above 600°C, analcime breaks down by the reaction:

$$2\,\text{analcime} = \text{albite} + \text{nepheline} + 2\,H_2O \qquad (11.4)$$

Analcime is decomposed at pressures higher than about 5 kbar by the following reaction (Fig. 11.1):

$$\overset{\text{analcime}}{NaAlSi_2O_6.H_2O} = \overset{\text{jadeite}}{NaAlSi_2O_6} + H_2O \qquad (11.5)$$

Figure 11.2 shows that under the condition of H_2O pressure equal to solid-phase pressure, stilbite, heulandite, laumontite and wairakite become stable successively in this order at higher temperature ranges (Liou 1971a). It should be noted that this order is generally in harmony with the observed orders of zeolite occurrence in the Taringatura Hills and the Tanzawa Mountains. This suggests that in these areas metamorphism occurred generally in the presence of a nearly pure H_2O phase.

At pressures higher than about 3 kbar, the lawsonite + quartz assemblage becomes stable in place of laumontite (Fig. 11.2) by the following reaction (Liou 1971a):

$$\underset{\text{laumontite}}{CaAl_2Si_4O_{12}.4\,H_2O} = \underset{\text{lawsonite}}{CaAl_2Si_2O_7(OH)_2.H_2O} + \underset{\text{quartz}}{2\,SiO_2} + 2\,H_2O \qquad (11.6)$$

For more experimental data related to stability relations of minerals in the zeolite and prehnite–pumpellyite facies, see Liou (1971a,b), A. B. Thompson (1971a,b), Zen & Thompson (1974), Winkler (1979) and Schiffman & Liou (1980).

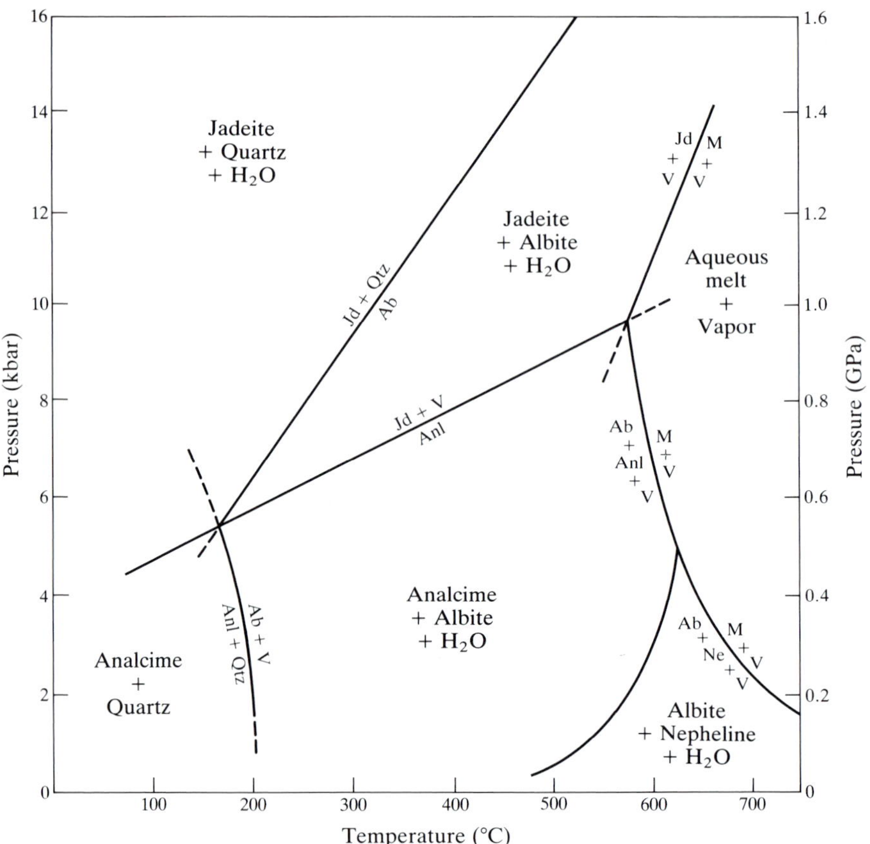

Figure 11.1 Stability fields of analcime, analcime + quartz, jadeite, and jadeite + quartz in the $NaAlSiO_4$–SiO_2–H_2O system in the presence of an aqueous vapor phase. V: aqueous vapor phase; M: melt. After Liou (1971b).

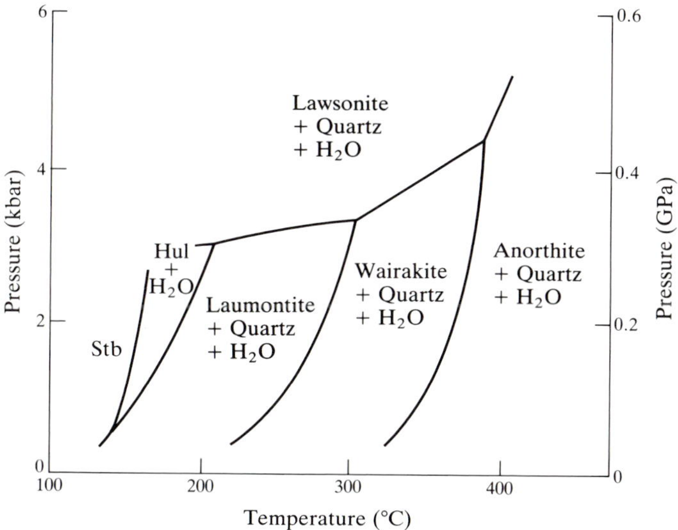

Figure 11.2 Stability fields of Ca-zeolites, lawsonite and anorthite in the presence of quartz and an aqueous vapor phase. After Liou (1971a) and Cho et al. (1987).

11.1.4 Factors, other than temperature and pressure, controlling zeolite formation

11.1.4.1 Relation of fluid pressure to the solid-phase pressure. At shallow depths, an intergranular fluid phase may be connected over a long distance through cracks and fissures, ultimately reaching the Earth's surface. If so, the fluid pressure is equal to the hydrostatic pressure, much smaller than the pressure of the adjacent solid phases (§4.1.2.3). At great depths, intergranular spaces may be disconnected from one another by cementation and metamorphic crystallization and each of them may be completely enclosed in solid rocks (§4.1.2.2). In equilibrium under this condition, the fluid pressure is equal to the pressure of the surrounding solid phases, and so the experimental data of Figures 11.1 and 11.2 are directly applicable if the fluid has pure H_2O composition. Thus, with increasing depth, the pressure of the fluid phase tends to increase from the hydrostatic to the lithostatic value. The actual depth at which fluid pressure becomes equal to solid-phase pressure must differ in different areas. Boles & Coombs (1977) have discussed how this factor could cause complications in zeolite formation.

11.1.4.2 Effects of CO_2. Ca-zeolites are isochemical with calcite + pelite mineral (or clay mineral) except for their H_2O and CO_2 contents, as shown for example by equations:

$$\text{heulandite} + CO_2 = \text{calcite} + \text{kaolinite} + 5\,\text{quartz} + 4\,H_2O \quad (11.7)$$
$$\text{heulandite} + CO_2 = \text{calcite} + \text{pyrophyllite} + 3\,\text{quartz} + 5\,H_2O \quad (11.8)$$
$$\text{laumontite} + CO_2 = \text{calcite} + \text{kaolinite} + 2\,\text{quartz} + 2\,H_2O \quad (11.9)$$

Thus, a higher chemical potential of CO_2 tends to decompose Ca-zeolites and produce carbonate-bearing pelitic mineral assemblages, whereas a higher chemical potential of H_2O tends to produce Ca-zeolites. In consequence, zeolite-bearing and zeolite-free mineral assemblages may form under the same temperature and pressure, depending on the values of the chemical potentials of CO_2 and H_2O (Zen 1961).

At a fluid pressure = solid-phase pressure = 2 kbar, for example, the laumontite side of reaction (11.9) is stable only for a mole fraction of CO_2 lower than about 0.0075 in the H_2O–CO_2 fluid, that is, in an extremely low CO_2 concentration only (A. B. Thompson 1971b).

11.1.4.3 Effects of other components in fluids. Some components, other than CO_2, in fluids must also have a large effect on mineral formation in the zeolite facies.

Fluids in quartz-free metamorphic rocks may commonly be unsaturated with SiO_2. Such a fluid may precipitate some particular kinds of zeolites, e.g. thomsonite, chabazite and natrolite (e.g. Miyashiro & Shido 1970). On the other hand, opal occurs commonly in zeolite-facies rocks, as in the Tanzawa Mountains. This suggests that metamorphic fluids oversaturated with SiO_2 are relatively common. If these oversaturated fluids participate in reactions, the equilibrium temperature range of heulandite, for example, increases in reaction (11.1), and that of albite increases in reaction (11.2).

Zeolites may form by reaction of glass and minerals with fluids. Depending on the original materials, the actual reactions differ, and various components will be taken from, or released into, the fluids. In a closed system, the composition of metamorphic fluids is controlled by the mineral assemblages of the system (§6.1.1). However, in rock complexes undergoing zeolite-facies metamorphism, and particularly in shallower parts of them, metamorphic fluids probably flow to a considerable extent, and flowing fluids may carry a variety of substances from outside the system.

11.1.5 Changes in the definitions of the zeolite facies

The concept of the zeolite facies has changed as its study has progressed.

(a) Coombs et al. (1959) defined the zeolite facies to include at least all the assemblages produced under physical conditions in which the following assemblages are commonly formed: analcime + quartz, heulandite +

quartz, and laumontite + quartz. Since in a silica-deficient environment certain zeolites such as analcime are stable in a much wider *P–T* range, the association of quartz with zeolite is regarded as essential.

(b) Some authors regarded the heulandite–analcime stage in the upper half of the depositional pile of the Taringatura Hills in the Southland Syncline as diagenetic, because the rocks contain too much uncrystallized material and are too far from equilibrium, and so regarded only the laumontite stage in the lower half as representing the zeolite facies (e.g. Fyfe et al. 1958, Winkler 1965). As stated above, however, it was later found that both heulandite and laumontite occur in virtually all horizons of the Southland Syncline, and so laumontite can no longer be regarded as diagnostic of a metamorphic origin with this meaning.

(c) In the Southland Syncline, analcime, heulandite and laumontite are the dominant zeolites, whereas in many other zeolite-bearing areas, such as the Tanzawa Mountains, other varieties of zeolites are also common and even show regular distribution patterns related to temperature in harmony with experimental data and theoretical considerations. Among them, stilbite + quartz and mordenite + quartz represent a lower temperature than heulandite (e.g. Cho et al. 1987, Miyashiro & Shido 1970: Figs 3, 8), whereas wairakite + quartz probably represents a higher temperature than the laumontite + quartz assemblage (Fig. 11.2, Seki 1966, Miyashiro & Shido 1970). Therefore, the total range of *P–T* conditions in which zeolite + quartz assemblages are stable, is probably considerably wider than was originally considered from data from the Taringatura Hills (Seki 1966, 1969). We may well expand the concept of the zeolite facies to cover all such zeolite + quartz assemblages.

(d) We made a postulate that metamorphic facies are defined in terms of temperature, pressure, and the chemical potential of H_2O (§7.4). This means that the effects of CO_2 and other components in fluid are ignored. If a fluid is a binary mixture of H_2O and CO_2, the chemical potential of CO_2 is a function of that of H_2O at a constant temperature and pressure, and so the above definition will cause no trouble. Since a fluid usually contains various other components, the chemical potential of CO_2 is independent of that of H_2O. The zeolite facies includes not only zeolite + quartz-bearing rocks but also spatially closely associated zeolite-free rocks. The latter rocks are free of zeolites, because of the absence of Na_2O and CaO in some cases and because of the high chemical potential of CO_2 in others.

11.1.6 Relation of the zeolite facies to metamorphic facies series

The zeolite facies represents not only very low temperature, but also low pressures. A zone of the zeolite facies occurs at the low-temperature end of many progressive metamorphic regions of the low-, medium- and high-*P/T* types. This probably means that the field *P–T* curves of these metamorphic regions usually have a positive slope.

A few examples are cited below. A zeolite-facies zone appears to be the lowest temperature part in low-*P/T* progressive metamorphic regions in several areas of Japan, e.g. the Tanzawa and Akaishi Mountains. In other areas, a zeolite-facies zone belongs to the medium-*P/T* progressive metamorphic sequence, e.g. probably the Southland Syncline. There is a zeolite-facies zone in the lowest-grade part of some glaucophane-schist belts, e.g. the Franciscan Complex in California and the western Alps.

11.2 Prehnite–pumpellyite facies

11.2.1 General characteristics

In the lower half of the Mesozoic sedimentary pile of the Southland Syncline described in the preceding section, prehnite and pumpellyite occur together with heulandite, laumontite, chlorite, calcite, albite and quartz. If this type of metamorphism occurs at somewhat higher temperatures, the zeolites + quartz assemblage will disappear and the resultant rocks will have prehnite and pumpellyite together with minerals common in the greenschist facies, such as chlorite, calcite, epidote, albite and quartz. Rocks of this type occur widely in Mesozoic metasediments of Canterbury in the South Island of New Zealand (e.g. Coombs 1960, Bishop 1972). For these rocks and associated metamorphic rocks, Coombs (1960, 1961) proposed the name of **prehnite–pumpellyite metagreywacke facies**. Later authors dropped metagreywacke to shorten the name, and this custom is accepted in this book. This facies is characterized by the occurrence of prehnite and/or pumpellyite, and the absence of zeolites, in quartz-bearing rocks.

Just as in the zeolite facies, metamorphic crystallization in the prehnite–pumpellyite facies usually takes place in volcanic and volcaniclastic rocks and greywackes containing volcanic materials. Associated pelitic rocks may be poorly crystallized, or may be chlorite-zone schists composed of quartz, albite, muscovite, chlorite, etc. Rocks of the prehnite–pumpellyite facies are non-schistose in some cases and schistose in others.

Prehnite–pumpellyite-facies metamorphism occurred in all geologic ages from the Archean (e.g. Jolly 1974) to the Cenozoic.

11.2.2 Examples of prehnite–pumpellyite-facies areas

In the Akaishi and Tanzawa Mountains of Japan, a zone of the prehnite–pumpellyite facies lies between zones of the zeolite facies and the greenschist facies (§11.1.2). In the Franciscan Complex of the Coast Ranges of California, a zone of the prehnite–pumpellyite facies passes into a zone of the blueschist facies with increasing metamorphic grade (Fig. 8.7). In other areas, an unmetamorphosed region grades directly into a zone of the prehnite–pumpellyite facies without an intervening zone of the zeolite facies. In the Paleozoic regional metamorphic complex of the northern Appalachians, for example, the lowest-grade zones belong to the prehnite–pumpellyite facies (e.g. Richter & Roy 1974, Zen 1974a,b).

Seki et al. (1964) discovered that in the progressive metamorphism of metavolcanic rocks in the Shimanto belt in central Kii Peninsula, Japan, metamorphic grade increases northward from an unmetamorphosed area through a prehnite–pumpellyite-facies zone and then a transitional zone to a greenschist-facies zone. Characteristic mineral assemblages are: prehnite + pumpellyite + epidote + chlorite in the prehnite–pumpellyite-facies zone, pumpellyite + epidote + chlorite + actinolite in the transitional zone, and epidote + chlorite + actinolite in the greenschist-facies zone. In other words, prehnite disappears at a lower grade than pumpellyite, and at the same time actinolite appears.

Hashimoto (1966) proposed the name of **pumpellyite–actinolite schist facies** for rocks of the above-mentioned type of transition zone, which is characterized by the assemblage pumpellyite + epidote + chlorite + actinolite. The P–T conditions of such a zone would lie between those of the typical prehnite–pumpellyite facies (free of actinolite), greenschist facies and blueschist facies. In this book, however, this facies is regarded as a part of the prehnite–pumpellyite facies, when it does not contain glaucophane. If an assemblage of this type is accompanied by glaucophane, it is treated as belonging to the blueschist facies.

11.2.3 Relevant experimental data

Schiffman & Liou (1980) synthesized Mg–Al pumpellyite and showed that the P–T stability fields of mineral assemblages of the prehnite–pumpellyite and greenschist facies vary considerably with pressure. Under H_2O pressures equal to solid-phase pressures of less than 5 kbar, pumpellyite decomposes at 300–360 °C, and prehnite at about 370–380 °C, whereas at 5–9 kbar, pumpellyite decomposes at about 370–400°C, and prehnite at a much lower temperature. Then, above about 380–400°C, metabasites come to show the tremolite (or actinolite) + chlorite + clinozoisite (or epidote) assemblage characteristic of the greenschist facies.

Prehnite and pumpellyite also tend to decompose under a relatively high chemical potential of CO_2.

11.3 Greenschist facies

11.3.1 General characteristics

The greenschist facies is characterized by quartz-bearing greenschists, that is, metabasites with the chlorite + actinolite + epidote + albite + quartz assemblage. Actinolite is absent in some greenschists.

Greenschist-facies rocks occur in the lowest-temperature zones in many regional metamorphic complexes, as originally considered by Eskola (1920, 1939). In some other regions, however, metamorphic crystallization reached beyond a zone of this facies to still lower temperatures, resulting in the formation of a zone or zones of the prehnite–pumpellyite, zeolite and/or blueschist facies (Fig. 8.3). In ordinary metapelitic regions, this facies is represented by chlorite and biotite zones (Figs 8.4, 8.5).

Prehnite, pumpellyite and glaucophane, which are characteristic of metamorphic facies lying on the low-temperature side of the greenschist facies, form readily in metabasites but not in typical metapelites. Hence, metapelites of the prehnite–pumpellyite and blueschist facies may be indistinguishable from those of the greenschist facies. The high-grade limit of the greenschist facies is marked by transitional states to the amphibolite facies that are characterized by the appearance of calcic oligoclase or more calcic plagioclase as well as of actinolite–hornblende coexistence.

Since the greenschist facies usually includes the chlorite and biotite zones for metapelites, a biotite-forming reaction generally occurs in the middle of the metamorphic grade range. The actual reaction differs from case to case, depending on the mineral assemblages of pelitic and semi-pelitic rocks present (§2.3.3, §10.1.1.2).

11.3.2 Observed mineral assemblages in the greenschist facies

Schematic *ACF* and *A'KF* diagrams for the greenschist facies are shown in Figure 11.3. As written in the accompanying figure captions, these diagrams cannot adequately represent the great diversity of mineral assemblages, because components ignored in these diagrams have essential effects on the mineralogical compositions (§5.3).

Metabasites. The common mineral assemblage of metabasites is actinolite + chlorite + epidote + albite (+ quartz + muscovite + sphene + calcite + opaque minerals). Either actinolite or chlorite may be absent in some rocks. In metabasites of the chlorite zone defined for metapelites, a small amount of muscovite (phengite) commonly coexists with Ca-amphibole, and biotite occurs very rarely, whereas in those of the biotite zone, biotite becomes more common and muscovite rarer. In all greenschist-facies rocks the An content of albite is usually 0–2%.

Metapelites. The most common mineral assemblage in the chlorite zone is chlorite + muscovite (phengite) + quartz. In the biotite zone, although this assemblage is still generally common, the biotite-bearing assemblage biotite + chlorite + muscovite (phengite) + quartz also occurs. In many cases, they are accompanied by albite, tourmaline, a graphite-like carbonaceous substance, an Fe-oxide (hematite or magnetite), a Ti-bearing mineral (sphene, rutile or ilmenite) and a carbonate (calcite or dolomite).

Biotite formation with increasing temperature in typical metapelites is accompanied by a decrease of the ($FeO + MgO)/Al_2O_3$ ratios of associated muscovite and chlorite (Ernst 1963b, Mather 1970, Miyashiro & Shido 1985).

Only pelitic rocks more aluminous than the above can contain highly aluminous minerals: chloritoid, pyrophyllite and/or paragonite, usually in association with chlorite and muscovite (Zen 1960, Tobschall 1969). Paragonite is highly aluminous as shown by the composition relation: paragonite+4 SiO_2 = albite + pyrophyllite. Rocks relatively low in Al_2O_3 and high in FeO/MgO may contain stilpnomelane, usually in association with chlorite and muscovite (e.g. Zen 1960).

Metagreywackes. Original clastic plagioclase changes into albite + epidote during metamorphism. Thus, the most common mineral assemblage of metagreywackes is muscovite + chlorite + epidote + albite + quartz with or without biotite. Biotite may begin to form in metagreywackes at a considerably lower temperature than in associated metapelites. In a southwestern part of the Scottish Highlands, Mather (1970) showed the biotite isograd for metagreywackes to lie about 5 km down grade as compared with the biotite isograd for metapelites.

Metagreywackes may contain stilpnomelane, actinolite, and/or microcline together with tourmaline. Stilpnomelane occurs more rarely and/or less plentifully in the biotite zone than in the chlorite zone because of consumption of stilpnomelane by reaction with muscovite (Brown 1971).

Calcareous rocks. Metamorphosed limestones are usually made up mainly of

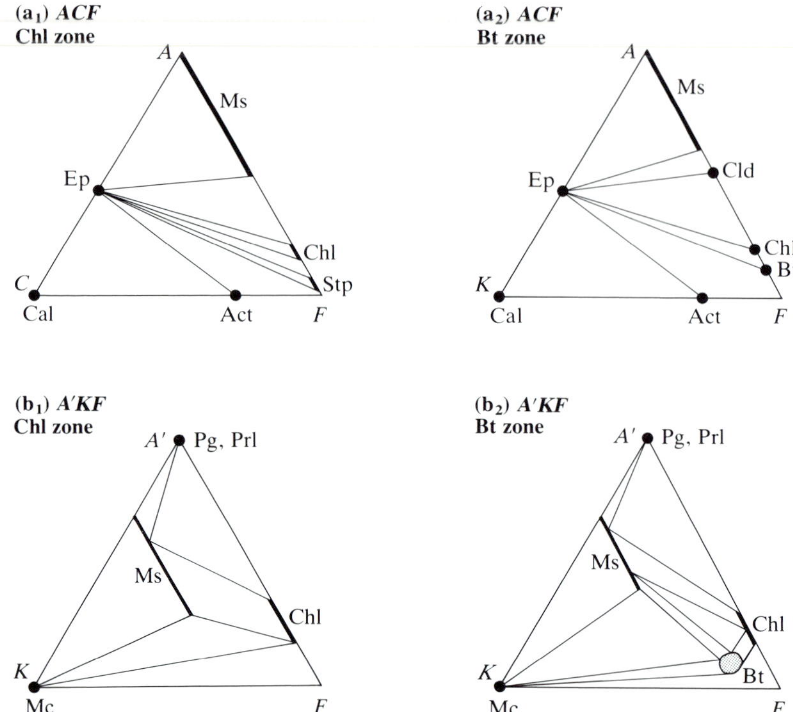

Figure 11.3 Schematic *ACF* and *A′KF* diagrams for the greenschist facies. (a_1) *ACF* diagram for the chlorite zone. The coexistence of muscovite + actinolite, of actinolite + chlorite, of calcite + chlorite, and of calcite + stilpnomelane are common, though not shown. Under high CO_2 activity, dolomite (ankerite) forms in the midpoint of the CF side instead of actinolite. (a_2) *ACF* diagram for the biotite zone. The coexistence of calcite + muscovite and of chlorite + biotite are common, though not shown. Under high CO_2 activity, dolomite (ankerite) forms instead of actinolite. (b_1) *A′KF* diagram for the chlorite zone. (b_2) *A′KF* diagram for the biotite zone. For symbols of mineral names see Appendix 1.

calcite accompanied by minor amounts of quartz, albite, microcline, muscovite and chlorite. Many such rocks also contain dolomite. In addition to the calcite + quartz assemblage, the dolomite + quartz assemblage, as represented by the left-hand side of the following equation, is usually stable instead of tremolite + calcite:

$$5\,\text{dolomite (or ankerite)} + 8\,\text{quartz} + H_2O = \text{tremolite (or actinolite)} + 3\,\text{calcite} + 7\,CO_2 \qquad (11.10)$$

Phlogopite occurs in some limestones.

11.3.3 Effects of CO_2 on greenschist-facies mineral assemblages

An increased chemical potential of CO_2 tends to decompose Ca-bearing silicates such as actinolite, epidote and sphene with the resultant formation of carbonates (for example, calcite, dolomite and ankerite). Among carbonates, dolomite and ankerite tend to form under a higher chemical potential of CO_2 than calcite. Quantitative stability relations pertaining to these changes are given by Will et al. (1990). These changes contribute to the mineralogical diversity in the greenschist facies, just as in the zeolite and prehnite–pumpellyite facies.

Graham et al. (1983) demonstrated how mineral assemblages of metabasites vary with the CO_2/H_2O ratio of intergranular fluids in the low-grade Dalradian metamorphic rocks in the Knapdale area in a southwestern promontory of the Scottish Highlands. Metasediments of this area include grits, conglomerates, psammites, phyllites, and siliceous and aluminous carbonate rocks. The metabasites were derived from doleritic and gabbroic sills, up to 140 m thick, together with basaltic volcanics. The area of research belongs to the greenschist facies and a transitional state to the amphibolite facies. They classified metabasites in the greenschist facies area into the following types on the basis of mineral assemblages.

- Type I: Ca-amphibole (actinolite or hornblende) + epidote + chlorite + quartz + albite + *sphene* (with or without muscovite, biotite and stilpnomelane). No carbonate.
- Type II: Ca-amphibole + epidote + chlorite + *calcite* + quartz + albite + sphene (with or without muscovite, biotite and stilpnomelane).
- Type III.1: epidote + chlorite + *calcite* + quartz + albite (with or without sphene, rutile, muscovite and biotite).
- Type III.2: chlorite + quartz + albite + *rutile* + muscovite (with or without dolomite (ankerite) and calcite).

These four assemblage types are stable at successively higher ranges of the chemical potential of CO_2 in fluid. The first three types are related to one another by the reaction:

$$3\,\text{tremolite} + 2\,\text{clinozoisite} + 8\,H_2O + 10\,CO_2 = 3\,\text{chlorite} + 10\,\text{calcite} + 21\,\text{quartz} \quad (11.11)$$

Types III.1 and III.2 are related to each other by the following reactions:

$$\text{biotite} + \text{chlorite} + 8\,\text{calcite} + 8\,CO_2 = \text{muscovite} + 8\,\text{dolomite} + 3\,\text{quartz} + 4\,H_2O \quad (11.12)$$

$$\text{sphene} + CO_2 = \text{calcite} + \text{rutile} + \text{quartz} \quad (11.13)$$

Metamorphosed mafic sills show zonal distributions of rock types, which have formed by infiltration and diffusion of relatively CO_2-rich fluid from the surrounding metasediments into the sill interiors during metamorphism. The interior of most massive mafic sills is made up of type I metabasites, whereas the schistose margins of the sills are made up of type III. Type II metabasites occur in intermediate zones. In thin, thoroughly schistose sills, type I and even type II may be lacking. It appears that infiltration was intense where schistosity was well developed. Textural relations indicate that an initial type I assemblage was later partly converted to type II, and then to a type III assemblage. These relations suggest that there was a gradual decrease of CO_2 concentration from the margin to the interior in these sills at the thermal peak.

Sphene is probably the most common Ti-rich mineral in metapelites, metagreywackes and metabasites in the greenschist facies. However, rutile or ilmenite occurs instead in many areas. Since these minerals have different chemical compositions, their stability relations are influenced by not only the chemical potential of CO_2 but also the associated minerals and bulk-rock chemical composition. If the intergranular fluid is a mixture of H_2O and CO_2 under greenschist facies conditions, sphene is stable only when the mole fraction of CO_2 is very low – less than 0.1 (Ernst 1972a, Ferry 1983a).

11.4 Transitional states between the greenschist and amphibolite facies

11.4.1 Mineralogical characteristics of the transitional states

The Ca-amphibole in metabasites is actinolite in the greenschist facies, and is hornblende in the amphibolite facies. The plagioclase in any rock of the greenschist facies is albite, usually with 0–2% An, whereas that in the amphibolite facies is generally close to, or more calcic than, An_{20}. Progressive metamorphic regions with zones of these two facies commonly have a transitional zone between them.

In regional metamorphic complexes, the transitional state is usually characterized by the coexistence of actinolite and hornblende in metabasites, and/or by the coexistence of albite and plagioclase more calcic than An_{20} in metabasic, metapelitic and metapsammitic rocks. A transitional state of this kind usually occurs in the high-biotite zone and/or the garnet zone as defined for metapelites (Figs 8.4, 8.5). In some contact metamorphic aureoles, on the other hand, a transitional state between the greenschist and amphibolite facies is characterized by occurrence of the actinolite + relatively calcic plagioclase assemblage.

The transitional states have been discussed by Apted & Liou (1983) and Maruyama et al. (1983) on the basis of their experimental data.

11.4.2 The compositional gap between actinolite and hornblende

11.4.2.1 Observed relations. Early workers supposed that the actinolite of greenschists changes into the hornblende of amphibolites continuously with increasing temperature. Shido & Miyashiro (1959), however, suggested the possible existence of a miscibility gap between the two Ca-amphiboles at relatively low temperatures, from their observations of metabasites showing coexistence of the two amphiboles in the transitional metamorphic grade. The two amphiboles occur together in the same metabasites, not only as zoned crystals but also as independent ones. When the two are in direct contact, there is a sharp boundary showing a Becke line between them. Some of later investigations gave accurate electron microprobe data showing the probable existence of a composition gap between coexisting actinolite and hornblende (e.g. Klein 1969, Brady 1974). These composition relations between actinolite and hornblende are illustrated in the composition–paragenesis diagram of Figure 5.17b.

A core of actinolite, rimmed by hornblende, probably crystallized earlier than the hornblende rim. This does not necessarily preclude the possibility of its equilibration by compositional adjustment after the formation of the hornblende rim. However, this textural relation is equivocal as evidence for equilibrium coexistence, although the boundary between them is sharp. Grapes & Graham (1978) claimed the relation to be ascribed not to a miscibility gap but to disequilibrium and kinetic factors.

Tagiri (1977) observed large crystals of originally igneous hornblende that had been changed into fine lamellar intergrowths of actinolite and hornblende in metabasite of the transitional-grade zone in the Abukuma Plateau. This supports the existence of a compositional gap.

Smelik et al. (1991) described, by means of the transmission and analytical electron microscopes, the occurrence of exsolution lamellae of actinolite and hornblende in a series of calcic amphiboles. This has given direct evidence for the existence of a miscibility gap between actinolite and hornblende.

11.4.2.2 Cause of the compositional gap. The composition of hornblende can be approximately derived from that of tremolite–actinolite $Ca_2(Mg,Fe)_5Si_8O_{22}$-$(OH)_2$ by the (Mg,Fe)Si → AlAl (Tschermak), Si → NaAl, and CaAl → NaSi substitutions (e.g. J. B. Thompson et al. 1982). Two hypotheses have been proposed for the origin of the compositional gap.

Solvus hypothesis. Many authors ascribed the observed compositional gap between actinolite and hornblende to the existence of a solvus at low temperatures. This view was supported by the hydrothermal synthetic experiments of Oba (1980) and Oba & Yagi (1987) on the tremolite–pargasite join, as well as on an analogous Fe-bearing system, where pargasite $NaCa_2Mg_4AlSi_6Al_2O_{22}$-

$(OH)_2$ was chosen as a representative Al-rich, Na-rich end-member of hornblende. A large miscibility gap due to a solvus was observed to exist on the join under 1 kbar at temperature below about 820°C.

Phase change loop hypothesis. If actinolite and hornblende are connected by a first-order phase transformation, they can coexist stably with a composition gap between them over a range of temperatures. This possibility was pointed out by Tagiri (1977), and was later advocated by Maruyama et al. (1983) from their observation of the progressive change of the compositional gap in metabasites in Kasuga (Japan) and the Yap Islands. They observed that, with increasing temperature, the gap becomes narrower and shifts toward the actinolite composition.

11.4.3 The peristerite gap in plagioclase

Becke (1903) was among the earliest observers who considered that plagioclase becomes more calcic with increasing metamorphic grade. Early workers unanimously considered the compositional change from albite to calcic plagioclase with increasing temperature to be continuous (e.g. Ramberg 1949, 1952: 50–54). The Ca and Al necessary for the formation of an additional An component were considered to be supplied generally by decomposition of associated epidote.

In the 1950s and the early 1960s, however, petrographic data indicating the occurrence of a very rapid or abrupt change in plagioclase composition near An_{10} were reported from many progressive metamorphic regions (e.g. Lyons 1955).

11.4.3.1 Peristerite solvus hypothesis. Some authors considered that what was regarded as a very rapid or abrupt change in plagioclase composition near An_{10} was probably actually a compositional gap that was caused by a solvus at low temperatures – the so-called peristerite solvus. This was soon supported by Evans's (1964) discovery of coexisting grains of albite (An_{0-2}) and oligoclase (An_{18-26}) in some samples from the Haast schists of New Zealand by means of the then-new instrument, the electron microprobe. Coexisting albite and oligoclase occur either side by side as separate grains or as zoned grains in which the boundary between an albite core and an oligoclase rim is sharp.

After Evans's discovery, Crawford (1966) and Cooper (1972) made more systematic observations on the peristerite gap in the Haast Schists of New Zealand. In the biotite zone defined for metagreywackes, albite ($An_{0.1-0.5}$) occurs in association with epidote. Oligoclase (An_{20-26}) begins to coexist with albite ($An_{0.2-0.4}$) and epidote approximately from the almandine isograd onwards. With a considerable further rise of temperature, albite disappears and oligoclase remains. Crawford (1966) found that quartzo-feldspathic mica schists in Vermont usually contain albite (An_{0-2}) together with carbonate (calcite and ankerite) in the low-grade part of the biotite zone, and albite (An_{2-7}) and oligoclase (An_{18-27}) in the higher-grade part of the biotite zone. In the almandine zone, usually only oligoclase (about An_{25}) occurs.

Maruyama et al. (1982, 1983) described the progressive mineral changes of metabasites in the contact aureole of Kasuga, central Japan, as well as in the metamorphosed basement of the Yap Islands in the western Pacific Ocean. In both areas, albite occurs in actinolite greenschists of the low-grade zone, and then the coexistence of oligoclase (An_{25}) with albite begins almost at the same grade as the coexistence of hornblende with actinolite. The compositions of the two coexisting plagioclases change with further rise of temperature toward the same value, An_{12}. In still-higher metamorphic grades, only one plagioclase occurs, which tends to become more calcic, with increasing grade from An_{12} to An_{60}. Thus, the closing of the peristerite solvus with increasing temperature appears really to have been observed. This gives strong support to the solvus hypothesis. The two metamorphic areas were probably metamorphosed at low pressures.

By comparison of the above data with observations in other metamorphic regions, Maruyama et al. (1982) concluded that, with an increase of pressure from 2 to 10 kbar, the temperature of the top of the solvus increases from about 420° to about 580 °C, with a concomitant shift of the top towards a more sodic composition.

11.4.3.2 Phase change loop hypothesis. Smith (e.g. 1975) proposed another hypothesis, which ascribes the compositional gap to the two-phase loop for the first-order phase transformation between low-albite and high-albite solid solution. This hypothesis was advocated by Orville (1974), Spear (1980a) and others. If high-albite solid solution shows unmixing on a solvus in the albite–oligoclase composition range, single-phase solid solutions on the high-temperature side of the solvus must show a large deviation from ideal mixing. However, Orville did not find a deviation of this kind in his experiment on exchange equilibria between plagioclase and hydrous chlorite solutions. Hence, denying the existence of the peristerite solvus, he ascribed the gap to a binary loop. He estimated the high–low transformation temperature for pure Ab composition to be about 575 °C, and the temperature rise with a pressure increase of 10 kbar to be of the order of 30 °C.

11.4.4 Diversity of the transitional state

The discontinuous progressive change from actinolite to hornblende, and that from albite to calcic oligoclase, occur in virtually the same temperature range in some metamorphic regions. In other metamorphic regions, however, one of the two progressive changes appears to occur at considerably lower temperatures than the other. We may consider two extreme cases on the basis of the temperatures of the two changes.

11.4.4.1 One extreme case: the albite–epidote-amphibolite zone. As one of the two extreme cases, we may consider metabasites in which the stable plagioclase is albite, and hornblende occurs without being accompanied by stable actinolite. Such metabasites show the mineral assemblage: albite + epidote + hornblende (possibly with quartz and garnet but without stable actinolite and oligoclase). This is a critical assemblage of what was called the *epidote–amphibolite facies* (albite–epidote–amphibolite facies) by Eskola and other authors. This facies was not included in Eskola's (1920) initial scheme of metamorphic facies, but was introduced as the facies intermediate between the greenschist and amphibolite facies in his final version (Eskola 1939).

However, later re-examination of metabasites once regarded as belonging to this facies revealed the presence of two coexisting Ca-amphiboles in many cases, and the presence of two distinct plagioclases in some. Therefore, typical metabasites of the epidote-amphibolite facies are not common, and hence this facies is not included in the major metamorphic facies but is treated as a zone in this book.

11.4.4.2 The other extreme case: the actinolite–calcic plagioclase zone. As the other extreme case of the transitional state, we consider metabasites in which plagioclase more calcic than An_{20} is accompanied by actinolite (with no albite and no hornblende, but possibly with quartz). This mineral assemblage is relatively common in contact aureoles, but not in regional metamorphic complexes (Miyashiro 1961: 306–307).

In typical contact aureoles composed of hornfelses, a zone of the greenschist facies is rare, whereas a zone characterized by metabasites with the assemblage actinolite + calcic plagioclase commonly occurs immediately on the low-temperature side of an amphibolite-facies zone. The plagioclase is labradorite or andesine. This assemblage was observed in the contact aureoles of the Iritono area (Shido 1958) and Arisu area (Seki 1961a), both in Japan.

In the Iritono aureole, the actinolite–calcic plagioclase zone represents the lowest temperature of metamorphic crystallization, whereas in the Arisu aureole, the actinolite–calcic plagioclase zone is followed on the low-temperature side by a zone of the greenschist facies. Thus, this area shows a facies series intermediate between those of the typical contact aureole of the Iritono area and low-*P/T* regional metamorphic areas where a zone of the greenschist facies is well developed. No Al_2SiO_5 minerals were found in the Iritono aureole. In the Arisu aureole, andalusite occurs in the middle grade and sillimanite in the high grade.

11.4.4.3 Cause of the diversity of the transitional state. Except for H_2O and CO_2 contents, the actinolite + calcic plagioclase assemblage is virtually isochemical with common greenschist assemblages such as chlorite + epidote + albite + quartz, and chlorite + calcite + albite + quartz, as is clear from equations such as:

$$9\,\text{actinolite} + 114\,\text{anorthite} + 64\,H_2O = 10\,\text{chlorite} + 66\,\text{epidote} + 77\,\text{quartz} \quad (11.14)$$

$$9\,\text{actinolite} + 15\,\text{anorthite} + 31\,H_2O + 33\,CO_2 = 10\,\text{chlorite} + 33\,\text{calcite} + 77\,\text{quartz} \quad (11.15)$$

(Here, the formula of chlorite is taken as $Mg_{4.5}Al_3Si_{2.5}O_{10}(OH)_8$.) Therefore, the occurrence of the assemblage actinolite + calcic plagioclase must depend mainly on temperature, pressure, and chemical potentials of H_2O and CO_2. The actinolite + calcic plagioclase assemblage tends to form under low chemical potentials of H_2O and CO_2. Since epidote usually contains much Fe_2O_3, oxidation conditions also have some effect.

The transitional states between greenschist and amphibolite have been experimentally investigated by Liou et al. (1974), Apted & Liou (1983) and others. Liou et al. as well as Maruyama et al. (1983) considered that the breakdown reaction curves of epidote have a gentler slope (dP/dT) than those of chlorites. So, as a transitional state between greenschist and amphibolite, albite–epidote amphibolite forms at relatively high pressures, whereas the actinolite + relatively calcic plagioclase assemblage forms at low pressures.

11.4.4.4 Relation to progressive mineral changes of associated metapelites. In progressive regional metamorphic complexes of the medium-*P/T* type, the discontinuous progressive changes in Ca-amphiboles and in plagioclase occur commonly in the garnet (almandine) zone defined for metapelites, and not infrequently in a somewhat wider range of metamorphic grades including the higher-grade part of the biotite zone.

The biotite and garnet isograds for metapelites are based on continuous reactions (§5.4.6). Moreover, there are several biotite-forming reactions (§2.3.3). Hence, the temperature ranges of the biotite and garnet zones differ from region to region, and this also must contribute to the diversity of the transitional state.

11.5 Amphibolite facies

11.5.1 General characteristics and partial melting

The amphibolite facies is characterized by quartz-bearing amphibolites, that is, metabasites with the assemblage plagioclase (An_{30} or more calcic) + hornblende + quartz, commonly accompanied by a small amount of epidote, biotite, garnet and/or salite (diopsidic pyroxene). With an increase in metamorphic temperature, epidote decreases and disappears. The formation of garnet is promoted by a higher FeO/MgO ratio of the rock together with relatively high pressure. The formation of salite is promoted by a higher CaO content of the rock.

In regions showing progressive metamorphism of pelitic rocks, the amphibolite facies usually includes the andalusite and the sillimanite–muscovite zone in the low-*P/T* type (Fig. 8.4), and the staurolite, the kyanite, and the sillimanite–muscovite zone in the medium-*P/T* type (Fig. 8.5). Eskola's (1914, 1915) type area for the amphibolite facies was the Orijärvi region in southwestern Finland, which belonged to the andalusite zone of low-*P/T* metamorphism.

The boundary between the amphibolite and granulite facies probably lies between 650° and 700°C (e.g. Fig. 7.3). In the presence of an aqueous fluid at pressures above 4 kbar, pelitic and granitic rocks begin to melt at temperatures between 600° and 650°C (e.g. Figs 1.1, 11.5, 11.6). Hence, partial melting probably begins in many rocks in the high-temperature part of the amphibolite facies in medium- and some low-*P/T* metamorphic complexes.

At pressures below about 4 kbar (that is, in a relatively low-pressure regime of low-*P/T* metamorphism) muscovite may undergo a dehydration reaction (dehydration-decomposition without melting) with the resultant formation of dry mineral assemblages characteristic of the granulite facies (Figs 11.5, 11.6).

If a melt of this kind is locally segregated from unmelted solid, the rock may become migmatitic. On the process of cooling following the thermal peak, the melt may crystallize back to form the same mineral assemblage as that which was present prior to the melting. If H_2O is removed during the melting, the newly crystallized rocks will show more dehydrated mineral assemblages.

The problem of melting is discussed in part in §1.2 and is treated again in relation to the granulite facies in the next section (§11.6.3).

11.5.2 Observed mineral assemblages in the amphibolite facies

ACF and *A'KF* diagrams for the amphibolite facies are given in Figure 11.4, although these diagrams cannot adequately show observed paragenetic relations, as the accompanying captions explain.

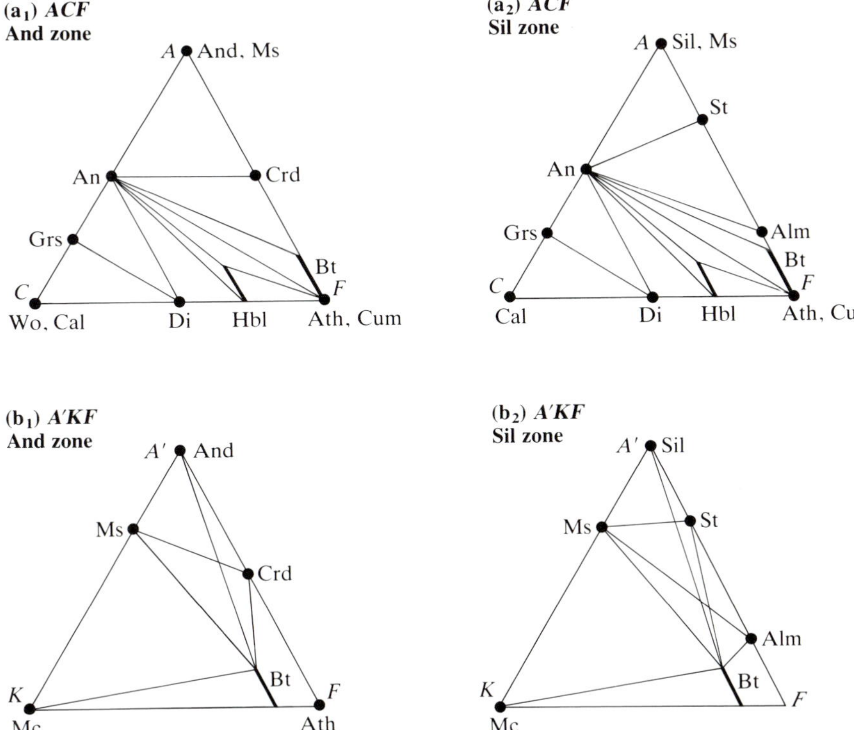

Figure 11.4 Schematic *ACF* and *A'KF* diagrams for the amphibolite facies. (a_1) *ACF* diagram for the andalusite zone of the low-*P/T* regional metamorphism. The coexistence of biotite + hornblende is common, though not shown. Andalusite, cordierite, anthophyllite and cummingtonite can occur only in K-feldspar-free rocks. (a_2) *ACF* diagram for the low-temperature part of the sillimanite–muscovite zone of the medium-*P/T* type. (b_1) *A'KF* diagram for the andalusite zone of the low-*P/T* regional metamorphism. The four-mineral assemblage muscovite + biotite + andalusite + cordierite is stable, though not adequately shown. (b_2) *A'KF* diagram for the low-temperature part of the sillimanite–muscovite zone of the medium-*P/T* type. Muscovite can coexist with sillimanite + staurolite + biotite, staurolite + biotite + garnet, and sillimanite + staurolite + garnet, though this is not adequately shown.

11.5.2.1 Metabasites. Metabasites of the amphibolite facies are usually amphibolites with mineral assemblages of the following types:

1: hornblende + plagioclase (possibly with biotite and quartz);

2: hornblende + plagioclase + garnet (possibly with cummingtonite and quartz);

3: hornblende + plagioclase + salite (or augite) (possibly with quartz).

Hornblende in metabasites of the transitional zone between the greenschist and amphibolite facies commonly shows a blue–green axial color for Z, whereas that in the amphibolite facies usually shows either a green or a brown axial color for Z. It is said that a higher content of Ti tends to produce a more brownish color.

In lower-grade parts of the facies, a small amount of epidote is commonly present, and amphibolites with relatively low FeO/MgO ratios may contain chlorite. Some amphibolites with a relatively low CaO contents contain, in addition to hornblende, cummingtonite and even gedrite and anthophyllite. The last two orthorhombic amphiboles are separated by a miscibility gap at relatively low temperature (Spear 1980b, 1982). Garnet (almandine) is common in amphibolites with relatively high FeO/MgO ratios of the medium-*P/T* type, but is absent in amphibolites of the low-*P/T* type.

11.5.2.2 Metapelites. The following assemblages are common in metapelites:

1: andalusite (or kyanite or sillimanite) + biotite + muscovite + plagioclase + quartz (possibly with cordierite);
2: staurolite + garnet + biotite + muscovite + plagioclase + quartz (possibly with kyanite);
3: sillimanite + garnet + biotite + muscovite + quartz (possibly with staurolite);
4: biotite + muscovite + plagioclase + quartz + K-feldspar.

K-feldspar is absent in typical metapelites. Metapelites commonly contain andalusite and cordierite in the low-pressure part, and kyanite (or sillimanite) and almandine in the high-pressure part, of the amphibolite facies. Common Ti-rich minerals are ilmenite and sphene in metapelites as well as in metabasites. In the higher-temperature part of the amphibolite facies, however, a large part of the Ti in rocks tends to be contained in biotite and hornblende. Rutile commonly occurs in some rocks instead of ilmenite and sphene.

11.5.2.3 Calcareous rocks. The following assemblages are common in metamorphosed calcareous rocks:

1: grossular + diopside + calcite (possibly with plagioclase and quartz);
2: grossular + diopside + wollastonite (+ calcite or quartz);
3: diopside + plagioclase + microcline + quartz.

11.5.3 Hornblende in amphibolites

Since typical amphibolites are composed of only two major minerals, plagioclase and hornblende, commonly with minor quartz, they have a high variance. The chemical composition of hornblende in these amphibolites varies not only with *P–T* condition and H_2O activity, but also with bulk-rock chemical composition (§5.2.6). In individual amphibolites, hornblende undergoes prograde compositional changes in relation to the formation, increase, decrease, and compositional change of associated minerals such as epidote, plagioclase, chlorite, garnet, quartz and sphene.

Spear (1980a, 1981), Blundy & Holland (1990) and others have discussed reactions that take place between components of plagioclase and hornblende in the presence of quartz; for example:

$$2\,\text{albite} + \text{tschermakite} = 2\,\text{anorthite} + \text{glaucophane} \quad (11.16)$$

$$\text{albite} + \text{tremolite} = \text{edenite} + 4\,\text{quartz} \quad (11.17)$$

$$28\,\text{tremolite} + 14\,\text{glaucophane} = 28\,\text{edenite} + 6\,\text{cummingtonite} + 92\,\text{quartz} + 8\,H_2O \quad (11.18)$$

The first of the three is an NaSi–CaAl exchange reaction between plagioclase and hornblende, whereas the second and third are net-transfer reactions. All these reactions proceed to the right with increasing temperature. The first two reactions, together with the final breakdown of epidote, tend to increase the An content of plagioclase with increasing temperature.

In the lowest-grade part of the amphibolite facies, amphibolites commonly contain small amounts of chlorite and epidote together with quartz, having a correspondingly smaller number of degrees of freedom. They resemble actinolite-bearing greenschists in their mineral assemblages. In metabasites of this kind, the composition of amphibole is under a considerable constraint imposed by the mineral assemblage. Holland & Richardson (1979) discussed chemical reactions controlling the composition of amphibole in such rocks. Laird (1980), Laird & Albee (1981a,b) and Maruyama et al. (1983) discussed progressive compositional changes in amphibole and other minerals in such rocks.

For the compositional variation and nomenclature of amphiboles in general, see Leake (1978) and Veblen (1981). For the atomic substitutions and stability relations of metamorphic amphiboles, see Veblen & Ribbe (1982) and J. B. Thompson et al. (1982).

11.5.4 Compositional gaps in plagioclase

At temperatures higher than the disappearance of the peristerite gap, that is, generally in the amphibolite facies, intermediate and calcic plagioclases occur. Some authors have reported finding gaps of various composition ranges in such plagioclase. Two phases of plagioclase of microscopic or macroscopic sizes with different An contents occur in the same rocks, and they may even be intergrown with a regular crystallographic orientation (e.g. Wenk & Wenk 1977, Spear 1980a, Grove et al. 1983). Wenk (1979) even reported coexisting albite and anorthite (with a gap ranging from An_5 to An_{90}) in amphibole schists that formed apparently under a condition transitional between the greenschist and amphibolite facies. These gaps might correspond to the Boggild and Huttenlocher gaps and other low-temperature immiscibilities in the plagioclase series.

11.6 Granulite facies

11.6.1 Low- and high-temperature subfacies of the granulite facies

11.6.1.1 Progressive mineralogical changes from the amphibolite to the granulite facies. If the temperature of metamorphism becomes higher than that of the amphibolite facies, the hornblende of amphibolites begins to decompose, with the resultant formation of coexisting ortho- and Ca-clinopyroxenes in the presence of quartz and plagioclase. This produces a mineral assemblage characteristic of the granulite facies. The decomposition reaction proceeds continuously over a considerable temperature range. Transitional metabasites, which still contain hornblende together with coexisting pyroxenes, occur in some high-temperature metamorphic areas of Phanerozoic orogenic belts and in extensive high-grade gneiss regions of the Precambrian. Complete disappearance of hornblende occurs in some cases, resulting in the formation of anhydrous mineral assemblages, such as orthopyroxene + Ca-clinopyroxene + garnet + plagioclase + quartz. The common occurrence of garnet of the almandine–pyrope series distinguishes the granulite facies from the pyroxene-hornfels facies, in which two pyroxenes occur but garnet of the above type is virtually absent.

In associated metapelites, muscovite decreases and then disappears with increasing temperature. Biotite also decreases. Complete disappearance of biotite occurs in quartz-bearing metapelites only in the highest-grade areas with the resultant formation of completely or virtually completely anhydrous mineral assemblages, such as sillimanite + cordierite + garnet + K-feldspar + plagioclase + quartz. Progressive changes of paragenetic relations in metapelites have already been shown in Figures 10.3 and 10.5.

A gradual transition of this kind from an amphibolite-facies to a granulite-facies area has been described in detail, for example, from Langöy, Norway (Heier 1960), southwestern Finland (Schreurs 1984, Schreurs & Westra 1986), southeastern Finland (Korsman 1977), southern India (Condie et al. 1982, Janardhan et al. 1982, Raase et al. 1986), Broken Hill, Australia (Phillips 1980), west Greenland (Wells 1979b), the Adirondack Mountains, New York (Engel & Engel 1962, Edwards & Essene 1988), and the northern Appalachians (Chamberlain & Lyons 1983).

11.6.1.2 Subdivisions of the granulite facies based on temperature. The boundary between the amphibolite and granulite facies is defined in this book as the *P–T* condition under which ortho- and Ca-clinopyroxenes begin to coexist with hornblende in some quartz-bearing metabasites with ordinary basaltic compositions. Even when only quartz-bearing metabasites are considered, the temperature range of the progressive decomposition reactions of hornblende into pyroxene-bearing assemblages varies greatly with the mineral assemblage (including the compositions of feldspars) and the bulk-rock FeO/MgO ratio. For this reason, usually in a transitional zone from the amphibolite to the granulite facies, amphibolites with only one pyroxene and pyroxene-free amphibolites occur closely mixed with two-pyroxene-bearing metabasites because of their differences in bulk-rock chemical composition.

Metabasites containing only one pyroxene (either clino- or orthopyroxene) are not diagnostic of the granulite facies. Diopsidic pyroxene occurs commonly in amphibolite-facies amphibolites relatively high in CaO content. In cummingtonite-bearing amphibolites, the decomposition of cummingtonite into orthopyroxene probably begins at a somewhat lower temperature than that of hornblende, with the resultant formation of orthopyroxene-bearing amphibolites under the *P–T* conditions of the higher amphibolite facies.

Eskola (1939) defined his granulite facies as being characterized by completely anhydrous mineral assemblages that form by the complete disappearance of amphiboles and micas in quartz-bearing rocks. However, areas of completely anhydrous granulite facies of this type are very rare and small. There are much wider regions of transitional states where, although many rocks show the two-pyroxene assemblage, all or most of them still contain hornblende and biotite. For this reason, it has become a custom in recent years to include all the transitional states in the granulite facies. The custom is accepted in this book. Thus, the granulite facies is divided into two: a low-temperature and a high-temperature subfacies, which correspond respectively to the transitional state and typical, virtually anhydrous granulite facies.

11.6.1.3 Temperature of, and melting in, the granulite facies. The determination of thermal-peak temperatures of granulite facies rocks is plagued by intrinsic difficulties caused by retrograde reactions as discussed later. The garnet–biotite Fe–Mg exchange, orthopyroxene–clinopyroxene and other geothermometers usually give temperatures in a range of 700–850°C for most rocks of the granulite facies. Some granulite facies regions show exceptionally high temperatures of 900–1000°C (e.g. Ellis 1980, Ellis et al. 1980, Sandiford et al. 1987, Bohlen 1991).

At such high temperatures, many rocks undergo partial or complete melting (Fig. 1.1). The temperature of melting, however, varies greatly with the amount and state of H_2O. So the amount and activity of H_2O is discussed in the next section, followed by the problem of melting in the granulite facies.

11.6.2 Low activity of H_2O and flow of CO_2-rich fluid

11.6.2.1 Observations suggesting low activity of H_2O. Since the mineralogical changes from the amphibolite to the granulite facies are caused by dehydration reactions, they are promoted not only by temperature increase, but also by decrease of the chemical potential or activity of H_2O in the environment. Indeed, many authors, including Wells (1979a), Phillips (1980), Bhattacharya & Sen (1986) and Lamb & Valley (1988), have estimated the H_2O activity roughly to be in the range 0.1–0.5 from their phase-equilibrium studies in granulite-facies regions in Greenland, Australia, India and the Adirondacks (New York), respectively.

It is known that fluid inclusions in granulite facies rocks commonly contain a large portion of CO_2 (§4.2). An increase of CO_2 content in intergranular fluid must cause a decrease of the proportion and activity of H_2O (Newton 1986a, 1989). However, fluid inclusions preserved in metamorphic rocks now exposed on the surface were generally trapped at stages after the thermal peak (§4.2). So CO_2-rich fluids in inclusions may differ in composition from the fluid that was present at the thermal peak (Lamb et al. 1991).

Granulite-facies metamorphism takes place at temperatures above about 650°C. At these temperatures, if an aqueous fluid is present between mineral grains, partial melting must be intensive in many or most of the quartzo-feldspathic rocks. However, although partial melting probably occurs, it does not appear to be very intensive in granulite-facies regions. This tends to support the idea that granulite-facies metamorphism occurs usually either in the absence of a fluid phase or in the presence of an H_2O-poor fluid.

11.6.2.2 Possible origins of the low activity of H_2O. For the origin of the low activity of H_2O, three main possibilities have been suggested by many authors. Each of the three may apply to some areas.

First, partial melting of pelitic and psammitic rocks probably usually begins in the high-temperature part of the amphibolite facies. The resultant melts can dissolve a large amount of H_2O but not CO_2, possibly leading to the formation of H_2O-poor, CO_2-rich fluid (e.g. Phillips 1980, Waters & Whales 1985, Bhattacharya & Sen 1986).

Secondly, low activities of H_2O may be realized in fluid-absent metamorphism. If rocks that have previously been dehydrated by high-temperature metamorphism are subjected to a second metamorphic event, there may be no fluid phase and rocks may show a very low activity of H_2O (e.g. Lamb & Valley 1984).

Thirdly, a large amount of CO_2 may form by decarbonation reactions of calcareous sediments, or may rise from the underlying mantle into the lower crust. The possibility of a massive rise of CO_2 from the mantle into the lower crust has been emphasized particularly by Newton et al. (1980).

The idea of granulite-facies metamorphism caused by infiltration of CO_2-rich fluids has been strongly advocated, particularly for the charnockite regions in southern India (e.g. Condie et al. 1982, Janardhan et al. 1982). In some areas there, the transition from amphibolite facies rocks to granulite-facies rocks was observed to take place in too short a distance (such as several centimeters) to be ascribed to a difference in temperature, so that a compositional variation in the fluid phase was suggested.

11.6.3 Partial melting and the effect of hydrous minerals

11.6.3.1 General remarks. It is widely believed that metamorphic rocks of the amphibolite facies usually have an intergranular aqueous fluid. If so, granitic and pelitic metamorphic rocks generally begin partial melting at some temperature between 600° and 650°C (Figs 1.1, 11.5). This fluid is probably usually so small in amount that it will disappear when the amount of melt reaches a few per cent of the rocks (§1.2.2). Thereafter, with increasing temperature the fluid becomes unsaturated with H_2O, and will show a progressively decreasing activity of H_2O. The decrease in H_2O activity tends to hinder the progress of melting.

Figure 1.1a shows that, if excess H_2O is present, some granitic rocks melt completely at a temperature around, or slightly above, 700°C. This complete melting, however, will not generally occur in regional metamorphism because of the very limited amount of H_2O present. If a large amount of CO_2 is added to the rock, either by decarbonation reactions, or by infiltration from outside, it will make an additional contribution to the decrease of the activity of H_2O.

Metapelitic and granitic rocks usually contain hydrous minerals, such as muscovite and biotite. In rocks with no aqueous fluid, even the small amount of H_2O in their crystal structures causes a large effect in decreasing the melting temperature of their host rocks, as discussed below.

With increasing temperature in metamorphism under pressures of less than about 4 kbar (i.e. in a relatively low-pressure regime of low-*P/T* regional metamorphism), muscovite undergoes dehydration reactions (**dehydration-decomposition**), producing anhydrous mineral assemblages of the granulite facies, as already shown in paragenetic changes from (d) to (e) and then to (f) in Figure 10.5. On the other hand, with increasing temperature in metamorphism at higher pressures (i.e. in a relatively high-pressure regime of low-*P/T* metamorphism and in medium-*P/T* metamorphism), muscovite undergoes simultaneous dehydration and melting (so-called **dehydration-melting**).

There is an extensive literature of theoretical and experimental investigations on the melting of metamorphic rocks (e.g. A. B. Thompson & Algor 1977, Atherton & Gribble 1983, Ashworth 1985, Le Breton & Thompson 1988, Vielzeuf & Holloway 1988).

In the next section, the characteristics of dehydration-decomposition and dehydration-melting are examined with examples involving muscovite. Then dehydration-melting of ordinary metapelites containing both muscovite and biotite is discussed.

11.6.3.2 Dehydration-decomposition and dehydration-melting of muscovite. Figure 11.5 shows a petrogenetic grid related to the decomposition and melting of muscovite. Here the phase assemblages are regarded as belonging, either to the five-component system K_2O–Na_2O–Al_2O_3–SiO_2–H_2O (free of CaO), or to its subsystems.

Q_1 is an invariant point belonging to the four-component system K_2O–Al_2O_3–SiO_2–H_2O, the K_2O end-member system of the above five-component system. Six phases (Ms, Kfs, Als, Qtz, V and M) coexist at the invariant point, and six univariant curves radiate therefrom.

Q_2 is an invariant point belonging to the above five-component system. Seven phases (Ms, Afs, Ab, Als, Qtz, V and M) coexist at the point, and seven univariant curves radiate therefrom. In this case, Ms, Afs and Ab are K–Na solid solutions. The K/(K + Na) ratios of these minerals are specific at a specific point on any of the curves. One of the univariant curves radiating from Q_2 reaches Q_1, where the K/(K + Na) ratio is unity. (This curve does not belong to the four-component system of Q_1.)

Q_4 is an invariant point of the three-component system $KAlSi_3O_8$–Al_2O_3–H_2O. At the point, five phases (Ms, Kfs, Crn, V and M) coexist. The dashed line that starts from Q_4 in a lower left direction represents dehydration-decomposition of muscovite by itself.

The grid shows that, at pressures lower than that of Q_2, dehydration-decomposition of muscovite occurs in the following sequence. In rocks containing albite and quartz, the decomposition begins by a reaction of the type (with coefficients neglected):

$$\text{muscovite} + \text{albite} + \text{quartz} = \text{potassic alkali feldspar} + \text{sillimanite} + H_2O \qquad (11.19)$$

This reaction represents a tieline-switching reaction from (d) to (e) in Figure 10.5. Then, final decomposition of muscovite occurs at a slightly higher temperature in albite-free, quartz-bearing rocks by continuous and discontinuous reactions of the type:

$$\text{muscovite} + \text{quartz} = \text{K-feldspar} + \text{sillimanite} + H_2O \qquad (11.20)$$

This reaction corresponds to the change from (e) to (f) in Figure 10.5.

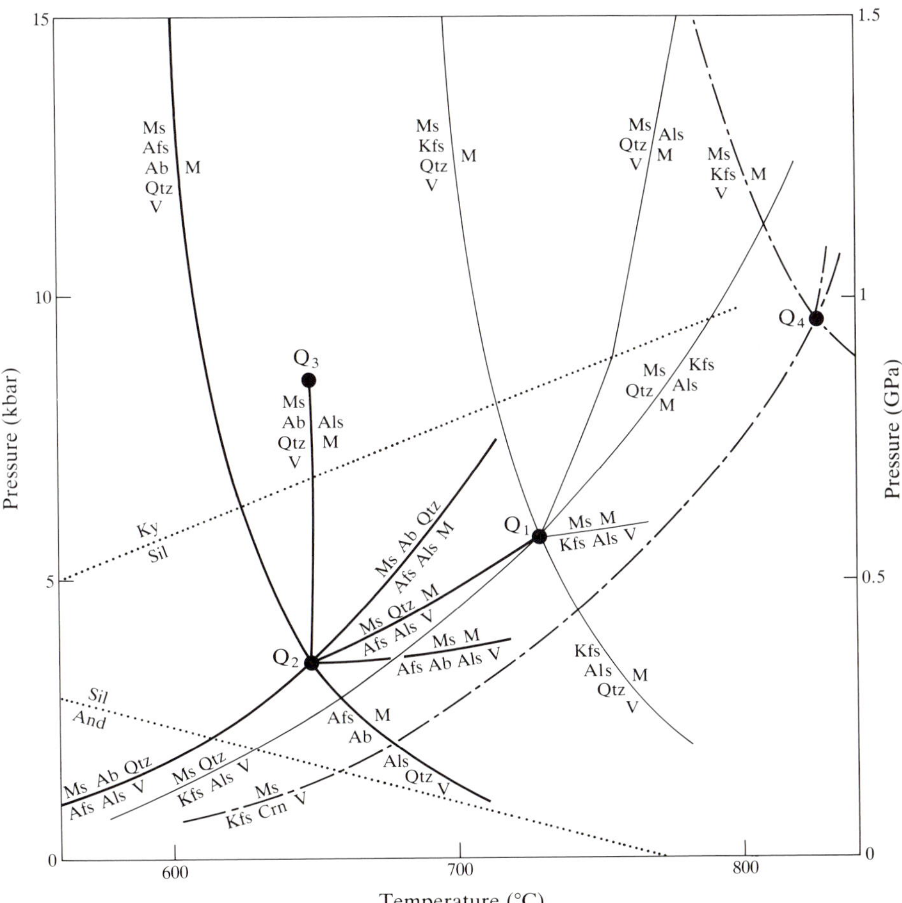

Figure 11.5 Petrogenetic grid showing dehydration-decomposition and dehydration-melting of muscovite. For the invariant points Q_1, Q_2 and Q_4, see text. Univariant curves radiating from Q_1, Q_2 and Q_4 are distinguished, respectively, by thin lines, thick lines and dashed lines. Q_3 is another invariant point of the five-component system K_2O–Na_2O–Al_2O_3–SiO_2–H_2O. At the point, seven phases (Pg, Ms, Ab, Als, Qtz, V and M) coexist, where Pg, Ms and Ab are K–Na solid solutions. Based on Huang & Wyllie (1974) and A. B. Thompson & Algor (1977). Dotted lines show stability boundaries between Al_2SiO_5 polymorphs (Holdaway 1971). V: aqueous vapor phase; M: melt. For mineral symbols, see Appendix 1.

In quartz-free rocks, final decomposition of muscovite occurs at a still-higher temperature by a reaction of the type:

$$\text{muscovite} = \text{K-feldspar} + \text{corundum} + H_2O \qquad (11.21)$$

At pressures higher than that of Q_2, however, melting is induced by dehydration of muscovite, and at the melting point muscovite coexists with melt. The temperature of dehydration-melting varies with the mineral assemblages of the rocks as shown in the grid. In the presence of potassic feldspar + sodic plagioclase + quartz, muscovite undergoes dehydration-melting at the lowest temperature. The assemblage muscovite + sodic plagioclase + quartz is a common mineral assemblage in metapelites. The grid indicates that this assemblage melts at about 650 °C in the presence of vapor and at a higher temperature in its absence. Generally speaking, melting of a mineral assemblage occurs at a considerably lower temperature in the presence of a vapor phase than in its absence. The temperature of melting is lowered by the presence of quartz and of albite.

On cooling, the melt may crystallize back into a mineral assemblage of the same system. It is likely that many rocks in granulite-facies regions underwent prograde partial melting of this type, followed by retrograde crystallization. If the rocks undergoing partial melting are kept for a long time under this high temperature, H_2O may gradually move out of the system, and then cooling may result in the production of mineral assemblages poor or lacking in muscovite.

11.6.3.3. Vapor-absent dehydration-melting of metapelites containing both muscovite and biotite. Ordinary metapelites contain both muscovite and biotite in association with relatively sodic plagioclase and quartz. In the ordinary pressure range of regional metamorphism, they begin to melt at about 650 °C in the presence of an aqueous vapor (Fig. 11.5).

In the absence of an aqueous vapor at pressures above about 4 kbar, these rocks undergo two-stage dehydration-melting, as shown in Figure 11.6 (Le Breton & Thompson 1988). First, muscovite reacts with plagioclase and quartz to form a melt by a reaction of the type:

$$\text{muscovite} + \text{plagioclase} + \text{quartz} = \text{K-feldspar} + Al_2SiO_5 + \text{melt} \qquad (11.22)$$

If the muscovite is phengitic in composition, the reaction products include cordierite, biotite or garnet, depending on pressure. Since an Al_2SiO_5 mineral is formed by the above reaction, a further temperature increase causes a dehydration-melting reaction of biotite of the type:

$$\text{biotite} + \text{plagioclase} + Al_2SiO_5 + \text{quartz} = \text{K-feldspar} + \text{garnet} + \text{melt} \qquad (11.23)$$

Since muscovite, biotite and plagioclase are solid-solution minerals, the temperatures of these reactions show a considerable variation with their compositions.

Figure 11.6 shows experimentally determined equilibrium curves (with esti-

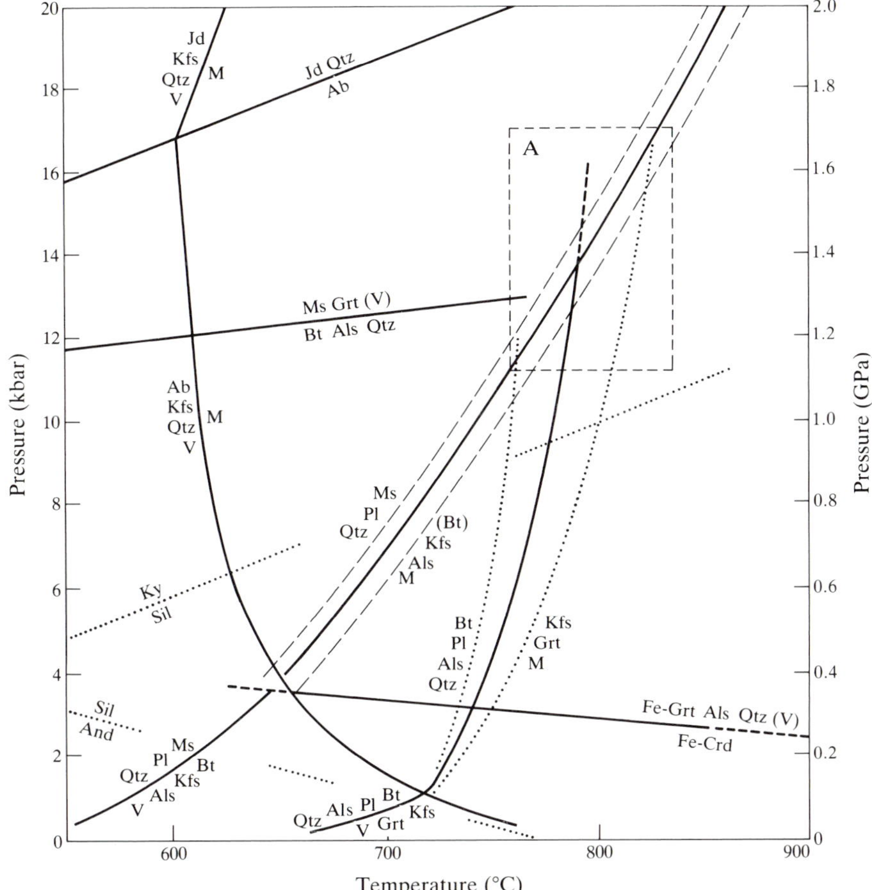

Figure 11.6 *P–T* conditions for the dehydration-melting of muscovite and biotite in metapelites in the absence of aqueous vapor. In addition, the diagram shows the solidus curve in the presence of vapor as well as dehydration-decomposition curves for muscovite and biotite in metapelites. Modified from A. B. Thompson & Algor (1977) and Le Breton & Thompson (1988). For mineral symbols, see Appendix 1. Box A is described in the text.

mated ranges of temperature variation) of these two reactions. The two curves intersect within box A at about 15 kbar. Thus, the two-stage dehydration-melting occurs over the pressure range of about 4–15 kbar.

At pressures higher than that of box A, muscovite and biotite undergo simultaneous dehydration-melting by a reaction of the type:

$$\text{muscovite} + \text{biotite} + \text{plagioclase} + \text{quartz} = \text{K-feldspar} + \text{garnet} + \text{melt} \qquad (11.24)$$

In regional metamorphism at pressures lower than about 4 kbar, dehydration-decomposition (not melting) of muscovite occurs first, and with a further temperature increase, the dehydration products undergo melting in the presence of vapor. At very low pressures, first muscovite, and then biotite, undergo dehydration-decomposition, and with a further temperature increase the dehydration products undergo melting in the presence of vapor.

11.6.4 Observed mineral assemblages and effects of increasing temperature

11.6.4.1 Metabasites. Metabasites show assemblages such as orthopyroxene + augite + garnet + plagioclase (commonly with hornblende, biotite and quartz)

Augite may contain exsolution lamellae of orthopyroxene and pigeonite (e.g. Jaffe et al. 1975). Decomposition of end-members of hornblende with increasing temperature may be expressed by reactions such as:

$$\text{actinolite} = 3\,\text{orthopyroxene} + 2\,\text{diopside} + \text{quartz} + 2\,H_2O \qquad (11.25)$$

$$\text{tschermakite} + \text{quartz} = 3\,\text{orthopyroxene} + 2\,\text{anorthite} + H_2O \qquad (11.26)$$

11.6.4.2 Metapelites. Decomposition and melting of muscovite are discussed on earlier pages. A detailed discussion of the paragenetic relations of muscovite, taking CaO into consideration also, is given by Evans & Guidotti (1966). Other minerals of metapelites are discussed below.

The stable polymorph of K-feldspar is microcline in the greenschist facies and low and middle grades of the amphibolite facies, but becomes orthoclase in high-amphibolite and granulite facies (Heier 1957, Shido 1958). However, microcline of retrograde origin is widespread in granulite-facies regions.

Biotite may decrease in amount with increasing temperature by some reactions with muscovite like:

$$\text{biotite} + \text{muscovite} + 3\,\text{quartz} = \text{almandine} + 2\,\text{K-feldspar} + 2\,H_2O \quad (11.27)$$

Because of this and other reactions including (11.19) and (11.20), K-feldspar can eventually coexist with Ca-free Al–Fe–Mg silicates such as sillimanite, garnet, cordierite and orthopyroxene. (At low temperatures up to the higher amphibolite facies, the assemblage muscovite + biotite forms in place of the assemblage K-feldspar + Al–Fe–Mg silicate.)

After muscovite disappears, biotite solid solutions may decompose by reactions like:

$$\text{biotite (phlogopite–annite)} + 3\,\text{quartz}$$
$$= 3\,\text{orthopyroxene} + \text{K-feldspar} + H_2O \qquad (11.28)$$

$$\text{biotite (phlogopite–annite)} + \text{sillimanite} + 2\,\text{quartz}$$
$$= \text{garnet} + \text{K-feldspar} + H_2O \qquad (11.29)$$

$$2\,\text{biotite (eastonite–siderophyllite)} + 6\,\text{quartz}$$
$$= 2\,\text{orthopyroxene} + \text{garnet} + 2\,\text{K-feldspar} + 2\,H_2O \qquad (11.30)$$

Since biotite has a lower FeO/MgO ratio than coexisting orthopyroxene and garnet, progress of these reactions decreases the FeO/MgO ratio of biotite. The usual TiO_2 content of biotite is a few per cent in amphibolite-facies metapelites and increases to about 5% in granulite-facies metapelites.

Garnet belongs to the almandine–pyrope series, and some garnets show a higher pyrope content than any garnet in the amphibolite facies. Metapelites with relatively low Al_2O_3 content and high FeO/MgO ratio may contain orthopyroxene. Some orthopyroxene in the granulite facies in metapelites shows an Al_2O_3 content as high as 10%. The stable Al_2SiO_5 mineral is usually sillimanite, but kyanite occurs very rarely (e.g. in the Granulitgebirge of Saxony, Germany).

Sapphirine and spinel are rare minerals which occur in quartz-free aluminous rocks in the lower part of the granulite facies. Under conditions where cordierite becomes unstable, sapphirine and even spinel form in quartz-bearing rocks as well (Ellis et al. 1980).

Osumilite is an aluminous mineral similar in composition to, but with a higher Mg/Fe ratio than, cordierite. It was originally discovered in volcanic rocks (Miyashiro 1956), but has been found to occur on a regional scale in metapelites from very high-grade parts of granulite-facies regions (e.g. Ellis 1980, Ellis et al. 1980, Kars et al. 1980, Grew 1982).

Figure 11.7 shows Lonker's (1980) interpretation of progressive changes of metapelites in the Frontenac axis of the Grenville province in Canada. The sillimanite–biotite tieline is stable in (a). With an increase of temperature, it is replaced by the cordierite–garnet tieline, as shown in (b). Then, the garnet–biotite tieline is replaced by the cordierite–orthopyroxene tieline, as shown in (c). With these changes, the composition field of biotite-bearing metapelites

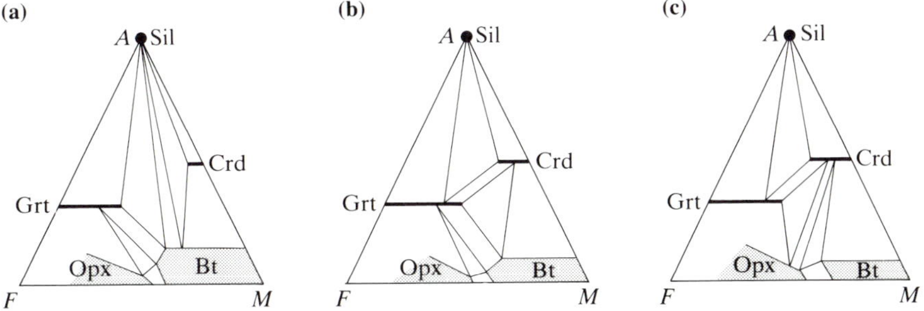

Figure 11.7 *AFM* diagrams showing progressive changes of paragenetic relations with increasing temperature in the order of (a)→(b)→(c) in K-feldspar-bearing metapelitic gneisses from the Frontenac Axis (Lonker, 1980). These diagrams are modified *AFM* projections through K-feldspar and quartz under a constant chemical potential of H_2O (§5.4.1).

becomes narrower, whereas those of cordierite-bearing and orthopyroxene-bearing metapelites become progressively wider.

11.6.4.3 Metagreywackes and quartzo-feldspathic rocks. Metagreywackes show assemblages such as:

orthopyroxene + garnet + plagioclase + orthoclase(perthitic) + quartz

Quartz and feldspars in granulite-facies rocks are whitish in some areas (e.g. the Granulitgebirge of Saxony and Lapland in Finland), but are dark-colored in many others. In the latter case, even quartzo-feldspathic rocks are dark-colored, sometimes looking like metabasites in hand specimen.

A well known example of such dark-colored rocks is **charnockite**, which occurs in the vast granulite-facies region of southern India. There, a large part of the region is made up of dark-colored hypersthene-bearing quartzo-feldspathic rocks with granitic composition. The rocks commonly contain small amounts of garnet and biotite (but by definition not highly aluminous minerals such as cordierite and sillimanite). Quartz is dark blue or bluish grey, and feldspars are dark blue, bluish green or brownish green. Charnockite appears to have formed from granitic and psammitic rocks by granulite-facies metamorphism (e.g. Ramberg 1952, Wendlandt 1981).

11.6.4.4 Calcareous rocks. Wollastonite occurs in some areas, whereas the calcite + quartz assemblage is stable in others. So metamorphosed calcareous rocks show assemblages such as:

1: wollastonite + diopside + calcite (possibly with sphene);
2: wollastonite + scapolite + calcite;
3: diopside + plagioclase + calcite (possibly with scapolite and quartz);
4: forsterite + calcite + dolomite (possibly with spinel, amphibole and phlogopite).

Ca-garnet may not be stable in quartz-bearing rocks in the low-pressure part of this facies (just as in the pyroxene-hornfels facies), but becomes stable in the high-pressure part.

In the granulite facies, Ti-rich minerals are usually rutile or ilmenite in metabasites as well as metapelites, but sphene occurs in calcareous rocks.

11.6.5 Changes of paragenetic relations with pressure in the granulite facies

11.6.5.1 Observed range of pressure. Some of the granulite-facies areas such as Uusimaa in southwestern Finland (Schreurs & Westra 1986) belong to a low-P/T facies series that has andalusite in the associated amphibolite-facies zone. The prevailing pressure was of the order of 3–5 kbar. Many other granulite-facies regions, such as those in southern India belong to a medium-P/T series with kyanite in the associated amphibolite-facies zone. The observed pressure usually ranges from 5 to 9 kbar (e.g. Raase et al. 1986, Janardhan et al. 1982, Newton 1983). Some other granulite-facies regions have given still higher pressures, such as 10–11 kbar (e.g. Newton 1983, Bohlen 1987).

Thus, the observed pressures range from about 3 to 11 kbar, roughly corresponding to 10–40 km in depth. These granulite-facies regions are usually underlain by continental-type crust with a thickness of 20–30 km. Hence, the present erosion surfaces of granulite-facies regions were located at a depth of about 10–40 km within a continental crust, 30–70 km thick, during metamorphism. In other words, the present erosion surfaces of granulite-facies regions represent not the lower crust but the middle crust, or even the lower part of the upper crust during metamorphism.

In the P–T diagram, on the low-pressure side of the granulite facies lies the pyroxene-hornfels facies (Fig. 7.1). Both of these facies are characterized by the coexistence of two pyroxenes in metabasites. Metapelites in the low-pressure part of the granulite facies commonly contain cordierite just as those in the pyroxene-hornfels facies do. On the high-pressure side of the granulite facies, lies the eclogite facies, which is characterized by the disappearance of plagioclase in metabasites (Fig. 7.3). Changes of paragenetic relations with pressure are discussed below.

11.6.5.2 Relation of cordierite to garnet. In metapelites the most important changes with increasing pressure are the decrease and disappearance of cordierite with the concomitant formation and increase of almandine–pyrope garnet.

Cordierite has a low density (d= 2.5–2.8), and many of the cordierite-producing reactions are accompanied by a large increase in the volume of the solid phases. So cordierite-bearing mineral assemblages are stable under low pressures and have a stability limit as pressure increases. They are widespread in pelitic rocks in contact aureoles and in low-P/T regional metamorphism, in the amphibolite and granulite facies (Fig. 8.4). In medium-P/T regional metamorphism, cordierite usually occurs only in the high-temperature part of the granulite facies (Fig. 8.5). Cordierite-bearing assemblages do not occur in the high-pressure part of the granulite facies.

Cordierite has an ordinary orthorhombic form as well as a high-temperature hexagonal form, named indialite. The orthorhombic form shows a variable degree of disorder (Miyashiro 1957, Wallace & Wenk 1980, Putnis & Holland 1986). H_2O molecules can be accommodated inside the open channel parallel to the c-axis of the cordierite structure. Anhydrous cordierite $(Mg,Fe)_2Al_4Si_5O_{18}$ is easily synthesized, but natural cordierite shows a variable H_2O content. The formula of hydrous cordierite is expressed as $(Mg,Fe)_2Al_4Si_5O_{18}.nH_2O$, in which *n* varies at least between 0.0 and 1.2, and is known to approach zero in the high granulite facies (Martignole & Sisi 1981, Lonker 1981). In many areas of granulite-facies metamorphism, intergranular fluid has a relatively high $CO_2/(H_2O + CO_2)$ ratio. Under these conditions, CO_2 molecules are also accommodated inside the open channel together with H_2O. Armbruster et al. (1982) discovered a cordierite with $CO_2/(H_2O + CO_2) = 0.75$ from a granulite-facies region of Lapland.

In granulite-facies metapelites, cordierite-bearing mineral assemblages are gradually replaced by garnet-bearing assemblages with a smaller volume with increasing pressure by various reactions such as:

$$3\,\text{cordierite} = 2\,\text{garnet} + 4\,\text{sillimanite} + 5\,\text{quartz} + 3n\,H_2O \quad (11.31)$$

This reaction corresponds to the cordierite–garnet–sillimanite triangle in Figure 10.3k and Figure 11.7. Reactions corresponding to other paragenesis triangles with cordierite and garnet at two corners in such diagrams are also possible.

Under anhydrous conditions, the Fe end-member reaction corresponding to reaction (11.31) occurs at about 2–4 kbar in the temperature range of 600–1000 °C, whereas the Mg end-member reaction requires a much higher pressure (about 6–12 kbar) (Martignole & Sisi 1981). Cordierite with a higher content of H_2O is stable to higher pressures than equivalent cordierite with a lower content of H_2O (Newton & Wood 1979, Martignole & Sisi 1981, Lonker 1981, Bhattacharya & Sen 1985).

Holdaway & Lee (1977) have shown that the stable coexistence of cordierite and garnet occurs only at temperatures higher than that of the breakdown reaction of muscovite in the presence of quartz, and hence, generally in the granulite facies. The direct change of cordierite-bearing assemblages into garnet-bearing assemblages occurs only under these conditions, although it is true that, in the amphibolite facies also, cordierite-bearing assemblages tend to form at lower pressures than garnet-bearing assemblages. Where the cordierite–garnet assemblage is accompanied either by sillimanite or orthopyroxene (e.g. Fig. 11.7c), the FeO/MgO ratios of cordierite and garnet, being functions of temperature and pressure, may be used as geobarometers, as attempted by Hensen & Green (1973). The ratios are sensitive to variation in pressure but not in temperature. In both the cordierite + garnet + sillimanite + quartz and cordierite + garnet + orthopyroxene + quartz assemblages, the FeO/MgO ratio of the minerals tends to increase with decreasing pressure.

11.6.5.3 Experimental determination of paragenetic relations of metabasites in the granulite and eclogite facies. Green & Ringwood (1967a) made an experimental study on the transformation of gabbro into eclogites under virtually anhydrous conditions. Plagioclase, pyroxene and possibly olivine are the main constituents of gabbro or pyroxene–hornfels-facies metabasite at low pressures. With an increase in pressure, garnet begins to occur and then increases in amount, whereas plagioclase and olivine decrease and disappear. Thus, the rock passes through granulite-facies conditions, and ultimately becomes a pyroxene–garnet rock, that is, eclogite.

The *P–T* relations of such a transformation for quartz–tholeiitic compositions are shown in Figure 11.8. For a medium value of the FeO/(MgO + FeO) ratio, the *P–T* field below curve B in the diagram represents a garnet-free granulite or gabbro assemblage. In this pressure range, the An component of plagioclase is increasingly incorporated into pyroxenes with increasing pressure. This results in the formation of highly aluminous pyroxenes containing Tschermak's compo nent: $(Ca,Mg)Al_2SiO_6$. In the pressure range between curves B and "Eclog",

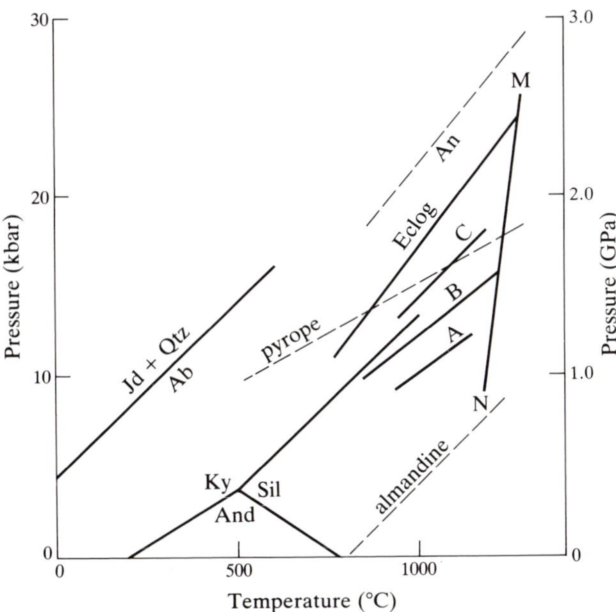

Figure 11.8 Granulite to eclogite transformation for quartz-tholeiitic compositions under a virtually anhydrous condition (Green & Ringwood 1967a). A, B and C are curves indicating the first appearance of garnet with increasing pressure for quartz-tholeiitic compositions with $Fe^{2+}/(Mg + Fe^{2+})$ = 0.9, 0.4 and 0.1, respectively. Curve "Eclog" indicates the disappearance of plagioclase with rising pressure. Line M–N is an approximate solidus under dry conditions. The low-pressure stability limits of pyrope and almandine end-members are shown by the curves designated as "pyrope" and "almandine", respectively. The high-pressure stability limit of anorthite is shown by the curve labeled as "An". The stability fields of kyanite, sillimanite and andalusite and the low-pressure stability limit of the jadeite + quartz assemblage are shown for comparison.

the reaction between orthopyroxene and plagioclase produces an increasing amount of garnet having an increasing pyrope/almandine ratio with rising pressure. At higher pressures within this range, the Ab component of plagioclase begins to change into the jadeite component of pyroxene, leading to the disappearance of plagioclase. The Tschermak's molecule of pyroxene decomposes with the resultant formation of additional garnet. Thus, typical eclogite with magnesian garnet and jadeite-rich clinopyroxene forms in the P–T field above the curve "Eclog".

For SiO_2-unsaturated alkali–olivine–basalt compositions, the appearance of garnet and the disappearance of plagioclase take place at pressures a few kilobar lower than those for quartz–tholeiite compositions. The higher the Ab content of plagioclase, the higher the pressure necessary for the disappearance of plagioclase (i.e. for the formation of eclogite). The higher the FeO/(MgO + FeO) ratio of the rock, the lower the pressure necessary for the formation of garnet.

It must be emphasized that the pressure necessary for reactions to form eclogites varies with the FeO/(MgO + FeO) ratio and the plagioclase composition; hence the transformation should be essentially gradual over a wide transitional P–T region.

Garnet begins to form at a lower pressure in quartz-free metabasites than in quartz-bearing metabasites. This relation may be understood by comparing the following two reactions in rocks without and with quartz.

$$\underset{\text{olivine}}{(Fe,Mg)_2SiO_4} + \underset{\text{anorthite}}{CaAl_2Si_2O_8} = \underset{\text{garnet}}{Ca(Fe,Mg)_2Al_2Si_3O_{12}} \quad (11.32)$$

$$\underset{\text{orthopyroxene}}{2\,(Fe,Mg)SiO_3} + \underset{\text{anorthite}}{CaAl_2Si_2O_8} = \underset{\text{garnet}}{Ca(Fe,Mg)_2Al_2Si_3O_{12}} + \underset{\text{quartz}}{SiO_2} \quad (11.33)$$

In reaction (11.32) the right-hand side is stable in the granulite and eclogite facies, whereas in reaction (11.33) the left-hand side is stable in the granulite facies and the right-hand side in the eclogite facies. These reactions were thermodynamically discussed by Wood (1975) and were later calibrated by Newton & Perkins (1982) and others.

11.6.6 Retrograde changes in granulite-facies rocks

Granulite-facies metamorphic rocks usually show characteristically strong retrograde mineralogical changes. The extent to which retrograde reactions take place depends on the amount of available H_2O and CO_2 in the retrograde stage as well as on the temperature and duration of the process. At the thermal peak in the granulite facies, rocks are usually partially melted, and a part of the H_2O that was liberated by prograde reactions is retained in the melt. When the melt is crystallized in the retrograde stage, the H_2O is liberated, and can react with thermal-peak minerals. The early part of the retrograde process occurs at high temperatures where rates of reactions are high. Hence, retrograde reactions usually occur extensively.

In granulite-facies rocks, net-transfer reactions in the retrograde stage produce a great variety of anhydrous and hydrous minerals, commonly as coronas and symplectites. Microprobe studies have clarified the widespread occurrence of retrograde exchange reactions between adjacent ferromagnesian minerals, which lead to the formation of chemical zoning in each of the minerals (Hess 1971, Lonker 1980, Schreurs 1985). Even virtually homogeneous minerals may have undergone compositional adjustment in the retrograde stage. Generally speaking, exchange reactions will continue at lower temperatures than net-transfer reactions. The widespread and intense retrograde reactions and melting cause great difficulty in geothermobarometry of granulite-facies rocks (Frost & Chacko 1989, Selverstone & Chamberlain 1990).

11.6.7 The relation of the granulite facies to crustal structure

Most of the wide granulite-facies regions occur in high-grade metamorphic regions of the Precambrian, and particularly of the Archean. However, recent progress of petrographic survey has clarified that granulite-facies rocks formed in a considerable number of Phanerozoic regions as well (Windley 1981, Pin & Vielzeuf 1983).

Bohlen & Mezger (1989) pointed out that the most common pressure range of regional-scale granulite-facies regions is 6–8 kbar, indicating that the present erosion surface was only 20–30 km in depth at the time of metamorphism. These regions are now usually underlain by 20–30 km of continental-type crust, suggesting that the original crustal thickness was about 50–60 km, and the present erosion surface was in the middle crust. On the other hand, granulite-facies xenoliths in volcanic pipes usually show a considerably higher crystallization pressure (9–12 kbar), and this suggests that these xenoliths, usually mafic in composition, were usually derived from the lower and lowermost continental crust.

Harley (1989) has given a comprehensive, petrologic review of granulite-facies regions of the world. He emphasized their diversity in origin and characteristics. The thermal peak is followed by a rapid decompression in some granulite-facies regions, but by a nearly isobaric cooling in others.

Bohlen & Mezger (1989) and Bohlen (1991) pointed out that many of the regional-scale granulite-facies complexes show anti-clockwise *P–T–t* paths like curve (d) in Figure 2.2. They ascribed the origin of these complexes to the heating effects of basaltic magmas that had been intruded from the mantle into the middle, lower and lowermost crust. These intrusions would have caused an increase of pressure and temperature in rocks in the middle and lower crust. The basaltic compositions of the intrusions would have caused an increase of the average density of the crust and hence little isostatic uplift and erosion on the surface, thus resulting in nearly isobaric cooling. They considered that these processes occur usually at depth in active continental margin arcs.

Granulite-facies rocks appear to show chemical characteristics distinct from lower-grade metamorphic rocks. Ramberg (1951, 1952) was the first to point out this possibility. He considered that granulite-facies rocks were basified during metamorphism, and he preferred chemical diffusion in the solid state to partial melting as a possible mechanism of basification. Eade & Fahrig (1971) calculated average chemical compositions of amphibolite- and granulite-facies regions in a part of the Canadian Shield. The granulite-facies regions are slightly higher in FeO and CaO, but slightly lower in SiO_2 and K_2O than the amphibolite-facies regions. The chemical migration in granulite-facies metamorphism, however, appears to be much more complicated than was once supposed, and varies with rock type and the mechanism involved (e.g. Barbey & Cuney 1982). More recent data are reviewed by Moorbath & Taylor (1986).

In the granulite facies, partial melting is probably widespread. Many authors ascribed the change in the bulk-rock chemical compositions to chemical transport by removal of the melts. Some authors considered, however, that the removal of melts cannot explain the observed chemical changes, and ascribed them to chemical transport by movement of fluids high in CO_2 content (Glassley 1983, Newton 1989).

12 Metamorphic facies characteristic of high-*P/T* regional metamorphism

12.1 A brief survey of the high-*P/T* metamorphic facies series

From the beginning of facies classification, Eskola (1920) recognized the existence of the **eclogite facies**, because eclogite, being basaltic in chemical composition, attracted wide attention for its remarkably peculiar mineralogical composition and for its possible presence as a major constituent of deep layers of the Earth. With further progress in his survey of the diversity of metamorphic rocks on the Earth, Eskola (1929, 1939) introduced the **glaucophane-schist facies**. Glaucophane, the most characteristic mineral of this facies, was known to occur most commonly in metabasites, although it also occurs in some other types of rocks. Thus, these two facies were characterized by the mineralogical compositions of metabasites. At that time it was not clear whether any metapelites and metapsammites existed in the eclogite facies. His idea of the glaucophane-schist facies was not widely accepted until the late 1950s.

In the early 1960s when the existence of the glaucophane-schist facies was already generally accepted, someone coined the name **blueschist facies** to replace it, without any justifiable reason. Although this is unfair to Eskola, this name has become so popular that it is very reluctantly accepted in this book.

Although Eskola regarded the blueschist and eclogite facies as being characterized by high pressures, he did not suspect the existence of a close genetic connection between the two. In his time, rocks of these two facies were known to occur in entirely different, and at least apparently unrelated places on the Earth.

In the 1980s our understanding of these two facies made a great advance through combined experimental and field studies. In particular, studies in the Alps and New Caledonia have clarified the eclogite-facies mineral assemblages in rocks of diverse chemical compositions, as well as the genetic and petrographic relationships between the blueschist and eclogite facies.

Both blueschist and zeolite facies are major metamorphic facies, rocks of which are widespread in the world. Between the P–T fields of these two facies a gap exists. As an example of metamorphic regions falling in this gap, Coombs (1960) discovered a zone in New Zealand that is characterized by the presence of lawsonite without being accompanied by glaucophanic amphibole and jadeite. Afterwards, more or less similar metamorphic tracts were found in a few other regions. In these tracts, the pressure of metamorphism was high enough to produce lawsonite, but not so high as to produce glaucophane or jadeite. A metamorphic facies representing the zones of this kind is included at the low-temperature, low-pressure end of the high-P/T facies series under the name **lawsonite–albite–chlorite facies**.

In the past a considerable number of other facies names have been proposed for metamorphic regions with glaucophanic amphiboles. In this book, only the three above are used, so that we get a simple scheme of facies classification in accordance with the general principle stated in §7.4.

Many high-P/T metamorphic regions include areas of the greenschist, prehnite–pumpellyite and zeolite facies (Fig. 8.3). Since these facies are described in the preceding chapter, they are not treated in this chapter. Thus, in this chapter only the above-mentioned three metamorphic facies are described in order of increasing temperature and pressure.

12.2 Lawsonite–albite–chlorite facies

12.2.1 Concept and history

In regional metamorphism at very low temperatures, the zeolite and prehnite–pumpellyite facies represent low pressures, whereas the blueschist facies represents high pressures. There exists a P–T range intermediate between these two cases, where rocks are crystallized at pressures high enough to form lawsonite and not zeolite (Fig. 11.2), but not so high as to form glaucophanic amphibole (glaucophane or crossite) or jadeitic pyroxene + quartz, or both (Fig. 12.1). Since the low-pressure limit for the formation of lawsonite is about 3 kbar at 200–350 °C (Liou 1971a), this intermediate field covers a pressure range from about 3 kbar to the pressure for the first formation of amphibole with a substantial content of glaucophane component. In this book, the metamorphic facies representing this P–T field is called the lawsonite–albite–chlorite facies after Turner (1981).

Prior to Turner, Winkler (1965) had proposed the name of **lawsonite–albite facies** for these rocks, and had defined it as being characterized by the presence of lawsonite and the absence of jadeitic pyroxene + quartz. He did not preclude the occurrence of glaucophanic amphibole in his facies. In this book, on the other hand, the occurrence of glaucophanic amphibole is precluded from our lawsonite–albite–chlorite facies, and the low-temperature, low-pressure limit of the blueschist facies is defined as being marked by the first occurrence of glaucophanic amphibole in rocks of ordinary compositions.

Coombs (1960) was the first to point out the existence of a metamorphic tract of the facies. His area was in Nelson at the northern end of the South Island in New Zealand, where lawsonite is widespread in metavolcanics as well as in metapelites, being accompanied by quartz, albite, chlorite, muscovite and calcite. No jadeite and no glaucophane have been found in this area. Similar metamorphic areas were subsequently found in California, Calabria (southern Italy) and some other places in the world (e.g. E. W. F. de Roever 1972, Turner 1981).

In areas of this facies, albite occurs in place of jadeitic pyroxene + quartz, and albite + chlorite occurs in place of glaucophane. In other words, the right-hand side of reaction (3.5) and the left-hand side of reaction (12.5) given later are stable because of relatively low pressure. Aegirine, riebeckite or magnesio-riebeckite may occur in some rocks, depending on the bulk-rock composition. The associated carbonate may be either calcite or aragonite, and pumpellyite and celadonite may occur.

This facies represents only a small range of P–T conditions and occurs in a small number of regions in the world. It is not clear whether it is appropriate to treat it as an independent major metamorphic facies on an equal footing with such facies as the blueschist and the amphibolite. This question is left for the future.

12.2.2 Examples of lawsonite–albite–chlorite-facies areas

Upper Wakatipu area, New Zealand. Kawachi (1974) mapped the Upper Wakatipu area in the southern part of the South Island in New Zealand. He described a progressive sequence of zones belonging to the prehnite–pumpellyite facies, lawsonite–albite–chlorite facies, pumpellyite–actinolite facies, and greenschist facies. The characteristic mineral assemblage of the lawsonite–albite–chlorite-facies zone is quartz + albite +lawsonite + pumpellyite + muscovite + chlorite + sphene.

The Franciscan Complex in the Pacheco Pass area, California. In this area, the high-grade zone defined for metagreywackes belongs to the typical blueschist facies with jadeite + quartz. Metagreywackes are composed of quartz, jadeite, glaucophane, lawsonite and aragonite. With decreasing metamorphic grade, jadeite and glaucophane decrease in amount and disappear, while albite and chlorite appear and increase. Thus, metagreywackes in the lowest grade zone show a lawsonite–albite–chlorite-facies mineral assemblage: quartz + albite + lawsonite + chlorite + aragonite +calcite. The calcite was regarded as having formed from aragonite by retrograde change (Maruyama et al. 1985).

12.3 Blueschist facies

12.3.1 History and definition

Eskola (1929, 1939) considered the blueschist facies (his glaucophane-schist facies) as representing a range of high pressures at low temperatures (Fig. 7.1). He mentioned not only glaucophane but also lawsonite as being critical for this facies. However, the great majority of petrologists at that time did not accept this idea and ascribed the origin of glaucophane to Na_2O-rich bulk-rock composition or metasomatic introduction of Na_2O from outside (e.g. Harker 1932, Turner 1948).

The existence of the blueschist facies was debated and ultimately came to be generally accepted in the late 1950s (for detail see Appendix III). A most important event that led to the general acceptance of the facies was W. P. de Roever's (1950, 1955a) discovery in Celebes (Sulawesi) in Indonesia that the jadeite + quartz assemblage is characteristic of the glaucophane schist group of metamorphic rocks, and that since jadeite + quartz is isochemical with albite, as shown by reaction (3.5), jadeite in quartz-bearing rock must have formed at a higher pressure than ordinary metamorphic rocks containing albite. This was confirmed by a thermochemical calculation (Adams 1953) as well as by synthetic experiments (Birch & LeComte 1960).

The jadeite + quartz assemblage occurs in the typical, high-pressure, low-temperature part of the *P–T* range of the blueschist facies. On the other hand, there is a wide range of conditions transitional from such a typical part to the prehnite–pumpellyite and greenschist facies of the medium-*P/T* facies series. Under these conditions, crossite containing a considerable amount of glaucophane component may form in rocks of appropriate compositions. In other words, this facies belongs partly to the jadeite–glaucophane type, and partly to the high-pressure transitional type. It was discovered in the 1980s that with increasing temperature and pressure the blueschist facies grades into the eclogite facies.

In short, the *P–T* field of the blueschist facies is bounded on the low-temperature, low-pressure side by those of the lawsonite–albite–chlorite, prehnite–pumpellyite and greenschist facies (Fig. 8.3). The boundary between these three facies and the blueschist facies is defined in this book as the P–T conditions where glaucophanic amphibole (amphibole containing a considerable amount of glaucophane component) begins to form in some rocks of ordinary basaltic compositions.

The P–T field of the blueschist facies is bounded on the high-temperature and high-pressure sides by that of the eclogite facies (Fig. 8.3). This book places the boundary from the blueschist to the eclogite facies at the P–T conditions where some metabasites in the ordinary range of chemical composition begin to become eclogites, that is, to show the assemblage omphacite + garnet, with increasing temperature and pressure. In this definition, glaucophanic amphiboles are still stable in many rocks in low- and medium-temperature parts of the eclogite facies.

The temperature of metamorphism of the blueschist facies appears to range from 170° to about 500°C.

12.3.2 Characteristic minerals of the blueschist facies

12.3.2.1 Glaucophane and related amphiboles. Glaucophane is stable only at high pressures. The low- and high-pressure limits of the stability field of the Mg end-member of glaucophane as determined by Carman & Gilbert (1983) are shown by thick lines marked as C—G in Figure 12.1.

The high-pressure limit for Mg-glaucophane is marked by the following solid–solid reaction (Carman & Gilbert 1983):

$$\underset{\text{Mg-glaucophane}}{Na_2Mg_3Al_2Si_8O_{22}(OH)_2} = 2\,\underset{\text{jadeite}}{NaAlSi_2O_6} + \underset{\text{talc}}{Mg_3Si_4O_{10}(OH)_2} \qquad (12.1)$$

The jadeite + talc assemblage on the high-pressure side has actually been discovered in some eclogite facies rocks.

On the other hand, the thick line marked M in Figure 12.1 indicates the equilibrium curve for the reaction: 25 glaucophane + 6 clinozoisite + 7 quartz + 14 H_2O = 6 tremolite + 9 chlorite + 50 albite, as determined by Maruyama et al. (1986). The glaucophane used in this study contains a considerable amount of FeO. This is a reaction marking the boundary between the blueschist and the greenschist facies.

Among glaucophane and other minerals there are stoichiometric relations such as:

$$\begin{aligned} &\text{glaucophane} + 3\ \text{calcite} + 3\ CO_2 \\ &\quad = 2\ \text{jadeite} + 4\ \text{quartz} + 3\ \text{dolomite} + H_2O \\ &\quad = 2\ \text{albite} + 2\ \text{quartz} + 3\ \text{dolomite} + H_2O \qquad (12.2) \end{aligned}$$

This illustrates that the formation of glaucophane is influenced by the activities of H_2O and CO_2.

Magnesioriebeckite $Na_2Mg_3Fe_2^{3+}Si_8O_{22}(OH)_2$, riebeckite $Na_2Fe_3^{2+}Fe_2^{3+}Si_8O_{22}$-

$(OH)_2$ and actinolite do not require particularly high pressure for their formation. Glaucophane forms a solid-solution series with magnesioriebeckite and riebeckite as well as with actinolite (e.g. Miyashiro & Banno 1958, Liou & Maruyama 1987). Intermediate members between glaucophane and riebeckite, called **crossite** in the following discussion, could form in rocks under relatively oxidized conditions and at pressures lower than Al_2O_3-rich glaucophane. In this book, the name **glaucophanic amphibole** is used to include both glaucophane

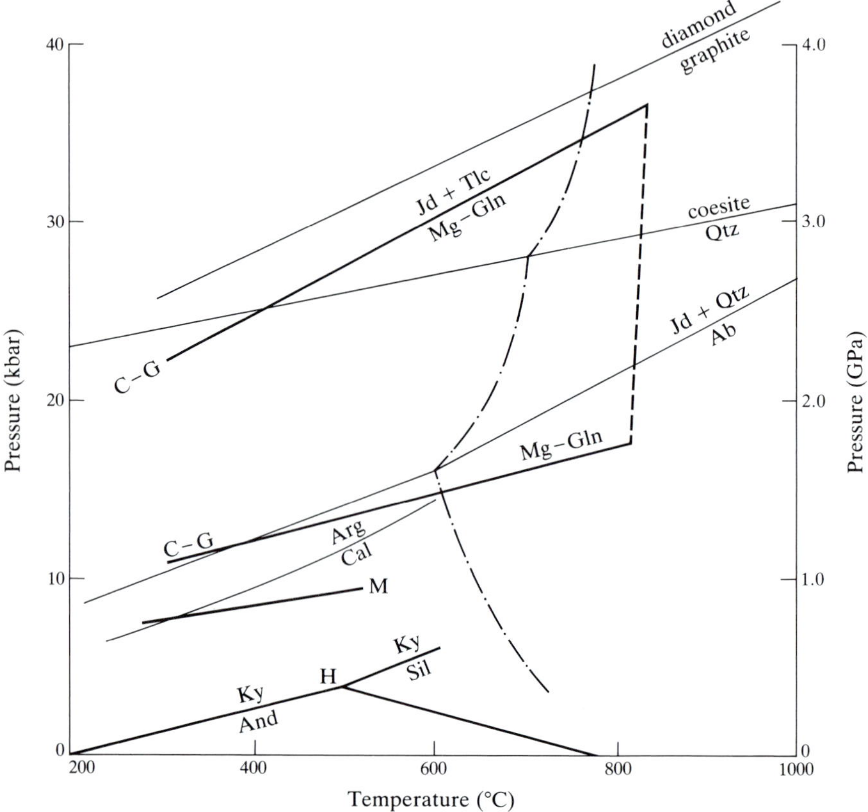

Figure 12.1 The solid-line segments of the enclosure C—G indicate the low-pressure and high-pressure limits for the stability field of Mg-glaucophane (Carman & Gilbert 1983). Line M represents the equilibrium curve for the reaction: 25 glaucophane + 6 clinozoisite + 7 quartz + 14 H_2O = 6 tremolite + 9 chlorite + 50 albite (Maruyama et al. 1986). Thin lines show equilibrium curves for the following reactions: aragonite = calcite, jadeite + quartz = albite, coesite = quartz, and diamond = graphite. The chain line shows the melting temperature of albite or jadeite in the presence of excess silica (quartz or coesite) and excess H_2O (Boettcher & Wyllie 1969). H indicates the triple point of Al_2SiO_5 minerals (Holdaway 1971).

and crossite. There is a miscibility gap between glaucophane and actinolite under some P–T conditions, while complete solid solutions form under others (Liou & Maruyama 1987). Thus, glaucophane and actinolite coexist in some cases, whereas intermediate members between the two minerals, called **winchite** (Leake 1978), occur instead in others.

12.3.2.2 Jadeite, omphacite and other Na–Ca clinopyroxenes. The equilibrium curve for the reaction: pure jadeite + quartz = albite, has been investigated by many experimental workers, as discussed in §3.4.3 (Figs 3.1, 12.1). The occurrence of the assemblage jadeitic pyroxene + quartz is characteristic of the relatively high-pressure part of the blueschist facies.

Natural jadeite is usually a solid solution containing aegirine, diopside and hedenbergite components. When it is accompanied by quartz, the pressure necessary for its formation is considerably lower than that for pure jadeite + quartz, and becomes lower with decreasing jadeite content, as shown in Figure 3.1.

Pyroxenes of the jadeite–aegirine series form a continuous solid-solution series as illustrated in Figure 12.2a. Under a specific P–T condition, pyroxenes of this series unaccompanied by quartz show a higher jadeite content than ones accompanied by quartz. Among proxenes accompanied by quartz, pyroxene accompanied by both quartz and albite gives the highest jadeite content. In the presence of albite and quartz, the pressure necessary for the formation of such a pyroxene solid solution becomes lower with decrease of jadeite component (Fig. 12.2b).

Clinopyroxenes of the system jadeite–aegirine–diopside–hedenbergite appear to have miscibility gaps between jadeite and augite (diopside + hedenbergite) at low temperature (Fig. 12.3a). The coexistence of two apparently stable Na–Ca clinopyroxenes is relatively common (e.g. Maruyama & Liou 1987).

Jadeite unaccompanied by quartz, being isochemical with albit + nepheline as shown by reaction (3.6), is stable down to a much lower pressure than jadeite + quartz (Fig. 3.1). In the presence of an aqueous fluid, however, analcime forms instead of jadeite at very low temperatures and pressures. For this reason, pure jadeite does not form at pressures below about 5 kbar in the presence of an aqueous fluid in rocks either with or without quartz (Fig. 11.1).

12.3.2.3 Lawsonite. Lawsonite is known to be stable at pressures down to about 3 kbar in the presence of H_2O fluid at 200–350°C (Liou 1971a) as shown in Figure 11.2. The high-temperature limit for the formation of lawsonite is shown in Figure 12.5b. It is near the boundary between the blueschist and the eclogite facies.

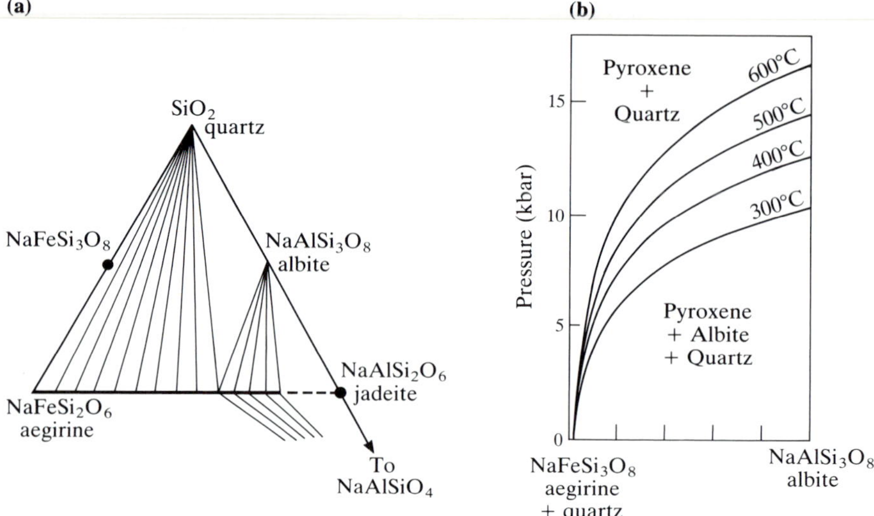

Figure 12.2 (a) Schematic isothermal, isobaric phase relations between albite, jadeite, aegirine and quartz. (b) Calculated compositions of pyroxene of the jadeite–aegirine series coexisting with quartz and albite based on the assumed ideality of pyroxene solid solutions. After Popp & Gilbert (1972).

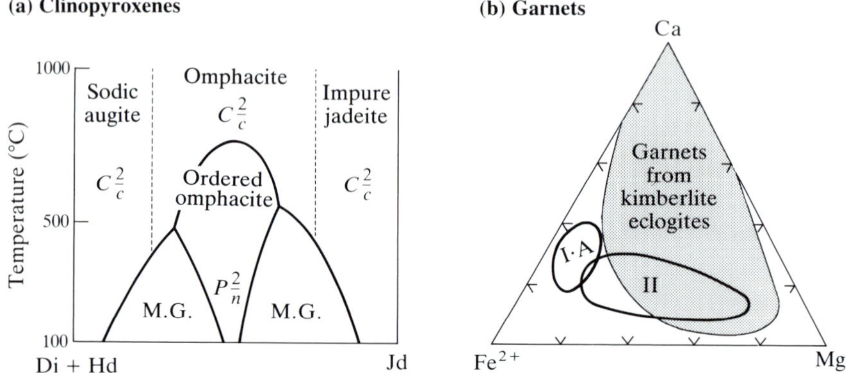

Figure 12.3 (a) Schematic phase diagram showing cation-ordering and miscibility gaps (M.G.) in the augite–jadeite series. Modified from Carpenter (1978) and Rossi (1988). (b) Composition ranges of garnets in eclogites showing various modes of occurrence. I.A: garnets from Sifnos and Franciscan eclogites (subgroup I.A). II: garnets from eclogites of western Norway (group II).

12.3.2.4 Aragonite. Aragonite is stable at high pressures. The boundary between the stability fields of calcite and aragonite lies at pressures considerably lower than the low-pressure limit of jadeite + quartz (Johannes & Puhan 1971), as shown in Figure 12.1. In most or many cases, however, aragonite once formed at high pressure appears to be lost by transformation into calcite at a later stage of the metamorphism.

12.3.2.5 Carpholite. Carpholite was known from the early 19th century as an Mn-rich mineral. The occurrence of Mn-poor, Mg–Fe-carpholite $(Mg,Fe)Al_2Si_2O_6(OH)_4$ (orthorhombic) in metapelites in blueschist-facies regions was established only recently by W. P. de Roever et al. (1967). Since then, it has been discovered from many localities in the Penninic zone of the western Alps. It is similar to actinolite in appearance, and similar to chloritoid in high $Al_2O_3/(FeO + MgO)$ ratio, but is more hydrated and so tends to form at a lower temperature than the latter (Chopin & Schreyer 1983). The composition relation may be shown by the equation:

$$\begin{aligned} &10\,\text{Mg–Fe-carpholite} \\ &\quad = 5\,\text{Mg–Fe-chloritoid} + 10\,\text{quartz} + 10\,H_2O \\ &\quad = 2\,\text{Mg–Fe-chlorite} + 8\,\text{kyanite} + 6\,\text{quartz} + 12\,H_2O \end{aligned} \qquad (12.3)$$

Hence, carpholite occurs only in highly aluminous rocks (Fig. 12.9). It has been shown experimentally that the stability field of Mg-carpholite has a low-pressure limit at several kbar (Schreyer 1988).

12.3.2.6 Deerite and other minerals. The hydrous iron silicates: deerite, howieite and zussmanite, occur in Fe-rich metasediments in blueschist-facies areas (Lattard & Schreyer 1981).

12.3.3 Low- and high-pressure parts of the blueschist facies

12.3.3.1 General remarks. In §8.1.2, the high-*P/T* type of metamorphism was subdivided into two divisions: the high-pressure transitional type and the jadeite–glaucophane type. The blueschist facies occurs in both divisions. Therefore, the blueschist facies is divided into two parts (subfacies) that correspond to the two divisions, as follows.

The *relatively low-pressure part* of the blueschist facies occurs in metamorphic regions of the high-pressure transitional type. Here the jadeitic pyroxene + quartz assemblage does not occur, and glaucophanic amphibole (usually crossite) occurs in metabasites but not in metapelites and metapsammites. The *relatively high-pressure part* occurs in metamorphic regions of the jadeite–glaucophane type. Here the jadeitic pyroxene + quartz assemblage occurs in rocks of appropriate chemical compositions, and glaucophane and crossite occur in almost all rock types, including metabasites, metapelites and metapsammites.

Brothers & Yokoyama (1982) published an elaborate description and discussion of this problem with special reference to the metamorphism in the Sanbagawa belt and northern New Caledonia.

12.3.3.2 The relatively low-pressure part of the blueschist facies: the Sanbagawa metamorphic belt of Japan. A well documented example of the relatively low-pressure part of the blueschist facies is seen in a middle-grade part of the Sanbagawa (Sambagawa) metamorphic belt in Japan (Fig. 9.2). It has been investigated for many years by Banno, Seki and their co-workers (e.g. Seki 1961b, Banno 1964, 1986, Ernst et al. 1970, Banno et al. 1978, Higashino et al. 1981, Banno & Sakai 1989). For the present situation of geologic studies, the reader is referred to the Special Issue on Sanbagawa Metamorphism of the *Journal of Metamorphic Geology* (Vol. 8(4), 1990).

The metamorphic region shows a progressive metamorphic sequence generally composed of a pumpellyite zone (free of prehnite), a crossite zone, and an epidote-amphibolite zone as defined for metabasites. Typical Al_2O_3-rich glaucophane hardly occurs (Brothers & Yokoyama 1982). Most of the glaucophanic amphiboles are crossites, which occur only in hematite-bearing (i.e. relatively well oxidized) metabasites in a very limited range of metamorphic grade. Table 12.1 summarizes detailed progressive mineralogical changes in ordinary metabasites, relatively well oxidized metabasites and metapelites.

As shown in the table, the occurrence of glaucophanic amphibole (crossite) is limited to metabasites in metamorphic grades ranging from the higher chlorite to the lower garnet zone defined for metapelites. Even within this range, metabasites are more frequently actinolite greenschist than crossite-bearing rocks. Hashimoto (1991) has shown that within this grade range bulk-rock $Fe_2O_3/(Fe_2O_3 + FeO)$ wt ratio is 0.15–0.51 in actinolite greenschists and 0.45–0.63 in crossite-bearing metabasites, whereas there is no systematic difference in Na_2O and CaO contents.

Metapelites in this belt are similar in mineral assemblages to those in the chlorite and biotite zones of medium-*P/T* metamorphism, except for that highly MnO-rich garnet becomes common at a considerably lower temperature than that of the biotite isograd. Hence, with respect to metapelites, this region may be divided into three progressive metamorphic zones characterized by the entrance of chlorite, garnet and biotite in order of increasing temperature. In the highest grade, oligoclase begins to occur (Enami 1983). It should be noted that glaucophanic amphibole occurs in some metacherts.

Rock type	Chlorite-bearing pelitic schists	Hematite-bearing basic schists	Hematite-free basic schists
↑ Metamorphic grade	Oligoclase–biotite zone	Oligoclase–hornblende zone	Oligoclase–hornblende zone
	Albite–biotite zone	Albite–hornblende zone	Albite–hornblende
	Garnet zone	Barroisite zone	Barroisite zone
		Crossite zone	Epidote–actinolite zone
	Chlorite zone	Winchite zone	Pumpellyite–epidote–actinolite zone
		Hematite–epidote–actinolite zone	
		Hematite–pumpellyite–actinolite zone	Pumpellyite–stilpnomelane–zone

12.3.3.3 Mineralogical changes with increasing pressure. Amphiboles intermediate in composition between glaucophane and actinolite and between glaucophane and riebeckite require pressures lower than typical glaucophane for their formation. So, in the transition from medium- to high-*P/T* states at low temperatures, crossite begins to occur at a lower pressure than typical glaucophane, and its occurrence is confined to hematite-bearing metabasites, because the riebeckite component in crossite requires a relatively well oxidized conditions.

Ernst (1963a) and Laird (1980) gave the following type of reaction between metabasites of the greenschist and blueschist facies:

$$\text{actinolite} + \text{chlorite} + \text{albite} = \text{glaucophane} + \text{chlorite} + \text{epidote} \qquad (12.4)$$

On the other hand, Brown (1977) gave the following reaction:

$$\text{actinolite} + \text{chlorite} + \text{albite} + \text{iron oxide} + H_2O = \text{crossite} + \text{epidote} + \text{quartz}.$$

The pressure necessary for the formation of glaucophane or crossite is considerably higher in metapelites than in metabasites. In either rock type, the occurrence of glaucophane appears to become increasingly common with increasing pressure. Brothers & Yokoyama (1982) concluded that with increasing pressure the frequency of occurrence of glaucophanic amphibole increases and the compositional change from crossite to typical glaucophane occurs by reactions like:

albite + chlorite + quartz = glaucophane + paragonite + H_2O (12.5)

albite + chlorite + actinolite = glaucophane + epidote + H_2O (12.6)

Thus, albite + chlorite decreases and disappears with a complementary appearance and increase of glaucophanic amphibole. In typical glaucophane schists, paragonite may occur even in rocks with an ordinary Al_2O_3 content because of composition relations such as equation (12.5).

It is empirically known that, in metamorphic regions of the jadeite–glaucophane type, a majority of jadeitic pyroxenes associated with quartz and albite show more than 70% jadeite component at least in some metamorphic grades (Compagnoni 1977, Maruyama et al. 1985, Yokoyama et al. 1986, Maruyama & Liou 1987, 1988). On the other hand, in glaucophane-schist belts of the high-pressure transitional type such as the Sanbagawa, some clinopyroxenes occur, but their jadeite content is lower than 60% when they are associated with quartz, although it may be as high as 90–100% where they are unaccompanied by quartz (Brothers & Yokoyama 1982).

12.3.3.4 The relatively high-pressure part of the blueschist facies. The relatively high-pressure part of the blueschist facies is characterized by the occurrence of jadeite solid solutions and lawsonite in various quartz-bearing rocks including metagreywackes. Glaucophanic amphiboles occur not only in metabasites but also in metapsammites and metapelites. The nearly pure jadeite + quartz assemblage may occur in rocks of appropriate compositions, such as metamorphosed quartzite or acidic igneous rocks.

An example of the relatively high-pressure part of the blueschist facies has been described by Brothers & Yokoyama (1982) from the blueschist–eclogite facies sequence in northern New Caledonia (Fig. 8.8). Jadeite occurs in metasediments and metamorphosed acidic igneous rocks in the higher-grade part of the blueschist-facies zone. The Franciscan metamorphic complex in the Coast Ranges of California also contains a zone representing the relatively high-pressure part of the blueschist facies (Fig. 8.7). Part of the blueschist facies zone shows a widespread occurrence of the jadeite + quartz assemblage and lawsonite.

12.4 Eclogite facies

12.4.1 History, definition and subdivision

12.4.1.1 The rock type eclogite. The rock name **eclogite** was coined in 1822 by Rene-Just Hauy of the Museum of Natural History in Paris. Although the original specification of the name was vague, its meaning was gradually settled through usage in later years, particularly in relation to the concept of the eclogite facies. According to the generally established usage in recent years, it means a rock of basic (mafic) composition mainly composed of omphacite and almandine–pyrope–grossular garnet. Omphacite is a clinopyroxene mainly composed of diopside, hedenbergite and jadeite components with minor aegirine. Twenty years ago, garnets of typical eclogites were usually considered to be MgO-rich. Now this limitation has been removed, because what are called eclogites of group I on a later page contain MgO-poor garnets, and yet it is convenient for genetic discussions to include these rocks in the eclogites.

Small amounts of quartz, coesite, kyanite, olivine, orthopyroxene, amphibole (winchite, barroisite, hornblende or glaucophane), phengite, paragonite, zoisite, epidote, rutile, graphite and diamond may be present. The absence of plagioclase is a most important characteristic of typical eclogite.

Eclogites range from quartz-tholeiitic to alkali-basaltic in chemical composition and also show much variation in FeO/MgO ratio. The above-mentioned variation in mineralogical composition results partly from such chemical variations and partly from differences in *P–T* condition of their formation.

12.4.1.2 Eskola's idea of the eclogite facies. Eclogite has a high density of about 3.4 $g\,cm^{-3}$ as compared with basalt (density 2.9) and amphibolite (density 3.0) in spite of its broadly basaltic chemical composition. In the early 20th century, this led to widespread speculation about the existence of an eclogite layer at great depths in the Earth. Basically following this line of thinking, Eskola (1920, 1939) proposed the idea of the **eclogite facies** as representing a *P–T* field characterized by the stability of vaguely defined eclogites. Not only eclogites themselves, but also associated rocks with other chemical compositions that were metamorphosed under the same physical conditions as eclogites should belong to the eclogite facies. However, Eskola had little reliable data on such associated rocks, because the eclogites known at that time occurred as inclusions in igneous rocks or as masses whose geologic relations to the surrounding rocks were not clear. Moreover, it was not known how rocks of various chemical compositions change their mineralogical composition in transition from the eclogite to other contiguous metamorphic facies. Thus, Eskola's idea was essentially incomplete.

12.4.1.3 Progress in the past 25 years. In recent years our knowledge and ideas about the eclogite facies have made marked progress, partly through experimental studies at very high pressures, and partly through field studies in coherent eclogite-facies metamorphic areas. Experimental studies have clarified not only the numerical values of temperature and pressure necessary for the formation of eclogites for various basic compositions, but also mineral assemblages of rocks of other chemical compositions formed under high pressures.

Field studies in the Alps, New Caledonia and other parts of the world have clarified the relatively common occurrence of coherent eclogite-facies areas that contain not only eclogites but also metapelitic, metapsammitic, calcareous, ultramafic and meta-acidic rocks. This finding has enabled us to understand the mineralogical variations of metamorphic rocks that crystallized under the same *P–T* conditions as eclogites. Although plagioclase does not occur in eclogites and most associated rocks, the possibility of occurrence of the mineral in some rocks of the eclogite facies is not precluded (Fig. 12.6a).

Moreover, progressive metamorphic regions showing a continuous change from the blueschist to the eclogite facies have been discovered in New Caledonia. Thus it has become possible to discuss the eclogite facies in a way analogous to other metamorphic facies. Carswell's (1990) book contains well written reviews of aspects of recent progress.

12.4.1.4 A more precise definition of the eclogite facies. The *P–T* field where metabasites become eclogite varies greatly with the bulk chemical composition of the rocks (Fig. 7.3). In future, the precise definition of the eclogite facies should be formulated so as to represent the *P–T* field where basic rocks with some precisely defined chemical composition become eclogite, or where metabasites show more precisely defined phase relations.

Boundary to the granulite facies. Some attempts have been made for a precise definition of the boundary between the eclogite and granulite facies. Green & Ringwood (1967a) showed in experiments under anhydrous conditions that the minimum pressure necessary for the formation of eclogite tends to become lower, as the metabasites become higher in FeO/MgO ratio and in the degree of unsaturation with SiO_2. They defined the eclogite facies as representing the *P–T* field where rocks of quartz-tholeiitic composition become eclogite. So, "eclogites" of alkali-basaltic composition could form not only in the thus-defined eclogite facies but also at lower pressures, that is, in the high-pressure part of the granulite facies. This discussion continues in §12.4.4.

Boundary to the blueschist facies. There is a wide transitional state between the blueschist and eclogite facies, where glaucophanic amphibole (glaucophane or crossite) is stable, but some metabasites become eclogitic, being mainly composed of omphacite, almandine-rich garnet and hydrous minerals. For example, Schliestedt (1986) described a beautiful example of an area belonging to such a transitional state from Sifnos in the Cyclades (Greece), where interbedded layers of eclogite (with or without glaucophane), garnet–glaucophane schist (with or without paragonite), and garnet–actinolite–glaucophane schist occur. All these rocks are broadly basaltic in chemical composition, and appear to have been metamorphosed under the same *P–T* conditions. In this book, the transitional state is included in the eclogite facies (§12.3.1).

12.4.1.5 Carswell's (1990) subdivision of the eclogite facies. Carswell (1990) proposed a subdivision of the eclogite facies based on temperature. As in Figure 12.4, the eclogite facies was divided into three divisions by boundaries at 550° and 900°C, as follows:

low-temperature division: about 450–550°C
medium-temperature division: 550–900°C, and
high-temperature division: 900 to about 1600°C.

In the next section, the modes of occurrence of eclogites are reviewed in relation to these divisions. The rock type eclogite shows diverse modes of occurrence. Correspondingly, eclogites vary in mineralogical composition and appear to have formed under a very wide range of *P–T* conditions.

12.4.2 Classification of eclogites based on their modes of occurrence

After the thermal peak in the eclogite facies, most eclogites underwent mineral changes at lower temperatures and/or lower pressures. In many cases it is hard to tell whether the later phase of crystallization took place in a declining stage immediately following the thermal peak, or in a separate, unrelated, later tectonothermal event. Moreover, it is hard to tell whether hydrous minerals in eclogites were in stable equilibrium with omphacite and garnet at the thermal peak, or formed in retrogressive or decompression stages. In the following description, only hydrous minerals that are usually regarded as having been stable at the thermal peak are named.

12.4.2.1 Eclogites of group I, which occur in eclogite-facies regions accompanied by blueschist-facies regions. Eclogites of this group are accompanied by a diversity of isophysical metamorphic rocks in many cases. In Sifnos and New Caledonia, such eclogites belong to Carswell's (1990) low-temperature division of the eclogite facies. In the western Alps, on the other hand, such eclogites show a very wide range of crystallization temperatures and pressures, covering both low- and medium-temperature divisions of the eclogite facies. These two types of group I eclogite regions are distinguished here by the names subgroups I.A and I.B, respectively.

Subgroups I.A: low-temperature eclogite regions. A typical eclogite-facies region of this type was described from the Island of Sifnos in the Cyclades in Greece (Schliestedt 1986, 1990, Schliestedt & Okrusch 1988). It represents the highest grade part of the wide glaucophane-bearing region of the Cyclades

Islands. Another typical example is the eclogite-facies zone in northern New Caledonia, which grades into a blueschist facies zone with decreasing temperature, as described in §12.4.5 (Brothers & Yokoyama 1982, Black et al. 1988). In the Franciscan of California, eclogites occur only as tectonic blocks in the blueschist-facies region, but are tentatively included in this subgroup on the assumption that a low-temperature eclogite-facies mass exists below the surface.

Eclogites of this subgroup, in addition to omphacite and garnet, generally contain quartz, zoisite, epidote, lawsonite (or pseudomorphs after lawsonite), glaucophanic amphibole, barroisite, phengite, paragonite, rutile, sphene and calcite. Not all metabasites are eclogites. Some of them are garnet–glaucophane schists, and others may be garnet–actinolite–glaucophane schists.

Geothermobarometric studies have given *P–T* values in the range of 450–550°C and 10–18 kbar (Fig. 12.4).

Metamorphosed acidic igneous rocks isophysical with such eclogites contain quartz, jadeite, garnet and phengite, possibly accompanied by K-feldspar, glaucophanic amphibole, paragonite and albite (Black et al. 1988). Isophysical metapelitic rocks may contain quartz, garnet, kyanite, phengite, paragonite, chloritoid, glaucophane and omphacite. Garnets in eclogites of subgroup I.A regions are almandine with a relatively small amount of MgO and CaO, as shown in Figure 12.3b.

Subgroup I.B: low- to medium-temperature eclogite regions. In the western Alps, a wide zone of the eclogite facies is accompanied by a genetically related zone of the blueschist facies on its west side (Fig. 12.8). In the eclogite-facies zone, eclogites and other rocks contain glaucophanic amphibole. Eclogites belong to the low-temperature division in parts of the zone, and to the medium-temperature division in other parts. Figure 12.4 shows the *P–T* conditions of eclogite formation in Gran Paradiso in a relatively low-grade part of the eclogite-facies zone and in Dora Maria in the highest grade part of the zone. Pressure of metamorphism was so high in the latter as to form coesite.

Eclogites in the Adula nappe in the central Alps show a gradual southward increase in crystallization temperature and pressure from the low-temperature to the medium-temperature division of the facies (Heinrich 1986).

Garnets in eclogites of subgroup I.B regions show a wide continuous composition field that covers the composition fields of both I.A and II in Figure 12.3b. This corresponds to the fact that, with respect to temperature of formation, the low-grade parts of subgroup I.B regions are similar to subgroup I.A regions, whereas the high-grade parts of subgroup I.B regions are similar to eclogites of group II described below.

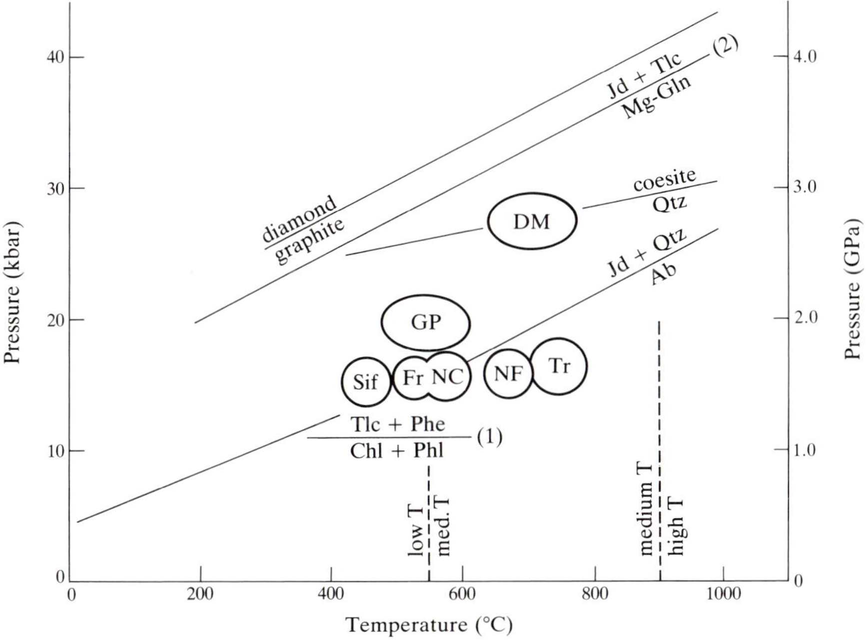

Figure 12.4 *P–T* conditions of formation of some eclogites. Subgroup I.A: Sif: Sifnos (Cyclades); Fr: Franciscan (California); NC: New Caledonia. Subgroup I.B: GP: Gran Paradiso; DM: Dora Maria, both in the western Alps. Group II: NF: Nordfjord; Tr: Tromso, both in Norway. Curve 1: low-pressure limit for the talc + phengite assemblage, that is, the equilibrium curve for reaction (12.11); curve 2: high-pressure limit for Mg-glaucophane, that is, the equilibrium curve for reaction (12.1). Carswell's (1990) low-, medium- and high-temperature divisions of the eclogite facies are shown.

12.4.2.2 Eclogites of group II, which occur in gneiss regions that belong to apparently amphibolite-like or granulite facies. Many eclogites that occur as bands or blocks in gneiss regions that seemingly belong to the amphibolite facies, probably belong to the true eclogite facies, and usually to the medium-temperature division of it. Besides omphacite and relatively MgO-rich garnet, these eclogites usually contain quartz, kyanite, zoisite, paragonite and amphibole (cummingtonite, hornblende, etc.). Examples are eclogites in western Norway, including the classical area of Nordfjord, originally described by Eskola (1921), and in the famous Glenelg area in the Scottish Precambrian.

So-called "external" eclogites in western Norway. Extensive investigations have been made on eclogites in western Norway in the past 15 years (Griffin 1987, Cuthbert & Carswell 1990), giving an entirely new insight into the origin of group II eclogites.

In the Caledonian orogen of western Norway, many masses of eclogites occur in autochthonous basements such as the so-called *Basal Gneiss Complex* (or *Western Gneiss Region*), situated between Trondheim and Bergen as well as in nappes. (For the Trondheim–Bergen region, refer to the map of Figure 2.3.) The eclogites are classified into two distinct groups. One group, called *internal eclogites*, occurs within garnet peridotites, and the other, called *external eclogites*, occurs in other types of rocks. These two appear to differ in age and origin. Only the external eclogites are reviewed below.

External eclogites commonly form pods and lenses ranging in length from decimeters to hundreds of meters, and occasionally larger bodies up to 6 km. Their margins are usually amphibolitized. The surrounding rocks are migmatitic orthogneisses, paragneisses, marble, quartzite, anorthosite, etc. Some external eclogites contain relict minerals that suggest that their protoliths were rocks of shallow or middle crustal origin, such as gabbros, dolerites and amphibolites. Very rarely glaucophane is also found in some eclogites.

The Basal Gneiss Complex was originally formed and metamorphosed in the Proterozoic, and was then deformed and metamorphosed again in the Caledonian orogeny (450–350 Ma), during which the external eclogites formed. It appears that during the Caledonian orogeny, the Baltic Shield collided with the Greenland continent. The continental crust of the Baltic shield was subducted beneath the overriding Greenland plate, and hence underwent eclogite-facies metamorphism (Krogh 1977, Cuthbert et al. 1983).

Measurements by the garnet–clinopyroxene Fe–Mg exchange geothermometer have shown that in the Basal Gneiss Complex temperatures of formation of external eclogites increase gradually over a distance of about 100 km northwestwards to the Atlantic coast from about 600° to about 800 °C. In other words, there exists a regional-scale regular increase of temperature in the direction of presumed continental subduction. This supports the view that eclogite-facies metamorphism occurred on a regional scale in the coherent Basal Gneiss Complex (Griffin 1987).

Pressures of eclogite formation appear to become generally higher with temperature from about 12 kbar to about 20 kbar. Smith (1984) discovered the occurrence of coesite in some eclogites, which appear to have formed at still higher pressures.

At present the rocks surrounding the external eclogites in the Basal Gneiss Complex generally show mineral assemblages in the amphibolite facies. However, there are some rocks having relict minerals and microstructures that suggest the former presence of granulite-facies and eclogite-facies mineral assemblages. It appears that not only the eclogites but also the surrounding rocks were once metamorphosed in the eclogite facies, and that only the surrounding rocks subsequently underwent mineralogical reconstitution in the amphibolite facies.

Granulite-facies eclogites. In many typical granulite-facies gneiss regions through the world, on the other hand, small masses of eclogites exist, which are considered to belong to the granulite and not to the eclogite facies. Examples are eclogites from South Harris in the Outer Hebrides, and eclogitic rocks in charnockite regions in southern India. Such eclogites contain no quartz or kyanite. Eclogites free of quartz and kyanite can form in both granulite and eclogite facies (§12.4.4).

12.4.2.3 Eclogites of group III, which occur as inclusions in kimberlite and basaltic rocks, or as layers in peridotites

Subgroup III.A: this subgroup includes eclogites in kimberlite. The primary accessory minerals of these eclogites include kyanite, corundum, coesite, sanidine, rutile, sphene, diamond, graphite and yagiite. These eclogites probably represent the highest-temperature, highest-pressure part of the eclogite facies. Geothermometers give temperatures in the range of 900–1400°C (Dawson & Carswell 1990). The presence of coesite and diamond indicates a particularly high pressure of formation at depths of 100 km or greater in the mantle (Fig. 12.1). Garnets in them show an unusually wide range of solid solution, continuous from pyrope to grossular (Fig. 12.3b).

These eclogites were originally solid rocks constituting the upper mantle and were brought up to the surface by rising magmas as inclusions. The eclogites may have formed by consolidation of basaltic magmas that had been generated by partial melting of upper mantle materials. Recently, however, many authors suspect that some of them may be fragments of oceanic crust that were subducted into, and metamorphosed in, the mantle (e.g. Schulze & Helmstaedt 1988).

Subgroup III.B: this subgroup includes eclogitic inclusions in basaltic rocks, such as those in Salt Lake Crater in Oahu, Hawaii, and at Delegate in New South Wales, Australia. At least some of these eclogites appear to have crystallized in the granulite facies. These eclogites may have formed in deeper parts of continental crusts.

Subgroup III.C: this subgroup includes eclogite layers in peridotitic rocks. An example is the "internal eclogites" within garnet peridotites in the Basal Gneiss Complex in western Norway. Rocks of this subgroup appear to be diverse in their origin, history and P–T conditions of formation.

12.4.3 Characteristic minerals of eclogites

Clinopyroxene. When analyses of clinopyroxenes are calculated in terms of the molecular percentages of end-members aegirine, jadeite, Ca-tschermak's component ($CaAl_2SiO_6$), hedenbergite and diopside in this order, clinopyroxenes in eclogites of the eclogite facies are mainly composed of diopside, hedenbergite and jadeite components. In the presence of albite and quartz, the maximum jadeite content tends to increase with prevailing pressure.

Clinopyroxenes from eclogites of the eclogite facies differ in composition from those from granulite-facies rocks (including granulite-facies eclogites). The molecular ratio of jadeite/Ca-tschermak's component is greater than 0.5 in the former, and smaller than 0.5 in the latter (e.g. White 1964).

Omphacite, which is an intermediate member between (diopside + hedenbergite) and jadeite, shows cation ordering at low temperatures. Presumably there are two miscibility gaps related to it, as shown in Figure 12.3a.

Garnet. Garnets in the eclogites of kimberlite pipes show a virtually complete series of solid solution between pyrope and Ca-garnet (grossular–andradite) with less than 50% almandine (e.g. Schulze & Helmstaedt 1988). On the other hand, garnets in eclogites of other groups are pyrope–almandine with less than 40% of Ca-garnet content (Fig. 12.3b). The pyrope contents are 70–15% in group II and below 20% in subgroup I.A.

On the basis of Ellis & Green's (1979) experimental work, the Fe/Mg distribution coefficient between garnet and clinopyroxene has been correlated with *P–T* conditions (Newton 1986b, Ghent 1988). The distribution coefficient is sensitive to temperature but not to pressure. Hence, the coefficient has been used as an indicator of temperature in eclogite, granulite and higher amphibolite facies. At lower temperatures, particularly in the blueschist facies, the relatively high Na_2O content and atomic ordering in clinopyroxene may greatly disturb this distribution coefficient. Ellis & Green's calibration was questioned by Pattison & Newton (1989), who gave temperatures 60–150° lower than Ellis & Green for the same value of the distribution coefficient.

Coesite and diamond. The presence of coesite suggests that the prevailing pressure was about 25 kbar or higher (Bohlen & Boettcher 1982), corresponding to depths of about 100 km or greater. The occurrence of diamond indicates still higher pressures (Fig. 12.1).

Chopin (1984) discovered the occurrence of coesite in a quartzite layer that contains pyrope, kyanite, phengite, jadeite, talc and rutile from the Dora Maria massif in the western Alps (also see Kienast et al. 1991). The coesite occurs as inclusions in pyrope. Chopin considers that the quartz of the quartzite was once transformed to coesite by unusually high pressure above about 25 kbar at 700–800°C that operated temporarily in the area, but was later changed back to quartz where it was not protected by enclosing pyrope. The rocks show the jadeite + talc assemblage instead of glaucophane (Fig. 12.1) and the jadeite + kyanite assemblage instead of paragonite (Fig. 12.5b). These assemblages are also stable at about 25 kbar or higher pressures, and belong to the eclogite facies (Schreyer et al. 1987).

Coesite was found also from eclogite (group II) in western Norway (Smith 1984) and China.

The occurrence of coesite in eclogite xenoliths in a kimberlite pipe was first discovered from the Roberts Victor Mine, South Africa (Smyth & Hatton 1977). Subsequently, the mineral was found in eclogite xenoliths from other kimberlites in South Africa, Siberia, and Colorado and Wyoming, USA (e.g. Schulze & Helmstaedt 1988).

12.4.4 Phase relations between the eclogite and granulite facies

12.4.4.1 Anhydrous experiments on basic rocks of the granulite and eclogite facies. Typical granulite- and eclogite-facies metabasites are virtually anhydrous. Increasing pressure causes changes in equilibrium mineral assemblages from the granulite to the eclogite facies essentially by solid–solid reactions, as discussed in §11.6.5.3 (Fig. 11.8).

The *P–T* curves indicating the appearance of garnet and the disappearance of plagioclase in metabasites of quartz-tholeiitic composition have been transferred from Figure 11.8 to Figure 12.5a (lines B and "Eclog", respectively). The *P–T* field above line "Eclog" represents the eclogite facies. The positions of these curves shift with the composition of eclogite, but are independent of the chemical potential of H_2O and the presence or absence of an aqueous fluid. With decrease in chemical potential of H_2O, the stable part of each curve extends to lower temperature, because the stability fields of hydrous mineral assemblages shrink.

Eclogite is the stable anhydrous metabasite in the range of pressure above line "Eclog" and up to about 100 kbar. Irifune et al. (1986) have shown that in rocks of basic compositions at pressures above 100 kbar, pyroxene decreases and garnet increases with increasing pressure, and that pyroxene-free garnetite (with or without stishovite) forms at about 150 kbar. This marks the high-pressure limit of the eclogite facies.

12.4.4.2 Differences in paragenetic relations between the granulite and eclogite facies. The formation of eclogite takes place at pressures a few kilobar lower in alkali-basaltic composition than in quartz-tholeiitic composition (§11.6.5.3). In an analogous way, it is empirically known that quartz-free eclogite occurs in some typical granulite facies areas (e.g. O'Hara 1961). So, we accept Green & Ringwood's (1967a) definition that the eclogite facies represents a *P–T* field above the line of disappearance of plagioclase for quartz-tholeiitic composition or quartz-bearing eclogite. In this case, quartz-free eclogite is not diagnostic of the eclogite facies, but may form in both the high-pressure part of the granulite facies and the eclogite facies. Green & Ringwood (1967a) proposed that the use of the rock name **eclogite** be confined to the eclogites of the thus-defined eclogite facies. If this nomenclature is accepted, the so-called eclogites of the granulite facies may well be called **garnet pyroxenites**.

The schematic *ACF* diagrams of Figure 12.6 indicate possible paragenetic relations in the eclogite facies under anhydrous conditions at two different pressures (i.e. below and above the high-pressure limit of anorthite). In Figure

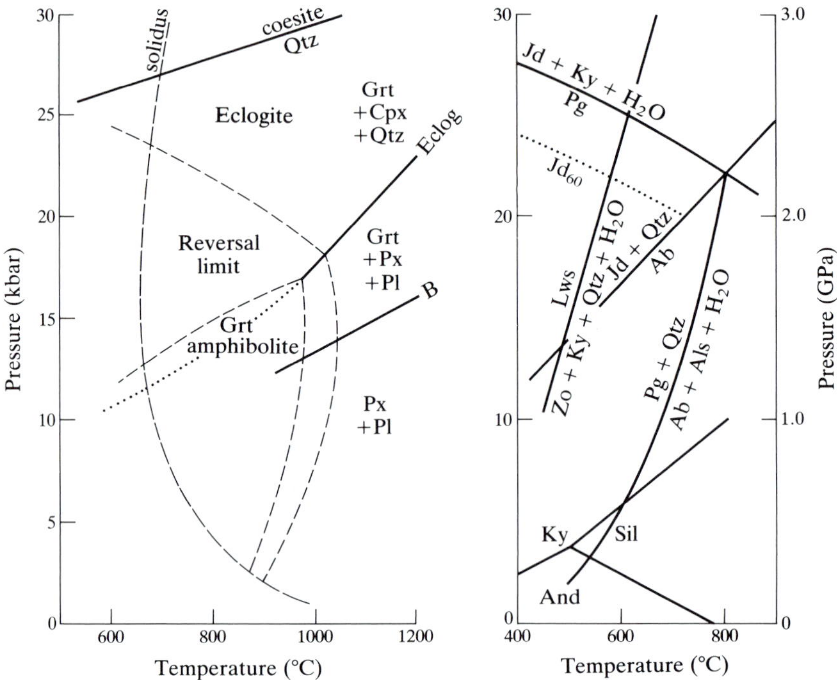

Figure 12.5 Phase relations of metabasites and some minerals relevant to the eclogite facies. (a) Mineralogical changes in rocks with quartz-tholeiitic compositions. Solid lines "B" and "Eclog" indicate the same mineral changes by solid–solid reactions as in Figure 11.8 (Green & Ringwood 1967a, Ringwood 1975). A possible extension of line "Eclog" is shown by a dotted line. Dashed lines indicate changes involving H_2O in the presence of an aqueous fluid (Essene et al. 1970).

(b) Decomposition of paragonite and lawsonite in the presence of an aqueous fluid. The following reactions are shown: paragonite = jadeite + kyanite + H_2O (Holland 1979a); paragonite + quartz = albite + Al_2SiO_5 + H_2O (Chatterjee 1972, Holland 1979a); jadeite + quartz = albite (Holland 1980); 4 lawsonite = 2 zoisite + kyanite + quartz + 7 H_2O (Newton & Kennedy 1963, Chatterjee et al. 1984). The dotted line marked as Jd_{60} shows equilibria for the reaction: paragonite = jadeite component + kyanite + H_2O for omphacite containing 60% jadeite.

12.6a, anorthite by itself is stable, but does not occur in rocks of ordinary basic compositions, whereas in Figure 12.6b, anorthite is not stable. Instead of eclogite in general, quartz-bearing eclogite and kyanite-bearing eclogite are diagnostic of the above-defined eclogite facies as distinguished from the granulite facies, if appropriate limits are fixed for the FeO/(FeO + MgO) and Na_2O/CaO ratios of the rocks. In the following two equations, the left-hand sides represent granulite-facies mineral assemblages, and the right-hand sides eclogite-facies ones.

$$\underset{\text{enstatite}}{4\,MgSiO_3} + \underset{\text{anorthite}}{CaAl_2Si_2O_8} = \underset{\text{pyrope}}{Mg_3Al_2Si_3O_{12}} + \underset{\text{diopside}}{CaMgSi_2O_6} + \underset{\text{quartz}}{SiO_2} \quad (12.7)$$

$$\underset{\text{enstatite}}{MgSiO_3} + \underset{\text{anorthite}}{CaAl_2Si_2O_8} = \underset{\text{diopside}}{CaMgSi_2O_6} + \underset{\text{kyanite}}{Al_2SiO_5} \quad (12.8)$$

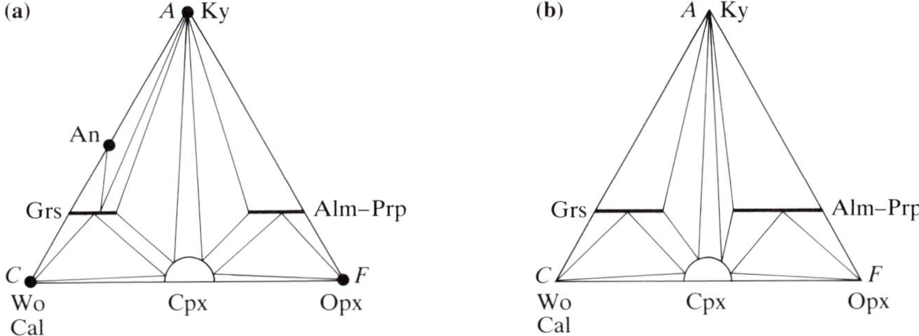

Figure 12.6 Two schematic *ACF* diagrams for the eclogite facies under anhydrous conditions. (a) For pressures lower than the breakdown pressure of anorthite. Modified from Yoder & Tilley (1962). (b) For pressures higher than the breakdown pressure of anorthite. Here, omphacite is shown as diopsidic clinopyroxene.

12.4.5 Progressive metamorphism from the blueschist to the eclogite facies in northern New Caledonia

A spectacular progressive sequence from the blueschist to the eclogite facies has been discovered in two regions. One is in northern New Caledonia, and the other is in the western Alps. The case in northern New Caledonia is reviewed below, whereas that in the Western Alps is discussed at the end of this chapter (§12.4.8), because studies of metapelites have played an essential rôle in the latter.

In northern New Caledonia an apparently coherent progressive metamorphic region shows a continuous sequence from the blueschist to the eclogite facies as shown in Figure 8.8 (Brothers & Yokoyama 1982, Yokoyama et al. 1986,

Black 1977, Black et al. 1988). Isograds generally trend in a northwest direction. The exposed metamorphic region is about 25 km wide as measured in a direction perpendicular to the isograds. The grade of metamorphism increases towards the northeast, that is, towards the Pacific Ocean. The original sediments were mainly pelitic and cherty with subordinate amounts of basaltic and calcareous rocks. Their ages of deposition range from Cretaceous to Eocene. The age of metamorphism is Oligocene.

Except for the incompletely crystallized lowest-grade part, glaucophanic amphibole occurs in metapelites and metabasites in all grades.

Lawsonite zone. In this zone, pelitic rocks are mainly composed of quartz, albite, phengite, lawsonite, chlorite and glaucophane with some paragonite, sphene and graphite, whereas metabasites are mainly composed of quartz, albite, phengite, actinolite, glaucophane and crossite, chlorite, lawsonite and pumpellyite sometimes with sphene and stilpnomelane.

Epidote zone. In this zone, lawsonite and pumpellyite disappear, whereas epidote and almandine appear in both metapelites and metabasites. In metabasites, omphacite begins to occur.

Omphacite zone. Metapelites and metabasites in this zone show similar mineral assemblages, both containing omphacite, glaucophane, hornblende (barroisite), epidote, almandine, phengite, paragonite, chlorite, albite and quartz with sphene and/or rutile. Phengite, albite and quartz are more common in metapelites, whereas zoisite, epidote and barroisite are more common in metabasites. Graphite is confined to metapelites. Some or all of the albite in the higher epidote and omphacite zones may be of retrograde origin. The mineral assemblages in the omphacite zone may be regarded as belonging to the eclogite facies.

Virtually pure jadeite in association with quartz was found in some metamorphosed acidic rocks in the epidote zone.

The P–T conditions of this metamorphism has been considered to be close to those of the equilibrium curve for the reaction: jadeite + quartz = albite (Brothers & Yokoyama 1982).

Part of the eclogite-facies P–T conditions appears to lie on the high-temperature side of the epidote-amphibolite facies (Fig. 8.3). Takasu (1984) reported an observation that suggests the formation of eclogite from epidote amphibolite by an increase of temperature from the Sebadani area in the Sanbagawa metamorphic region (high-pressure transitional type) in central Shikoku, western Japan.

12.4.6 Phase relations between the eclogite and blueschist facies

12.4.6.1 Experimentally investigated relations of the blueschist and amphibolite facies to the eclogite facies in the presence of an aqueous fluid. Under the pressures of the eclogite facies (> 10 kbar), the equilibrium curves of many dehydration reactions show a negative slope in the presence of an aqueous fluid (Fig. 3.3). Thus, not only increasing temperature but also increasing pressure tends to cause dehydration reactions even in the presence of an aqueous fluid. The change of hydrous mineral assemblages characteristic of the blueschist and amphibolite facies to less hydrous or anhydrous mineral assemblages of the eclogite facies may be caused or promoted by increasing pressure in the presence of an aqueous fluid (Fry & Fyfe 1969, Newton 1986b). This relation is illustrated in Figure 12.5.

The dashed lines in Figure 12.5a show that in rocks of quartz-tholeiitic composition in the presence of an aqueous fluid, the *P–T* field of formation of eclogites lies on the high-pressure side of that of the formation of amphibolites. An analogous relation holds for rocks of alkali-olivine-basaltic composition as well.

Figure 12.5b shows the equilibrium curves for the decomposition of lawsonite and paragonite in the presence of an aqueous fluid at high pressures. These two minerals are widespread in blueschist-facies metabasites but absent in typical eclogites. Decomposition of paragonite by itself in response to increasing pressure produces jadeite + kyanite + H_2O. However, omphacite usually forms in metabasites in place of jadeite. The equilibrium curve for the formation of omphacite + kyanite + H_2O from paragonite lies at lower pressures compared with the corresponding curve for jadeite +kyanite + H_2O. In Figure 12.5b, the dotted line indicates such an equilibrium curve for omphacite containing 60% jadeite component.

However, the disappearance with increasing metamorphic grade of lawsonite, paragonite and other hydrous minerals commonly occurs not by decomposition reactions of these minerals by themselves, but by reactions with associated minerals at much lower temperatures and pressures.

12.4.6.2 Effects of the chemical potential of H_2O. Field work on some eclogites associated with glaucophane schists has provided evidence suggesting the existence of an aqueous fluid during metamorphism (e.g. Holland 1979b, Heinrich 1986, Pognante 1991). However, it is also possible that some eclogites formed under conditions of lower chemical potential of H_2O. The lower chemical potential of H_2O should cause the shrinkage of the stability fields of hydrous minerals and consequently the enlargement of the stability field of eclogites to lower temperatures and pressures (Fry & Fyfe 1969). Under these conditions, the extension of line "Eclog" to lower temperatures in Figure 12.5a represents the low-pressure limit of the eclogite facies at lower temperatures. The slope of the curve "Eclog" would become smaller in the extended part (Ringwood 1975), as shown by a dotted line in the diagram.

12.4.7 Metapelites of the eclogite facies

12.4.7.1 Significance of the assemblage talc + phengite. In the study of metapelites of the eclogite facies, the most remarkable breakthrough occurred with Schreyer's discovery of the high-pressure stability of the assemblage talc + phengite, followed by Chopin's discovery of the widespread occurrence of this assemblage in the western Alps. Phengite is a muscovite solid solution containing relatively large amounts of Fe + Mg and Si through Tschermak substitution AlAl → (Fe,Mg)Si (Fig. 5.13).

Abraham & Schreyer (1976) pointed out the possibility that the assemblage talc + phengite may be stable only at high pressures. The validity of this idea was experimentally confirmed by Schreyer & Baller (1977). Through an elaborate experimental study, Massonne and Schreyer (1989) clearly depicted the P–T fields where this and related assemblages are stable. The talc + phengite assemblage is stable at pressures above about 11 kbar in the presence of an aqueous fluid (Fig. 12.7a).

Chopin (1981) discovered the widespread occurrence of the assemblage talc + phengite in metapelites from many places in and around the Gran Paradiso massif in the western Alps (Fig. 12.8). This assemblage forms in metapelites in place of the assemblage biotite + chlorite that is widespread in medium-P/T metamorphic regions. The crucial factor leading to the formation of the assemblage talc + phengite is the advancing shrinkage with increasing pressure of the bulk-rock composition field where chlorite is stable. In this respect, the formation of such talc may be visualized by simplified reactions such as:

$$\text{chlorite} + \text{quartz} = \text{talc} + \text{garnet} + H_2O \qquad (12.9)$$

$$\text{chlorite} + \text{quartz} = \text{talc} + \text{chloritoid} + H_2O \qquad (12.10)$$

Goffé & Chopin (1986) found that the assemblage talc + phengite occurs in a zonal area including the Gran Paradiso massif, which represents a certain range of metamorphic grade in the progressive-like high-P/T metamorphic sequence of the western Alps. In the same zonal area, eclogites also occur (Fig. 12.8). This zone may be assigned to the low-temperature part of the eclogite facies.

12.4.7.2 Experimentally determined phase relations. Figure 12.7a shows the experimentally determined P–T stability field of the assemblage talc + phengite in the presence of an aqueous fluid in the Al_2O_3–MgO–SiO_2–K_2O–H_2O system (Massonne & Schreyer 1989).

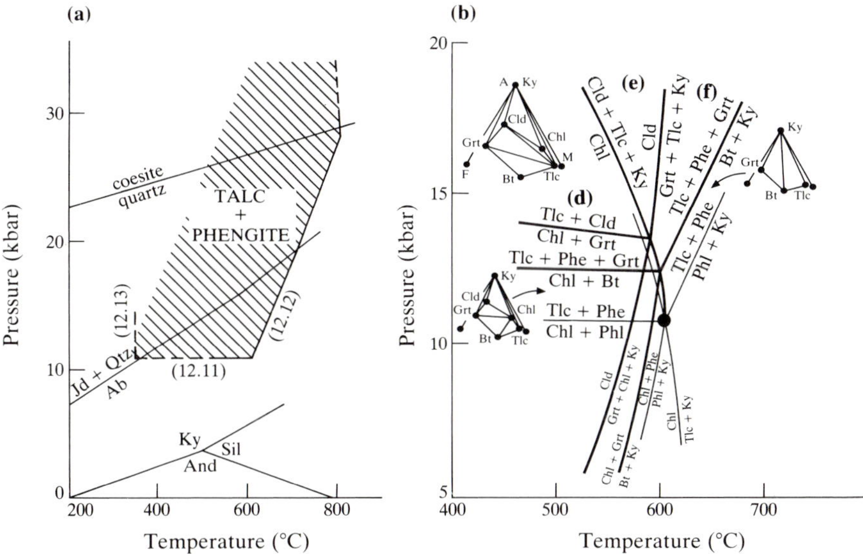

Figure 12.7 (a) Stability field of the talc + phengite assemblage in the presence of quartz and an aqueous fluid. The numbers of the reaction equations defining the stability limits are shown. (b) Thick lines indicate estimated phase relations in the six-component system Al_2O_3–FeO–MgO–SiO_2–K_2O–H_2O in the presence of quartz, phengite and an aqueous fluid (solid lines). In the *AFM* diagrams, the relative positions of Fe–Mg solid-solution minerals are shown schematically by dots. Thin lines indicate univariant curves in the five-component system Al_2O_3–MgO–SiO_2–K_2O–H_2O in the presence of quartz, phengite and an aqueous fluid. Staurolite is ignored. After Massonne & Schreyer (1989).

The low-pressure limit and the high-temperature limit of the stability field are represented by the following reactions (with coefficients ignored), respectively:

$$\text{chlorite} + \text{phlogopite} + \text{quartz} = \text{talc} + \text{phengite} \pm H_2O \qquad (12.11)$$

$$\text{talc} + \text{phengite} = \text{phlogopite} + \text{kyanite} + \text{quartz} + H_2O \qquad (12.12)$$

The junction of the equilibrium curves for reactions (12.11) and (12.12) makes an invariant point at about 10.8 kbar and 610 °C. The low-temperature limit of the stability field is inferred to be represented by:

$$\text{chlorite} + \text{K-feldspar} + \text{quartz} = \text{talc} + \text{phengite} + H_2O \qquad (12.13)$$

The thin lines in Figure 12.7b show some univariant curves in the five-component system Al_2O_3–MgO–SiO_2–K_2O–H_2O in the presence of quartz, phengite and an aqueous fluid. In actual metapelites, rocks contain FeO in addition and so must be treated as belonging to the six-component system Al_2O_3–FeO–MgO–SiO_2–K_2O–H_2O, and moreover talc + phengite (muscovite) is commonly accompanied by other minerals of the same six-component system, such as chlorite, garnet, chloritoid and kyanite. Because of the increase of the variance by addition of FeO, the mineral assemblage that is stable at the invariant point of the five-component system becomes stable along a line starting from the invariant point in the six-component system, and the mineral assemblage that is stable on a univariant curve of the five-component system becomes stable in an area in the six-component system (§10.4). The assemblage talc + phengite + another mineral of the six-component system is stable in a part of the *P–T* stability field of the talc + phengite assemblage by itself. Thick lines in Figure 12.7b show estimated stability fields of such assemblages in the six-component system Al_2O_3–FeO–MgO–SiO_2–K_2O–H_2O. On the univariant curves that separate stability fields, a discontinuous reaction (tieline-switching or terminal) occurs so that the topology of the *AFM* diagram changes.

Experimentally determined stability fields of some other relevant high-pressure minerals are summarized by Schreyer (1988). The stability of talc in metapelites at very high pressures and low to moderate temperatures is well shown by the petrogenetic grids by Koons & Thompson (1985), Massonne & Schreyer (1989), Spear & Cheney (1989) and Guiraud et al. (1990).

12.4.8 Progressive high-P/T metamorphism in the western Alps

12.4.8.1 Zonal mapping. The study of early Alpine high-*P/T* metamorphism in the western Alps along the French/Italian border has proved very difficult due to the occurrence of pre-Alpine metamorphic rocks, tectonic displacements caused by post-metamorphic nappe formation, and the overprinting of late Alpine metamorphism. Nevertheless, Bearth (1962) showed the existence of a zone of lawsonite and glaucophane, and on the eastern side of it another zone characterized by the disappearance of lawsonite and the appearance of chloritoid and garnet in addition to glaucophane.

Goffé & Chopin (1986) published a metamorphic map shown in Figure 12.8, which indicates that lawsonite and Mg–Fe-carpholite occur in the western, low-temperature part of the metamorphic region, whereas the talc + phengite assemblage together with eclogite occurs in a zone that lies to the east of the above-mentioned zone, and includes the Gran Paradiso and Monte Rosa massifs. Coesite-bearing rocks, representing the highest temperature and pressures, occur in the Dora Maria massif in the easternmost zone (Chopin 1984, Kienast et al. 1991). All these zones belong to the Penninic domain (Fig. 9.9).

The grade of metamorphism changes discontinuously at nappe boundaries. Although grade tends to increase eastwards over the whole high-*P/T* region of the western Alps, the increase is not monotonic (Pognante 1991).

On the east side of these zones lies the Sesia–Lanzo zone, belonging to the Austro-Alpine domain (Fig. 9.9). This zone is metamorphosed partly in the

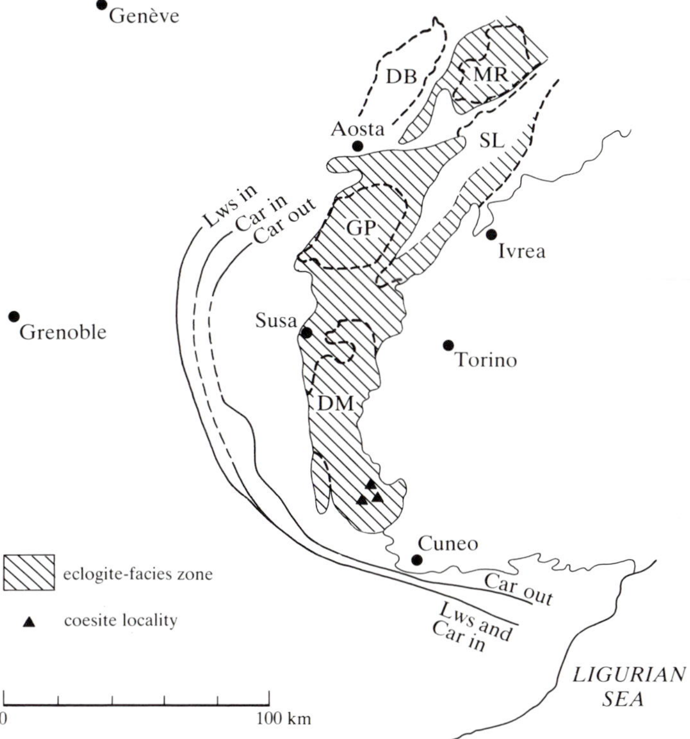

Figure 12.8 High-*P/T* metamorphic zones in the western Alps, mainly based on Goffé & Chopin (1986) and Pognante (1991). The map shows the isograds indicating the entrance of lawsonite and of carpholite as well as that indicating the disappearance of carpholite, all in the blueschist-facies zone. The eclogite-facies zone is shown by a ruled pattern. MR, GP and DM indicate, respectivley, the Monte Rosa, Gran Paradiso and Dora Maria massifs in the Penninic domain. DB and SL indicate, respectively, the Dent Blanche nappe and Sesia–Lanzo zone of the Austro-Alpine domain (Fig. 9.9).

eclogite facies and partly in the blueschist facies, a fact which attracted wide attention among Italian geologists from the middle 1970s (Compagnoni 1977, Compagnoni et al. 1977, Pognante 1991). On the west side of the lawsonite and glaucophane zone, there is a wide zone of the prehnite–pumpellyite and zeolite facies within the Helvetic domain (Fig. 9.9). This zone appears to reach the vicinity of Grenoble (Frey et al. 1974).

12.4.8.2 Progressive changes of paragenetic relations. The inferred changes of paragenetic relations with increasing temperature (and hence with advancing dehydration) are shown in diagrams (a) to (f) in Figure 12.9. Muscovite (phengite) + quartz (or coesite) may accompany any assemblage in the diagrams. (Different diagrams in this figure are connected by reactions that do not involve phengite and so the diagrams may be regarded as belonging to a K_2O-free five-component system.) Diagrams (d), (e) and (f) correspond to fields (d), (e) and (f), respectively in Figure 12.7b.

Some talcs in metapelites contain a considerable amount of FeO with FeO/(FeO + MgO) ratios up to 0.15. Some chloritoids in metapelites are magnesian with MgO/(MgO + FeO) ratios up to 0.74. Neither albite nor jadeite occurs in talc-bearing rocks, although the jadeite + quartz assemblage has been found in associated metagranitoids. Biotite (phlogopite) has been found not in metapelites but in associated dolomitic marble.

12.4.8.3 Geologic relations. The observed ages of metamorphism tend to increase eastwards (Ernst 1975, Goffé & Chopin 1986). In other words, the metamorphism of the whole region is diachronous-progressive. The metamorphism in the lawsonite–carpholite zone occurred at about 40 Ma, whereas that in the high-grade zones appears to be of 90–130 Ma (late and middle Cretaceous).

The metamorphism appears to have been caused by the eastward subduction of the European continental plate beneath another continental mass (the Apulian continent). The subducted plate reached a depth of about 100 km at least, so as to produce coesite. The major parts of the metamorphic region, lying in the Penninic domain, were derived from sediments and underlying continental-type basements in the small ocean between the two continents. The Sesia–Lanzo zone represents the part of the margin of the Apulian continent that was sheared off and subducted together with the tectonically underlying European plate and the small ocean basin (Ernst 1975, 1977, Goffé & Chopin 1986).

For a general review of Alpine metamorphism, refer to Frey et al. (1974). For a review of glaucophane schists in the Alps and other Mediterranean regions, refer to Okay (1989).

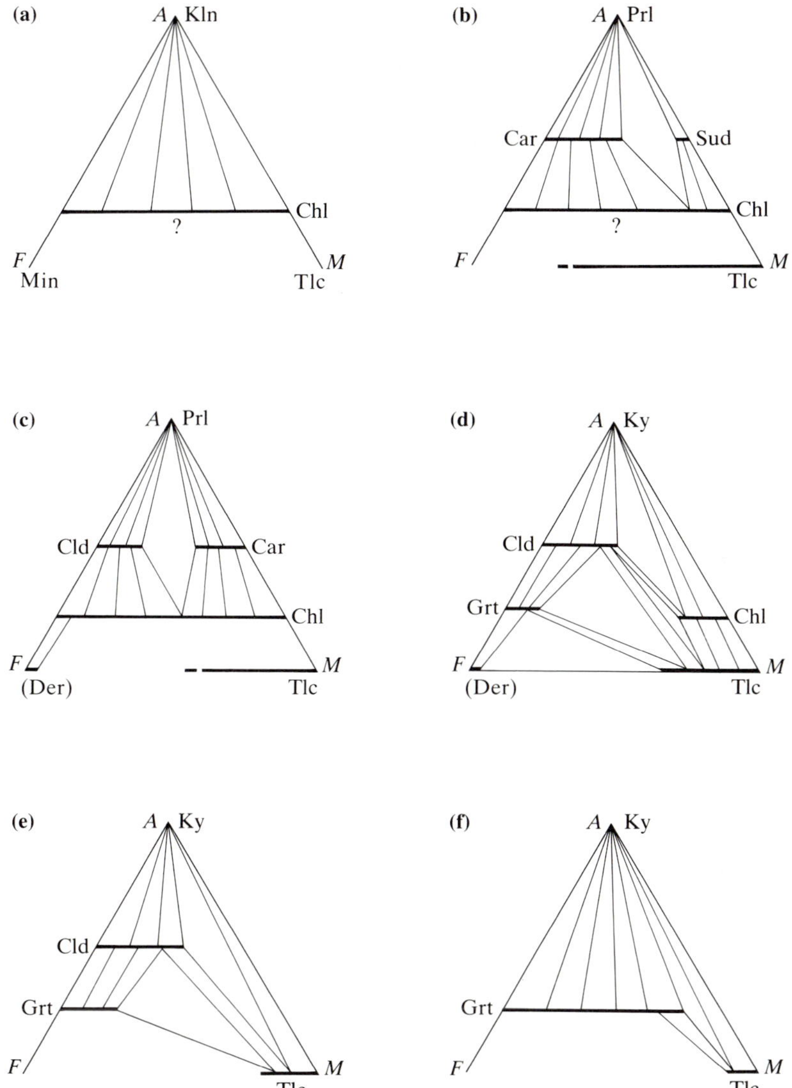

Figure 12.9 *AFM* diagrams showing progressive changes of pelitic rocks (based on Goffée & Chopin 1986). Temperature increases from (a) to (f). (a) Pelitic protoliths; (b) and (c) blueschist-facies metamorphic rocks in the carpholite zone (about 300 °C, 4–8 kbar); (d) and (e) eclogite-facies metapelites in zones with talc and chloritoid (500–600 °C, about 15 kbar); (f) eclogite-facies metamorphic rocks in a zone with coesite after the breakdown of chloritoid (700–800 °C, above 27 kbar). Additional phases are quartz and phengite in (a) to (e), and coesite and phengite in (f).

13 Metamorphic facies characteristic of contact metamorphism

13.1 Diversity of contact metamorphism

Contact metamorphism occurs around igneous intrusions as a consequence of their heating effects. Corresponding to the diversity of igneous intrusions in terms of temperature, depth and tectonic conditions, the resulting contact metamorphism is very diverse with respect to the temperature at which noticeable metamorphic crystallization begins, P–T conditions, metamorphic facies series and the width of the contact aureole. The diversity was described in detail in my former book (Miyashiro 1973: 115–19, 266–92, 302–303), and such detailed descriptions and discussions are not repeated here. Only a short survey is given below for the convenience of the reader.

The metamorphic facies observed in contact aureoles include the greenschist facies, amphibolite facies with andalusite, kyanite and sillimanite, pyroxene-hornfels facies, and granulite facies, among which the amphibolite facies with andalusite is probably most widespread.

13.1.1 Progressive contact metamorphism of metabasites

Metabasites in the outermost zone of a contact aureole commonly show mineral assemblages in the greenschist facies, although schistosity may be lacking. Towards the igneous intrusion, this zone is followed by a zone of a transitional state between the greenschist and amphibolite facies, and then by a zone of the amphibolite facies. In some areas, the transitional state is represented by what is called the **actinolite-calcic plagioclase zone** (§11.4.4). The occurrence of this type of zone is confined to contact-metamorphic aureoles.

Metabasites in the innermost zone, which is in direct contact with the plutonic mass, are commonly amphibolites without garnet, and are virtually the same as amphibolites of the low-P/T regional-metamorphic complexes. The innermost zone relatively rarely reaches still higher temperatures, and shows the breakdown of amphiboles. The resulting mineral assemblages of metabasites are orthopyroxene + clinopyroxene + plagioclase + quartz without garnet. The metamorphic facies characterized by this assemblage is called the **pyroxene-hornfels facies**.

The highest temperature part of contact aureoles is usually in the amphibolite facies around the common granitic masses of orogenic belts, and in the pyroxene-hornfels facies around some dioritic and gabbroic masses, and around alkalic intrusions, e.g. alkali granite and nepheline syenite.

13.1.2 Progressive contact metamorphism of metapelites

In pelitic areas, the outermost zone of a contact aureole is a chlorite zone (without biotite) in relatively rare cases. More commonly a chlorite zone of this kind is lacking and the outermost zone is characterized by the occurrence of biotite together with muscovite and chlorite. In other words, noticeable metamorphic crystallization usually begins at a temperature where chlorite and biotite are both stable. Metamorphic crystallization advances towards the plutonic intrusion in the core of the contact aureole. With some increase of temperature, cordierite usually begins to form. Furthermore andalusite may also appear, if highly aluminous rocks are present.

In metapelites of contact aureoles of the pyroxene-hornfels facies, muscovite is no longer stable, and K-feldspar (orthoclase) instead becomes widespread, while biotite is still present. Garnet is virtually absent.

In many contact aureoles, andalusite is stable up to the highest grade. In other contact aureoles, however, andalusite occurs in lower temperature zones, and sillimanite in higher temperature zones, just as in low-P/T regional metamorphism. This difference could be ascribed to higher pressures in the latter than in the former.

The contact aureoles around some shallow minor intrusions show mineralogical characteristics of still higher temperatures including a large extent of partial fusion in metapelites. These conditions are called **sanidinite facies**.

In aureoles around anorthosite–gabbroic complexes in Precambrian regions of Canada (Berg & Wheeler 1976, Arima & Gower 1991) and Norway (Tobi et al. 1985), contact metamorphism in the granulite facies has been observed, and osumilite has been found to occur in high MgO/FeO metapelites in the highest grade zone.

13.1.3 Facies names for contact metamorphism

Turner (1968, 1981) proposed a series of new names for metamorphic facies of contact metamorphism in spite of their close similarity in composition–paragenesis relations to some metamorphic facies of regional metamorphism. Thus, he named the greenschist facies of contact aureoles the albite–epidote-hornfels facies, and the amphibolite facies of contact aureoles the hornblende-hornfels facies. This nomenclature, however, is a violation of the fundamental principle of metamorphic facies. A metamorphic facies should be defined in terms of composition–paragenesis relations, and not in terms of geologic modes of occurrence. All the facies that essentially show the same composition–paragenesis relations should be called by one and the same facies name (§7.1.2). For this reason, Turner's facies names for contact metamorphism are rejected in this book.

Since the greenschist, amphibolite and granulite facies have already been described in Chapter 11, and the actinolite–calcic plagioclase zone in §11.4.4, only the two higher temperature metamorphic facies characteristic of contact aureoles: the pyroxene-hornfels and the sanidinite facies, are described below.

13.2 Pyroxene-hornfels facies

This facies is characterized by the occurrence of ortho- and clinopyroxene in metabasites in association with plagioclase and quartz (without being accompanied by garnet). The P–T conditions of this facies are on the high-temperature side of the amphibolite facies. In the transitional zone from the amphibolite facies to this facies, hornblende accompanies two pyroxenes in metabasites, although the degree of metamorphic crystallization is often very poor.

In metapelites, muscovite has already decomposed, but biotite is still stable and usually common. The decomposition of muscovite results in the common occurrence of orthoclase. The stable Al_2SiO_5 mineral is usually andalusite, but sillimanite may occur in some cases (Fig. 13.1a). Almandine–pyrope does not occur in rocks of ordinary compositions. In this respect, the pyroxene-hornfels facies contrasts with the granulite facies, in which almandine–pyrope garnet is very widespread in metapelites as well as in metabasites. It should be noted that pure almandine by itself is stable even at very low pressures (below 1 kbar) if the O_2 fugacity is very low. So almandine may occur in some rocks with unusual chemical compositions, e.g. very high FeO/MgO ratios. Spessartine can easily be synthesized at atmospheric pressure, and so occurs in Mn-rich contact-metamorphic rocks.

(a) Pyroxene-hornfels facies

A And, Sil
An Crd
Bt
C F
Wo Cpx Opx

(b) Sanidinite facies

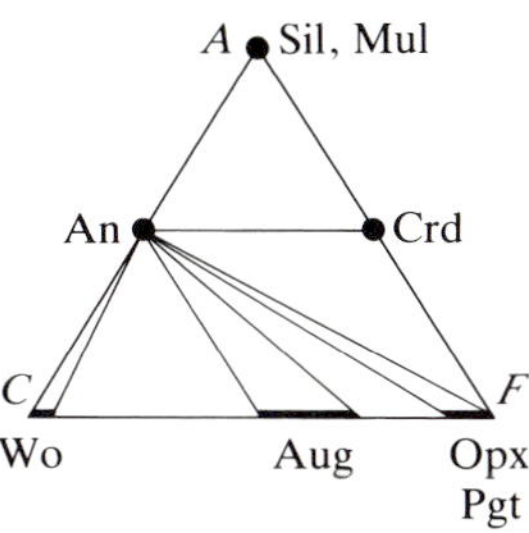

Figure 13.1 *ACF* diagrams for (a) the pyroxene-hornfels and (b) the sanidinite facies. Mul: mullite.

In calcareous rocks, wollastonite, anorthite and diopside are stable. Although traditionally Ca-garnet has been shown in the *ACF* diagram for the pyroxene-hornfels facies, the occurrence of this mineral in quartz-bearing rocks in this facies may be seriously doubted. Ca-garnet by itself is stable to a high temperature at low pressures, corresponding to this facies, e.g. up to 850°C at 1 kbar for grossular. In the presence of quartz, however, its stability is limited to much lower temperatures under a low activity of CO_2, e.g. up to 570°C at 1 kbar for grossular. In place of Ca-garnet + quartz, the wollastonite + anorthite assemblage is probably stable in this facies (Miyashiro 1973: 275, 279). It is likely that confusion of quartz-bearing and quartz-free rocks has occurred in past descriptions of calcareous rocks of this facies.

This facies is developed in the inner zone of some contact aureoles, e.g. the classical Oslo (Kristiania) aureoles around alkalic plutonic intrusions in Norway (Goldschmidt 1911). Experimental data of the phase relations at low pressures are reviewed by Schreyer (1976) and Winkler (1979).

13.3 Sanidinite facies

The sanidinite facies is characterized by the occurrence of minerals indicating very high temperatures at very low pressures, comparable with volcanic conditions, e.g. tridymite, cristobalite, sanidine, anorthoclase, high-temperature plagioclase and glass. However, quartz is commonly the stable silica mineral. Although cordierite (orthorhombic), sillimanite, orthopyroxene and ordinary wollastonite form in many cases, some rocks may contain instead hexagonal cordierite (indialite), mullite, pigeonite and Ca–Fe-wollastonite, respectively. Cordierite may exhibit a considerable K_2O content (Schreyer et al. 1990).

Osumilite may also occur. Because of the high temperatures, a complete series of solid solution forms between K- and Na-feldspars, and partial melting occurs in rocks of pelitic and psammitic compositions. Metamorphism of the sanidinite facies is sometimes called pyrometamorphism.

Since this facies represents a very wide range of temperatures, the paragenetic relations vary considerably with temperature. Figure 13.1b shows only greatly simplified relations. Most rocks of this facies did not reach chemical equilibrium, and this makes determination of equilibrium paragenetic relations very difficult.

The rocks of this facies occur in inclusions within volcanic rocks and some minor intrusions as well as in country rocks adjacent to some mafic minor intrusions. In P–T conditions, this facies is probably on the high-temperature side of the pyroxene-hornfels facies, and represents very low pressures, lower than any regional metamorphism.

A continuous series of solid solution forms between K- and Na-feldspars at temperatures above about 650 °C at 1 kbar. The temperature of the solvus crest increases with pressure at a rate of about 20° $kbar^{-1}$ (Parsons 1981). It may be due to relatively high pressures that a continuous series of alkali feldspar solid solution does not form in regional metamorphism, even in metamorphic facies higher in temperature than about 650 °C, e.g. the granulite facies.

However, some of the pyroxene-hornfels facies zones could be above 650 °C at pressures as low as those of the sanidinite facies. Thus, we arrive at the question as to why a continuous series of alkali feldspar solid solution does not form in the pyroxene-hornfels facies. It is conceivable that in the pyroxene-hornfels facies the rate of cooling is sufficiently slow as to cause retrograde unmixing and other adjustments of alkali feldspars. Any one metamorphic facies should represent a definite range of P, T, and the chemical potential of H_2O, and not of the rate of cooling in the retrograde stage. Because the natural occurrence of continuous Na–K-feldspar solid solution appears to depend seriously on the rate of cooling, this mineral should be excluded from the definition of the sanidinite facies.

APPENDICES

APPENDIX I
Symbols for mineral names used in this book

Ab albite
Act actinolite
Afs alkali feldspar(N)
Alm almandine
Als Al_2SiO_5 mineral(N)
Amph amphibole(N)
An anorthite
And andalusite
Anl analcime
Ann annite
Arg aragonite
Ath anthophyllite
Aug augite
Bt biotite
Cal calcite
Cam Ca-clinoamphibole
Car carpholite(N)
Chl chlorite
Cld chloritoid
Coe coesite(N)
Cpx Ca-clinopyroxene
Crd cordierite
Crn corundum
Cum cummingtonite
Czo clinozoisite
Der deerite(N)
Di diopside
Dol dolomite
En enstatite
Ep epidote
Fa fayalite
Fo forsterite
Fs ferrosilite
Gln glaucophane
Grs grossular
Grt garnet
Hbl hornblende
Hd hedenbergite
Hem hematite
Hul heulandite
Ilm ilmenite
Jd jadeite
Kfs K-feldspar
Kln kaolinite
Ky kyanite
Lct leucite
Lmt laumontite
Lws lawsonite
Mag magnetite
Mc microcline
Min minnesotaite(N)
Ms muscovite
Ne nepheline
Ol olivine
Opx orthopyroxene
Or orthoclase
Pg paragonite
Pgt pigeonite
Phe phengite(N)
Phl phlogopite
Pl plagioclase
Pmp pumpellyite
Prh prehnite
Prl pyrophyllite
Prp pyrope
Px pyroxene(N)
Qtz quartz
Rt rutile
Sil sillimanite
Spl spinel

Spn	sphene
Sps	spessartine
St	staurolite
Stb	stilbite
Stp	stilpnomelane
Sud	sudoite(N)
Tlc	talc
Tr	tremolite
Ts	tschermakite
Wo	wollastonite
Zeo	zeolite(N)
Zo	zoisite

Note: Symbols for mineral names have been taken from Kretz (1983) so far as possible. In the above list the mineral names that are not found in Kretz's table are marked as (N). In a few figures, symbols that are not listed above are used. In such cases the symbols are defined in the pertinent figure captions. Commonly used symbols other than the above are: V, vapor (usually aqueous vapor); M, melt (silicate melt or liquid).

APPENDIX II
Glossary of metamorphic petrogenesis

additive components Conventional components of minerals as expressed by ordinary chemical formulas, such as $FeSiO_3$ in pyroxene and Fe_2SiO_4 in olivine (§3.1.4). Coined by J. B. Thompson (1982a). This name is used in combination with **exchange components**.

andalusite–sillimanite type (or **series**) See ***P/T* ratio types**.

Barrovian zones A sequence of index-mineral zones that are characterized by the successive occurrence of chlorite, biotite, almandine garnet, staurolite, kyanite and sillimanite with increasing temperature in a progressive metapelitic region (§1.6.3 and Figs 1.4 and 1.5). Named in honor of George Barrow (1912), who discovered such a sequence in the Scottish Dalradian. The zones form in medium-*P/T* metamorphic regions where pelitic rocks have bulk-rock compositions plotting below the chlorite–garnet line in the *AFM* diagram (§10.1.1). See **progressive metamorphism** and ***P/T* ratio types**.

buffering See **external buffering** and **internal buffering**. In many cases, the term buffering is used to indicate only internal buffering (Greenwood 1975).

closed system A chemical system that can exchange heat and work but not chemical substances with the surroundings. cf. **open system**.

composition–paragenesis diagram A general name for any diagram that shows the relationships between the bulk-rock chemical composition and mineral assemblages in a group of metamorphic rocks under a definite set of externally controlled conditions, such as *ACF*, *A'KF* and *AFM* diagrams (§5.2).

continuous reaction A reaction that takes place continuously over a range of temperatures under a given pressure in the presence of an intergranular fluid with a constant composition (§2.3.2, §5.6.4). This term is commonly applied to the idealized 6-component metapelitic system: SiO_2–Al_2O_3–FeO–MgO–K_2O–H_2O, in which any mineral assemblage with three *AFM* phases (+ muscovite + quartz) represents a continuous reaction if the relevant ferromagnesian minerals are regarded as binary Fe–Mg solid solutions (§5.4.4). The temperature of an isograd based on a continuous reaction depends on the FeO/MgO ratio of the metapelites exposed in the region even under a given pressure and a given chemical potential of H_2O. The idea (not the name) of continuous reaction came from J. B. Thompson (1957) and J. B. Thompson & Norton (1968). Cf. **discontinuous reaction**.

critical mineral assemblage The mineral assemblage that characterizes a zone of a specific range of metamorphic grade in a progressive metamorphic region. The assemblage is created on the isograd at the low-temperature limit of the zone, and so characterizes the isograd as well (§2.3). cf. **index mineral**, and **metamorphic grade**.

decarbonation reaction A reaction that liberates CO_2. H_2O and other volatiles may or may not participate in it. If they do not, the reaction is called a **simple decarbonation reaction** in this book (§3.2.1).

dehydration reaction A reaction that liberates H_2O (§2.1). CO_2 and other volatiles may or may not participate in it. If they do not, the reaction is called a **simple dehydration reaction** in this book (§3.2.1, §3.5.1).

dehydration-reaction isograd An isograd based on a definite dehydration reaction. In the common pressure range of regional metamorphism and in the presence of a virtually pure aqueous fluid, a dehydration-reaction isograd approximates to a thermal-peak isotherm (§2.2.3).

devolatilization reactions Usually include dehydration and decarbonation reactions, but may also include reactions liberating oxygen (§3.2.1).

diachronous-progressive metamorphism In this book, this term is used for metamorphism that shows not only an apparently progressive increase of thermal-peak temperature in a certain direction, but also an increase of the age of metamorphism in the same direction, as in the Franciscan (§8.2, §9.3.2). It is not progressive metamorphism in the proper sense. Each diachronous-progressive metamorphic region appears to be made up of a number of metamorphic age provinces. cf. **progressive metamorphism**.

discontinuous reaction Reaction that takes place at a specific temperature under a given pressure in the presence of an intergranular fluid with a constant composition (§2.3.2, §5.4.5). This term is commonly applied to the idealized 6-component metapelitic system: SiO_2–Al_2O_3–FeO–MgO–K_2O–H_2O, in which any mineral assemblage with four *AFM* phases (+ muscovite + quartz) represents a discontinuous reaction if the relevant ferromagnesian minerals are regarded as binary Fe–Mg solid solutions. The temperature of an isograd based on a discontinuous reaction is independent of the bulk composition of the metapelites exposed in the region. Such discontinuous reactions may be subdivided into **tieline-switching reactions** and **terminal reactions** (§5.4.5). The idea (not the name) of discontinuous reaction came from J. B. Thompson (1957) and J. B. Thompson & Norton (1968). cf. **continuous reaction**.

exchange components Components of solid-solution minerals that express the extents of isomorphous replacements such as $FeMg_{-1}$ (§3.1.4). Proposed by J. B. Thompson (1982a). Used in combination with the term **additive components**.

exchange reaction A reaction that shows exchange of atoms such as Mg and Fe between coexisting minerals (§3.2, §3.7). Exchange reactions do not change the amounts (numbers of moles) of relevant minerals, and cause little change in the volume of the system and so are insensitive to pressure. cf. **net-transfer reaction**.

external buffering Maintaining the chemical potential of a component in a rock at a constant or nearly constant value through influence from outside the rock. Usually the constancy of the chemical potential is maintained through a metamorphic area (§6.1). cf. **internal buffering**.

field *P–T* curve In this book, this term means such a curve on the *P–T* diagram that shows the course of variation of thermal-peak *P–T* condition along a line perpendicular to thermal-peak isobars (and so usually approximately perpendicular to dehydration-reaction isograds) on the ground surface of a metamorphic region (§2.2.6). An example is curve A′—B′—C′—D′ in Figure 2.6. Commonly (but not always) the curve has a positive slope. A field *P–T* curve is polychronous, and does not generally match any part of the *P–T–t* paths of relevant rocks (Fig. 2.6). See **progressive metamorphism**.

geothermal gradient Rate of increase of temperature with depth (degree km^{-1}), measured vertically downward in the crust (§2.6.2). The gradient varies with depth. For average geothermal gradient, see Figures 2.6 and 8.2. cf. **temperature gradient** and **metamorphic thermal gradient**.

grade of metamorphism See **metamorphic grade**.

index mineral The mineral that is used to characterize a zone of a specific range of metamorphic grade in a progressive metamorphic region (§2.2.2). In some cases use of an index mineral is misleading. Instead, a critical mineral assemblage may well be used (§2.3.1). cf. **critical mineral assemblage** and **metamorphic grade**.

infiltration Pervasive flow of fluid into a metamorphic rock.

instantaneous isobars Isobars (isopleths) showing the distribution of pressure within the crust at a specific time (§2.2.3). cf. **thermal-peak isobars**.

instantaneous isotherms Isotherms (isopleths) showing the distribution of temperature within the crust at a specific time (§2.2.3). cf. **thermal-peak isotherms**.

internal buffering Maintaining the chemical potential of a component in a rock at a constant or nearly constant value through chemical reaction with minerals present in the rock itself (§6.1). Sometimes it is simply called buffering (Greenwood 1975). cf. **external buffering**.

inverted metamorphic zones A higher grade index-mineral zone overlying a lower-grade one in the present-day crust (§2.6.4). Such a structure forms either by an originally inverted distribution (i.e. upward increase) of thermal-peak temperature or by post-metamorphic deformations.

isograd Boundary between two adjacent zones that have different index minerals or critical mineral assemblages. An isograd that has been mapped in routine petrographic work and is characterized by an index mineral may be called a **tentative isograd**. Such an isograd is a true isograd (called a **reaction isograd**), if it has been shown that a specific reaction or a specific set of reactions occurs on the isograd because of a continuous spatial variation in *P–T* condition or in chemical potential of H_2O and/or CO_2, resulting in a change in mineral assemblage. A reaction isograd is polychronous (§2.5). An isograd is the intersection of an isogradic surface with the erosion surface of the Earth. The idea and use of isograd began with Barrow (1893). Although the term isograd was coined by Tilley (1924), his definition was inappropriate (Carmichael 1978: 770). cf. **tentative isograd**, **reaction isograd**, **index mineral**, **critical mineral assemblage** and **metamorphic grade**.

kyanite–sillimanite type (or **series**) See ***P/T* ratio types**.

metamorphic crystallization General term that indicates any processes of mineral change and new mineral formation by any mechanisms during metamorphism. Petrologists have traditionally used the term **recrystallization** in this sense. In recent years, however, the term recrystallization has been used by structural geologists in a much narrower sense. To avoid confusion, the term recrystallization is not used in this book.

metamorphic facies Eskola (1915) defined a metamorphic facies as including all the metamorphic rocks that are supposed to have crystallized under a certain definite range of *P–T* condition. This book defines a metamorphic facies as including all the metamorphic rocks that are supposed to have crystallized in such a definite volume in *P–T*–(chemical potential of H_2O) space that is characterized by the stability of a critical mineral assemblage. In addition it is explicitly stated that the metamorphic rocks that coexist in a sufficiently small metamorphic area and can be regarded as representing equilibria within a small volume in *P–T*–(chemical potential of H_2O space), are treated as belonging to the same metamorphic facies. In this definition, the effects of the chemical potential of CO_2 are not taken into consideration (§7.1.2, §7.4). Table 7.1 gives a system of metamorphic facies names and their critical mineral assemblages as used in this book.

metamorphic facies series A series of metamorphic facies observed in a progressive metamorphic region. See ***P/T* ratio types**.

metamorphic grade Although this term has been widely used in the literature, it has been ill defined and ambiguous. In this book, the term is used as a scale for the advance of mineralogical changes with increasing temperature across a progressive metamorphic region. An isograd (reaction isograd) is a line indicating a constant metamorphic grade within a metamorphic region. So far as such mineralogical changes are mainly controlled by temperature (thermal-peak temperature), the term metamorphic grade approximately indicates temperature, and thus low and high grades mean approximately low and high temperatures of metamorphism. See **progressive metamorphism**, **isograd** and **reaction isograd**.

metamorphic thermal gradient The metamorphic thermal gradient at a point within the crust is the rate of increase of thermal-peak temperature measured in a direction perpendicular to the isogradic surface that is based on a dehydration reaction and passes through the point (De Yoreo et al. 1991). Although such an isogradic surface approximately represents a constant temperature, it is generally polychronous (§2.6.2). cf.

temperature gradient and **geothermal gradient**.

mineralogical phase rules The versions of the Gibbs phase rule that are adapted to certain models of metamorphic rock systems. **Goldschmidt's mineralogical phase rule** applies to a closed system model of metamorphic rocks, whereas **Korzhinskii's mineralogical phase rule** applies to metamorphic rock systems externally buffered with respect to certain components (§5.1).

net-transfer reaction Reaction that causes a change in the amounts (numbers of moles) of minerals in a rock (§3.2). They include all reactions other than exchange reactions. Proposed by J. B. Thompson (1982b). A reaction that causes a change in the amount of minerals and in addition involves exchange of atoms between coexisting minerals, is classed as a net-transfer reaction. cf. **exchange reaction**.

open system A chemical system which can exchange not only heat and work but also chemical substances with the surroundings. External buffering may occur in such a system. cf. **closed system**.

paired metamorphic belts Paired metamorphic belts are composed of a high-*P/T* and a lower *P/T* ratio metamorphic belt, which run parallel with each other generally along a continental margin, with the high-*P/T* belt lying on the ocean side (§9.1). Proposed by Miyashiro (1961). For examples, see Table 9.1 and Figure 9.1.

phase components Chemical components that are used to express the chemical composition of minerals, such as Fe_2SiO_4 and Mg_2SiO_4 in olivine, and SiO_2 in quartz (§3.1.2). cf. **system components**.

plurifacial metamorphic rock A rock that contains groups of metamorphic minerals which formed at two or more different times under different *P–T* conditions. The two or more different times (phases) of metamorphic crystallization belong to one and the same orogeny or thermal event in some cases (e.g. the thermal peak and an immediately following retrograde phase), and to two or more different orogenies in other cases (i.e. polymetamorphic rocks). See §2.1.5. Coined by W. P. de Roever & Nijhuis (1963). cf. **polymetamorphic rock**.

polymetamorphic rock A rock that suffered two or more genetically unrelated metamorphic events (Read 1957: 306). The two or more events may belong to different orogenies, but may well be merely different phases separated by a period or periods of marked cooling in the course of one orogeny. In cases where the existence of such a cooling-down period is not clear, the uncommitted term plurifacial should be used instead (§2.1.5). cf. **plurifacial metamorphic rock**.

prograde metamorphism Metamorphic crystallization that takes place in a rock when temperature is temporally increasing (i.e. increasing with time) prior to the thermal peak (§2.1). An example is crystallization between points B and C in Figure 2.1. In prograde metamorphism, the temporal change of *P–T* condition of each rock is expressed by a *P–T–t* path. cf. **progressive metamorphism** and ***P–T–t* path**.

progressive metamorphism Metamorphism in a region where thermal-peak temperature increases continuously and regularly in a certain direction on the surface (§2.2.1). In other words, the term **progressive** is used in this book to refer to increase of temperature (thermal-peak temperature) **spatially** across a region in contrast to the word **prograde** which refers to temporal increase of temperature (i.e. temperature increase with time) in individual rocks. The term progressive metamorphism became popular through Harker's (1932) book, in which, however, the term was used to include both prograde and progressive metamorphism of this book. In progressive metamorphism, the spatial variation of *P–T* condition at the thermal peak of each rock across the metamorphic region is expressed by field *P–T* curves. cf. **prograde metamorphism**, **metamorphic grade**, **diachronous-progressive metamorphism** and **field *P/T* curve**.

***P/T* ratio types of regional metamorphism (or metamorphic regions)** Miyashiro (1961) proposed a three-fold classification of regional metamorphism, metamorphic regions and metamorphic facies series based on *P/T* ratios. The **low-*P/T* type (andalusite–sillimanite series)** is characterized by the occurrence of andalusite in a lower grade and sillimanite in a higher grade. The **medium-*P/T* type (kyanite–sillimanite series)** is characterized by the occurrence of kyanite in a lower grade and sillimanite in a higher grade. The **high-*P/T* type (glaucophanic metamorphism)** is characterized by the occurrence of glaucophane. The last-named type is further divided into: **high-pressure transitional type** (with glaucophane but without jadeite + quartz), and **jadeite–glaucophane type** (with both glaucophane and jadeite + quartz). See Chapter 8.

***P–T–t* path** The course of temporal change of *P–T* conditions of a rock, or a curve showing the course in a *P–T* diagram (§2.1.4).

reaction isograd A true isograd, on which a specific reaction or a specific set of reactions occurs owing to a spatially continuous change in *P–T* conditions and/or in chemical potential of H_2O and/or CO_2 (§2.2, §2.4), resulting in the creation of a critical mineral assemblage (§2.3). Proposed by Winkler (1979). The term reaction in the above context does not connote a process but refers only to a mass-balance relationship between mineral assemblages on the opposite sides of the isograd (Carmichael 1978: 770). A reaction isograd shows the distribution of mineral assemblages that formed at the thermal peak of each rock in a region. Since the thermal peak is polychronous within a region, a reaction isograd is also generally polychronous. Reactions that occur on isograds are either continuous or discontinuous. cf. **tentative isograd**, **continuous reaction**, **discontinuous reaction** and **critical mineral assemblage**.

recrystallization See **metamorphic crystallization**.

regional metamorphism Metamorphism that produces regional-metamorphic complexes, which are exposed on a regional scale (i.e. in a large region) in regions with a continental-type crust (i.e. continents and island arcs), and which show a temperature distribution unrelated to individual igneous intrusions (§1.3). Regional-metamorphic complexes appear to have formed in a number of different tectonic settings (§1.3.2). Regional metamorphism commonly but not always produces foliated metamorphic rocks.

retrograde metamorphism Metamorphic crystallization that occurs in a rock with decreasing temperature after the thermal peak, that is, for example between points C and D in Figure 2.1 (§2.1). cf. **retrogressive metamorphism**.

retrogressive metamorphism This term has traditionally been used for two different processes: (a) retrograde metamorphism as defined above, and (b) a younger event of metamorphic crystallization that is lower in temperature than the older one(s), where the younger and older events belong to different orogenies (§2.1.5). This term has a practical convenience in descriptive work, because commonly it is not easy to determine which of the above two processes occurred in actual metamorphic rocks.

simple decarbonation reaction A reaction of the type: B = D + CO_2, where B and D denote one or more minerals. See Figure 3.5.

simple dehydration reaction. A reaction of the type: B = D + H_2O, where B and D denote one or more minerals. See Figure 3.5.

solid–solid reaction A reaction that involves solid minerals alone, that is, with no participation of a fluid phase or volatile substances in grain-boundary adsorption films (§3.2, §3.4). The equilibrium curves of most solid–solid reactions have a gentle slope as compared with those of dehydration reactions in the presence of an aqueous fluid at pressures above a few kilobar. cf. **dehydration reaction** and **decarbonation reaction**.

system components Components that appear in the phase rule (§3.1.2). For example, metapelites are commonly simplified as composed of six system components: SiO_2, Al_2O_3, FeO, MgO, K_2O and H_2O. cf. **phase components**.

temperature gradient The temperature gradient at a point within the crust is the rate of increase of temperature measured in a direction perpendicular to the instantaneous isothermal surface that passes through the point at the time under consideration (§2.6.2). cf. **geothermal gradient** and **metamorphic thermal gradient**.

tentative isograd A tentatively drawn isograd that marks the appearance or disappearance of a mineral (called an index mineral) in a progressive metamorphic region. It is either a reaction isograd (i.e. a true isograd), or a false isograd (§2.3.1). See **isograd**, **reaction isograd**, **index mineral**, **critical mineral assemblage**, **continuous reaction**, and **discontinuous reaction**.

terminal reaction A reaction that causes the appearance or disappearance of a mineral on the *AFM* diagram (§5.3.5). See **discontinuous reaction**.

thermal peak The state of highest temperature that a metamorphic rock experiences in the course of metamorphism, such as point C in Figure 2.1. In other words, the point of the highest temperature in the *P–T–t* path of a rock, such as the point marked with an open circle on *P–T–t* paths in Figure 2.2. Metamorphic rocks tend to preserve the mineral assemblages that equilibrated at or near the thermal peak. The thermal peak is not synchronous through a metamorphic region.

thermal-peak isobars Isobars (isopleths) showing the distribution of thermal-peak pressure in a metamorphic region (§2.2.3). A thermal-peak isobar is polychronous. An isograd based on a solid–solid reaction that causes little change in volume approximates to a thermal-peak isobar. cf. **instantaneous isobars**, **reaction isograd** and **solid–solid reaction**.

thermal-peak isotherms Isotherms (isopleths) showing the distribution of thermal-peak temperature in a metamorphic region (§2.2.3). A thermal-peak isotherm is polychronous. In the common pressure range of regional metamorphism and in the presence of a virtually pure aqueous fluid, a dehydration-reaction isograd approximates to a thermal-peak isotherm. cf. **instantaneous isotherms**, **reaction isograd** and **dehydration-reaction isograd**.

thermal-peak pressure The pressure of a rock at the time of its thermal peak. Usually this is the pressure recorded in the mineral assemblage of the rock.

thermal-peak temperature The temperature of a rock at the time of its thermal peak. Usually this is the temperature recorded in the mineral assemblage of the rock.

tieline-switching reaction A reaction that causes a change in a tieline connecting minerals on the *AFM* diagram (§5.4.5). See **discontinuous reaction**.

APPENDIX III
A history of metamorphic petrology

III.1 Descriptive studies and increasingly improved understanding of genetic relations of metamorphic rocks and regions

III.1.1 The advent of microscopic petrography and Grubenmann's classification of crystalline schists

Classification and nomenclature of metamorphic rock types began with the science of geology in the late 18th century. The magnifying lens helped the identification of minerals, leading to the development of a system of classification of rock types based on their mineral compositions in the early 19th century. Chemical analysis of rocks began in the mid-19th century.

The introduction of the polarizing microscope in the 1870s revolutionized the description and classification of igneous and metamorphic rocks. Immediately after its introduction, the new tool was thought to be useful for the observation of textures and fluid inclusions. Shortly, however, its usefulness in the identification of minerals was realized. Thus, detailed classification based on the mineral compositions became possible. In the succeeding 30 years, a new science called **petrography** developed in Germany and France.

Students went to these countries from all over the world to learn petrography. On their return, they described igneous and metamorphic rocks in their own countries. In this way a rough petrographic survey spread across the world. Some ideas, even though very incomplete, on the distribution of rock types on the Earth were formed in the early 20th century. This is seen, for example, in Harker's (1909, fig. 22) map showing the distribution of alkalic and subalkalic igneous rocks in the world, as well as in Daly's (1914) book dealing with the origin of igneous rocks in relation to the Earth's structure.

Since the early petrographers preferred to study igneous rocks rather than metamorphic rocks, systematization of the descriptive petrography of metamorphic rocks was delayed until the publication of *Die kristallinen schiefer* (1904–06) by Ulrich Grubenmann (1850–1924) of the Swiss Federal Institute of Technology in Zurich. He classified crystalline schists (meaning regional-metamorphic rocks) into three imaginary **depth-zones**, called epi-, meso- and kata-zones in order of presumed increasing depth of their origin in the crust. Rocks of each depth-zone were divided into 12 groups according to their chemical compositions. Prior to Grubenmann, the idea of depth-zones itself had been initiated by J. J. Sederholm and by F. Becke.

For a more detailed history of early days of petrography, see Miyashiro (1973: 429–40).

III.1.2 The search for more realistic viewpoints of classification

When it was first proposed, Grubenmann's depth-zone classification of regional-metamorphic rocks represented a marked advance in metamorphic petrography. Unfortunately it continued to dominate most of the European continent for the unduly long time of 40 years, mainly because it was advocated by Paul Niggli, his successor in Zurich (Grubenmann & Niggli 1924), and so became an obstacle to further progress. Harker and Eskola, however, thought it to be too speculative, too rigid and too static. They formulated their own views to some extent by criticizing Grubenmann's.

Harker's and Eskola's views are discussed later in detail in relation to the progress of paradigms about the factors controlling metamorphic minerals and mineral assemblages (Appendix III.2). In the following section their views are dealt with only in relation to the classification of metamorphic rocks.

Alfred Harker (1918, 1932) of the University of Cambridge criticized Grubenmann as being too static, and considered that the mineralogical changes of metamorphic rocks with increasing temperature could be traced by systematic description of progressive metamorphic regions. He himself produced a beautiful systematic description of progressive metamorphism of the Barrovian sequence in a part of the Scottish Highlands (Fig. 1.5).

Pentti Eskola (1920) of the University of Helsinki criticized Grubenmann's depth-zones as being too speculative and too rigid. Instead, he proposed a system of metamorphic facies, each of which represented a specific range of P–T conditions and was capable of being determined empirically from the mineral assemblages of metamorphic rocks (Fig. 7.1). Thus, he classified metamorphic rocks according to the metamorphic facies they belonged to.

III.1.3 The increasingly improved perspective of regional metamorphism

When the first systems of metamorphic petrology were organized as stated above in Britain and Scandinavia before the Second World War, they were naturally mainly based on data obtained in these regions and particularly from relatively civilized parts of them. For this reason they were strongly biased. Since each region on the Earth has its own characteristics, each geologic problem can be studied more easily in some regions than in others. Hence, the progress of not only petrology but also all fields of geology has been connected with the geographical expansion of mankind's activity on the Earth.

In the late 1940s and the 1950s, systematic descriptive study of metamorphic rocks was extended to most regions of the world. This caused great changes not only in our descriptive knowledge of metamorphic rocks but also in our petrogenetic thinking, as will be outlined below.

III.1.3.1 Genetic significance of the granulite facies. Eskola's (1939) proposal of the granulite facies (Fig. 7.1) was based mainly on his knowledge of the vast granulite-facies region of Lappland in the remote northern part of Scandinavia. He did not know of any granulite facies areas that showed a progressive metamorphic relation to other facies. Granulite-facies regions showing a progressive metamorphic relation with the amphibolite facies in western Norway and other places in Europe were not known at that time. So Eskola had no clear idea about the origin of this facies and its relation to other facies.

The major granulite-facies regions of the world are in Archean regions far from the civilization centers of the early part of the 20th century. During and immediately after the Second World War, progress of geologic work in India, Africa and Greenland clarified the existence of regional-metamorphic regions showing a progressive transition from the amphibolite to the granulite facies. Thus, it came to be accepted that the granulite facies represents the highest- temperature part of ordinary regional metamorphism, and lies on the immediate high-temperature side of the amphibolite facies.

Petrographic studies of granulite-facies areas such as the charnockite regions of India and other places became a fashionable subject in the 1950s.

III.1.3.2 Discovery of the zeolite and prehnite–pumpellyite facies. Large regions of low-temperature metamorphism are exposed in the South Island of New Zealand. Douglas S. Coombs of the University of Otago investigated these regions, leading to the establishment of the **zeolite, prehnite–pumpellyite** and **lawsonite–albite–chlorite facies**, and thereby contributed to the extension of our understanding of mineral formation towards lower temperatures (Coombs 1954, 1960, 1961). Thereafter, metamorphic rocks of these three facies were found to occur in many places all over the world.

III.1.3.3 Andalusite-producing and glaucophane-producing regional metamorphism. Before 1950, the Barrovian sequence (Fig. 1.5) and similar types of metamorphism were virtually the only well documented progressive regional metamorphism in the world. So Harker (1932) gave this type of metamorphism the name of "normal regional metamorphism". At that time, there were fragmental petrographic data suggesting the possible existence of regional metamorphism that was clearly different from the Barrovian type. Andalusite-bearing schists and glaucophane schists in particular, which do not appear in the Barrovian sequence, were known to occur in central Europe and the Alps respectively. Andalusite-bearing schists were known to occur even in part of the Scottish Highlands. Harker explained away such rocks by *ad hoc* assumptions (Appendix III.2).

In the 1950s, systematic descriptive studies of progressive metamorphism were extended to Japan, where a number of regional-metamorphic belts different from the Barrovian type occur. The Ryoke and other metamorphic regions of regional scale have andalusite in place of kyanite, whereas the Sanbagawa and other belts have glaucophane in some rocks. Miyashiro (1961) regarded all the diversity of such metamorphic regions as representing different types of regional metamorphism that took place under different *P–T* conditions. This led to his proposal of a three-fold classification of progressive regional metamorphism: andalusite–sillimanite, kyanite–sillimanite and jadeite–glaucophane types. The metamorphism of the Barrovian sequence was thought to represent a special case of the kyanite–sillimanite type. This expanded the concept of regional metamorphism.

It should be noted that in Japan, and probably in other countries also, recognizing the existence of the types of regional metamorphism that produce andalusite or glaucophane was not a result of a mere increase of petrologic data but was directly connected with a change of thinking in metamorphic petrology, a change that is described as the change from the first to the second paradigm in Appendix III.2.

III.1.3.4 Significance of the eclogite facies. Eskola (1920) paid special attention to eclogites because of their peculiar mineralogical compositions, leading to the proposal of the eclogite facies (Fig. 7.1). However, the genetic and tectonic significance of this facies was not clear. It was not even known whether this facies was stable in the Earth's crust.

Abrupt and remarkable progress occurred in the 1980s. Progressive metamorphic regions showing a transition from the glaucophane-schist facies to the eclogite facies were found in New Caledonia and the Alps. Brothers and his co-workers in particular discovered a beautiful case of such progressive metamorphism in northern New Caledonia (e.g. Brothers & Yokoyama 1982). Chopin and other people discovered such a progressive sequence with eclogite-facies metapelites showing extremely high-pressure mineral assemblages in the western Alps (e.g. Goffé & Chopin 1986). These findings settled the position of the eclogite facies in crustal metamorphism (e.g. Carswell 1990).

III.1.4 Studies of rock-forming minerals

III.1.4.1 Optical mineralogy Physicists began crystal optics in the early 19th century. It was adapted to and developed for mineral identification in petrographic–microscopic work. These studies led to a better understanding of the relationships among rock-forming minerals.

Prior to the introduction of the microscope, for example, it had been believed that feldspars were a family composed of a large number of discrete minerals, called albite, oligoclase, andesine and so on, which have similar but slightly different fixed chemical compositions. Combined microscopic, optical and chemical investigations in the late 19th century showed this idea to be false. It was found that actually feldspars were continuous series of solid solution, commonly showing zonal growth of different members.

In the late 19th and early 20th century, petrologists were enthusiastic about optical studies, including determination of optical constants of rock-forming minerals as well as the development of optical-microscopic techniques. The early editions (1927, 1933) of A. N. Winchell's *Elements of optical mineralogy*, Part II, were most important handbooks for petrologists to have at their elbow.

The optical constants of all important rock-forming minerals had been measured and correlated with their chemical compositions by about 1930. The enthusiasm of petrologists for optical investigations at that time was based on their expectation that the determination of optical constants would serve as a substitute for chemical analysis which was more difficult and troublesome. This expectation was fulfilled in binary solid solutions such as plagioclase, but not in complicated solid solutions such as amphiboles and biotites. The usefulness of optical constants was very limited. The heyday of optical mineralogy ended in the 1930s.

III.1.4.2 Early chemical and X-ray studies. A great number of chemical analyses of minerals were published in the late 19th and the early 20th century, and were compiled in Doelter & Leitmeier's (1912–31) huge series of books, *Handbuch der Mineralchemie*. However, these analyses were made mostly on easily obtainable, large specimens, and so were of very limited use in metamorphic studies.

In Doelter's time, the chemical formulas of minerals were calculated from chemical analyses on the wrong principle of grouping elements according to their valencies. In many silicates, it was found that the atomic ratios between such valency groups of elements could not be represented by small integers. This led to a difficulty in the derivation of chemical formulas.

On the other hand, the crystal chemistry formulated on the basis of X-ray crystallographic studies by Bragg and Goldschmidt in the 1920s provided the new principle that ionic substitutions occurred between ions with similar sizes irrespective of their valencies. Thus, it was realized that the formulas of silicate minerals should be calculated by grouping elements according to their ionic radii and not to their valencies. A sweeping revision of the chemical formulas of silicates occurred in the 1930s, as summarized by Berman (1937) and Strunz (1941).

X-ray crystallography gave a further new insight into the nature of some complex minerals. For example, most authors had regarded amphiboles as anhydrous minerals like pyroxenes or even polymorphous with pyroxenes, until Warren (1930a) showed that hydroxyl ions must be essential in the tremolite structure, and that in the structure there existed a large vacant site which could accommodate an Na ion in some cases. On the basis of these findings, Warren (1930b) showed that all monoclinic amphiboles had essentially the same crystal structure as tremolite, and so their chemical formulas should be derived from that of tremolite by isomorphous substitutions between cations with similar radii. This led to a drastic change, not only in the formulas of amphibole members, but in the method of compilation of chemical and optical data for the amphibole group (e.g. Hallimond 1943).

III.1.4.3 Structural transformations. Because a large number of detailed optical–chemical studies were made on feldspars from the late 19th century, in the 1930s mineralogists came to feel that almost all the important problems of the feldspars had already been solved. This confidence was shattered by Koehler's (1941) discovery that the optic properties of plagioclase varied greatly with the temperature of its formation, and that plagioclases showing high-temperature optics occurred in volcanic rocks, whereas plagioclases showing low-temperature optics occurred in plutonic and metamorphic rocks. Koehler's work was performed in Vienna during the Second World War. Because of the abnormal wartime and post-wartime conditions, the Austrian journal that published this discovery was not well circulated. Rumors of his discovery and then his papers spread slowly over the years (Koehler 1949). His discovery evidently showed the existence of unknown kinds of petrologically important structural transformations in plagioclase.

The news caused a serious reconsideration of the state of our knowledge not only of feldspars but also of all rock-forming minerals. It was suspected that such transformations could exist in other minerals as well. Just in time the X-ray diffractometer was invented, and so X-ray experiments on minerals became simple and precise. Thus in the very late 1940s, a great surge of interest occurred in the detailed crystallographic investigation of rock-forming minerals in combination with synthetic and optic studies. *The Journal of Geology* published a special issue on the early results of such investigations on feldspars in September 1950. The first page of the issue began with words of appreciation to Koehler.

During the course of these studies, our understanding of the nature of subtle structural changes has greatly advanced. Although the idea of order–disorder transformation was introduced into the study of rock-forming minerals at least as early as 1934 by Tom Barth, it attracted little attention at the time. When the above-mentioned surge of interest in crystallographic study was beginning, Buerger (1948) delivered a timely address on the order–disorder transformation and its significance. It exerted a great influence on mineralogists and petrologists. Shortly, various structural conversions were discovered or suspected, e.g. in the cordierite group of minerals (Miyashiro 1957). Early results of such studies were reviewed in detail by Eitel (1958).

Since the 1960s, great technological innovations have occurred in X-ray work combined with high-temperature and high-pressure experimental studies. However, the direct impact of crystallographic work to metamorphic studies appears to have decreased.

III.1.4.4 Introduction of the electron probe microanalyzer. The advent of the electron probe microanalyzer in the mid-1960s opened a new age in the study of composition relations of equilibrium assemblages in metamorphic rocks. Among the most remarkable results obtained by the new tool in the 1960s were the confirmation of the existence of the peristerite gap in metamorphic plagioclase by Evans (1964), and the discovery of the widespread occurrence of zoned garnets in metamorphic rocks and genetical discussions on them by Hollister (1966) and others.

III.1.5 Radiometric dating of metamorphic events

III.1.5.1 Advent of K–Ar dating. Radiometric dating by the U–Pb and U–He methods began in the early 20th century, and provided a new idea of the length of geologic time. These early methods, however, were hard to apply to metamorphic rocks. The age determination of ordinary metamorphic rocks became possible by the introduction of the K–Ar method in the 1950s. Dating of common K-bearing minerals such as hornblende, muscovite and biotite by this method led to a drastic change in our understanding of age relations of metamorphism. Introduction of the $^{40}Ar/^{39}Ar$ method in the 1960s has increased the applicability and reliability of K–Ar dates.

There are many orogenic belts that suffered two or more independent metamorphic events. Before the use of K–Ar dating, associated metamorphic rocks of two distinct ages had commonly been confused and treated as belonging to a single genetic group or a single sequence of progressive metamorphism. For example, when a series of progressive metamorphic zones was for the first time delineated on a map of the central Alps by Niggli (1960), rocks formed by Cretaceous high-P/T metamorphism and by Eocene medium-P/T metamorphism were lumped together.

III.1.5.2 Dating of Precambrian metamorphism. Most drastic changes of ideas caused by K–Ar dating occurred in the study of Precambrian orogenies and metamorphisms. Prior to K–Ar dating, it had generally been speculated that magmatism and metamorphism were very intense in the Archean, but became weak in the Proterozoic. From this idea, high-grade metamorphic regions of the Precambrian were regarded as Archean, whereas the weakly metamorphosed or unmetamorphosed Precambrian regions were thought to be Proterozoic. In the late 1950s and early 1960s, K–Ar dating showed the falsehood of this practice. For example, the Grenville province in Canada and the Svecofennian orogenic belt in Scandinavia had traditionally been regarded as Archean because of their high-grade metamorphism, but were found to be middle and early Proterozoic, respectively.

Dating showed the existence of a greater number of orogenic events in the Precambrian than had been expected. Starting in 1959, the Geochronology Laboratory of the Geological Survey of Canada contributed greatly to the elucidation of Precambrian orogenies (e.g. Stockwell 1968). It was soon realized that a K–Ar date does not indicate the true time of crystallization of the mineral, but only the time since the mineral last cooled below its argon diffusion closure temperature (about 530°C for hornblende, 350°C for muscovite, and 310°C for biotite). Thus, instead of K–Ar dates, the whole-rock Rb–Sr isochron date and the U–Pb date of zircon came to be used to indicate the true age of the thermal peak of metamorphism.

On the other hand, K–Ar dates came to be used as a clue to the cooling history of metamorphic regions, which is usually related to that of the post-metamorphic uplift and erosion.

III.1.6 Tectonics of metamorphic belts

Miyashiro (1961) noted the occurrence of the so-called paired metamorphic belts in various parts of the circum-Pacific regions, and supposed that they had formed in a trench zone and the associated arc zone, respectively (§9.1).

The formulation of plate tectonics in the 1960s and its introduction into geology around 1970 enabled tectonic discussion of regional-metamorphic belts (e.g. Miyashiro 1973). For detailed historical reviews of this problem, refer to Miyashiro et al. (1982) and Miyashiro (1991).

III.2 A history of paradigms about the petrological factors controlling metamorphism

III.2.1 Paradigms and research programs in science

A brief general comment on the paradigm in science is inserted here, because the concept of a paradigm is not widely understood among geologists.

It is widely believed that the advance of science is a process of continual addition of new techniques, data, and ideas to the existing stockpile of knowledge. If this is the case, the growth of science should be accumulative, and a matter of piling up new materials on top of the existing system. Indeed, the researches of most scientists are of this type. Thomas S. Kuhn (1962) called such scientific work **normal science**.

In a period of normal science, there is a way of thinking which a whole body of related scientists believes to be correct. These scientists interpret their data, devise explanations, modify existing hypotheses, and write textbooks, all in accordance with this way of thinking. Kuhn gave the name **paradigm** to such a way of thinking (or a set of beliefs), which becomes the framework for the research and educational activities of all related scientists. On the basis of the paradigm, an individual scientist carries out detailed research (normal scientific research) into a certain narrow range of problems. During the course of it, data, explanations, techniques, and technical terms accumulate, with the result that in due course his writings become **esoteric**, that is, incomprehensible to the nonspecialists.

Kuhn stressed that really fundamental changes in science, such as that from Ptolemy's to Copernicus' astronomy, the formation of Newtonian physics, Einstein's proposal of relativity theory, and the chemical revolution of Lavoisier, were not developments piled up on top of older systems, but were **revolutions**. A revolution in science shatters the paradigm which forms the foundation of the age's normal science, and creates a new one. So, the normal science of the preceding period dies out, science enters a new stage of development, and a new normal science begins. There have also been smaller revolutions which have not had as wide an influence as the four examples cited above, but which have still overthrown a normal science in a restricted field.

Kuhn's view of the development of science was formulated on the basis of his analysis of the history of astronomy, physics and chemistry. Geology differs greatly from these sciences in the nature of basic concepts and of systems of knowledge and the methods of research. However, it appears that geological sciences also show a style of historical development as was formulated by Kuhn. This was first pointed out by Tuzo Wilson (1968a,b,c, 1971), who claimed that the advent of plate tectonics was a revolution in Earth sciences and that plate tectonics was a new paradigm in this sense. This view was supported by many later authors (see Cohen 1985).

I believe that the study of metamorphism has also shown this style of historical development. In other words, it has experienced two revolutions in this sense.

If we look at various instances of scientific revolutions in the history of science, it is not usually the case that an old paradigm is proven to be false and then a new paradigm is born and comes to be widely accepted. Even facts that are brilliantly explained by the new paradigm may be explained somehow or other, if old theories are modified and patched up repeatedly within the framework of the outdated paradigm. A revolution occurs when the great majority of scientists come to feel that a new paradigm is more plausible than an old one, and when they adopt the new one as the framework for their research activities. In other words, a scientific revolution occurs not through a proof in the true sense, but rather through persuasion and conversion. Wilson has pointed out that plate tectonics was successful in attracting the great majority of Earth scientists to cause a revolution in this sense during the first half of the 1970s.

In short, a paradigm is an internally consistent system of science based on a theory and philosophy, the validity of which may not have been proven. It forms the framework of scientific researches and education in a period of time. In this section, I describe the history of paradigms about the petrological factors controlling metamorphism.

It should be noted that besides Kuhn's idea of paradigms, there exists another successful attempt to explain the historical development and decline of scientific theories as well as scientific revolutions. This is Lakatos's (1970) idea of **research programs**. A research program is composed of a sequence of successive, logically connected theories which have some basic assumptions in common (called the **hard core** of the program). The sequence represents successive modifications of theories, keeping the basic assumptions intact. In the same period there could exist two or more rival research programs that deal with the same subject but have different basic assumptions and so are incompatible with each other.

Kuhn's and Lakatos's ideas appear to be complementary to each other. In the following pages, I describe three paradigms about the petrological factors controlling metamorphism from Kuhn's viewpoint. However, these three paradigms may well be regarded as two research programs from Lakatos's viewpoint; what is called the first paradigm represents one research program, while what are called the second and third paradigms together represent the other research program. Although some features of historical development and decline of petrological theories appear to be better understood in terms of research programs, this issue is not discussed in the following pages mainly due to considerations of space.

III.2.2 The first paradigm, characterized by the concepts of stress minerals and normal regional metamorphism

Although various genetic views on metamorphic rocks were expressed in the 19th century, they were ill defined and usually fragmentary, and did not constitute a paradigm. Kuhn (1962) pointed out that there is a sort of scientific research without paradigms, and that acquisition of a paradigm and of the esoteric type of research that it permits is a sign of maturity in the development of any given scientific field. In this sense, the study of metamorphism was immature in the 19th century and the early 20th century. Some other branches of geology may still be in such an immature state even today. Although Grubenmann began a general consideration of the factors controlling metamorphism, the first paradigm was formulated by Harker (1918, 1932) and was expanded by Turner (1948).

III.2.2.1 Grubenmann's depth-zone hypothesis. Ulrich Grubenmann classified regional-metamorphic rocks into three depth-zones, named epi-, meso- and kata-zones in order of presumably increasing depth of their metamorphism (Appendix III.1.1). In his view, both temperature and pressure increased in the order of epi-, meso- and kata-zones, whereas shearing stress was strong in the epi-zone, very strong in the meso-zone, and weaker in the kata-zone.

The problem of shearing stress aside, Grubenmann's view was based on a thermal model of the crust in which both temperature and pressure are functions of, and increase with, depth, and hence isothermal surfaces within the crust are horizontal and stationary, as illustrated in Figure 2.11a. In such a case, temperature is a function of pressure. Thus, the *P–T* conditions of all regional metamorphisms of the world are represented by a single curve with a positive slope in the *P–T* plane (like N–E–F in Figure 2.11b).

III.2.2.2 Harker's formulation of the first paradigm. Alfred Harker accepted the essential feature of the Grubenmann model, and built a greatly refined system of metamorphic petrology on it (Harker 1918, 1932). He thought that in the crust pressure increases downward with temperature, and that pressure is a function of temperature. He further thought that the maximum value of shearing stress that a rock can maintain decreases with increasing temperature. In the "normal" case, regional metamorphism takes place under the maximum shearing stress possible at the pertinent temperature. Hence, shearing stress is also a function of temperature in "normal regional metamorphism". In other words, temperature may usually be regarded as the only independent variable that controls regional metamorphism. He thought that the regional metamorphism of the Barrovian zones in the Scottish Highlands (Fig. 1.5) was a typical example of such "normal regional metamorphism".

Harker considered that shearing stress is absent in contact metamorphism, and ascribed the observed differences in mineral composition between normal regional metamorphism and contact metamorphism to the presence and absence of shearing stress, respectively. He considered that strong shearing stress in normal regional metamorphism produces such characteristic minerals as chlorite, almandine, staurolite and kyanite, which he named **stress minerals**. On the other hand, the absence of shearing stress in contact metamorphism leads to the production of such characteristic minerals as andalusite and cordierite. He named them **anti-stress minerals**.

Harker formulated his view in the 1910s and 1920s, when the theory of geosyncline and tectonic cycle was becoming dominant in geology. He therefore considered that regional metamorphism takes place in the thick sedimentary pile in a slowly subsiding geosyncline. With advancing accumulation of sediments, the older parts of the pile are depressed and enter deeper places, which are at higher temperature and pressure. Since this geosynclinal sinking and sedimentation are very slow processes, thermal conduction approximately keeps step with it, maintaining nearly horizontal isothermal surfaces in the sedimentary pile. In the depths of the geosyncline, thermal expansion of rocks takes place, producing great lateral pressure, folding of rocks, shearing stress, and regional metamorphism.

In his view, pressure and shearing stress are definite functions of temperature. So the course of temporal change of pressure and shearing stress which individual metamorphic rocks experience during increase of temperature must be identical to the variation of pressure and shearing stress that is observed in the progressive sequence of zones across the metamorphic region.

Harker was aware of the existence of some regional-metamorphic rocks containing andalusite and cordierite. He thought that such rocks form by abnormal regional metamorphism in which shearing stress is smaller than the maximum value possible at the pertinent temperature.

This first paradigm dominated metamorphic petrology in the 1930s and 1940s. The most expanded form of this paradigm was given by Turner (1948), as discussed later.

III.2.3 The second paradigm, characterized by the concept of chemical equilibrium controlled by temperature and pressure, and of the diversity of regional metamorphism due to differences in pressure

III.2.3.1 Start of the second paradigm by Goldschmidt and Eskola. Victor Moritz Goldschmidt (1888–1947) of the University of Kristiania (Oslo) initiated the idea that the mineral assemblages of metamorphic rocks were controlled by temperature and pressure in combination with the bulk chemical composition of individual rocks, and that they should follow the laws of chemical thermodynamics. This was the starting point of the second paradigm.

Goldschmidt (1911, 1912a,b) investigated contact-metamorphic aureoles around diverse alkalic intrusions in the Oslo region, and showed that the mineral assemblages of metamorphic rocks could be interpreted as having reached a near-equilibrium state under a limited range of temperatures and pressures. This meant the possibility of application of chemical thermodynamics to metamorphic mineral assemblages. At that time chemical thermodynamics was a new, rapidly developing field of science. Goldschmidt attempted to apply the phase rule and some chemical thermodynamic calculations to metamorphic rocks (Appendix III.3.1).

Then, Pentti Eskola of the University of Helsinki generalized this viewpoint by application of it to all metamorphic rocks on Earth. He proposed a classification of metamorphic rocks based on the concept of metamorphic facies, each of which represented a specific range of temperature and pressure (Eskola 1915, 1920, 1939). He considered that temperature and pressure were independent variables of metamorphic conditions, that both temperature and pressure had great effects on mineral assemblages, and that glaucophane schists and eclogites were typical high-pressure rocks.

Eskola differed from Harker in the respect that he did not regard shearing stress as controlling the mineral composition of metamorphic rocks, that he attempted to explain mineral assemblages of metamorphic rocks by application of equilibrium thermodynamics, that he regarded pressure as a variable independent of temperature in metamorphism, and that he emphasized the effect of pressure on mineral assemblages.

These differences led to crucial differences of opinion between them about the genesis of the glaucophane schists and of metamorphic complexes containing "stress and anti-stress minerals". Eskola regarded glaucophane schists as high-pressure rocks, whereas since Harker did not accept the mineralogical effects of high pressures, he ascribed the formation of glaucophane schists to the high Na content of the rocks, which might have been either a chemical feature inherited from their original rocks or a result of Na-metasomatism during the metamorphism. Harker explained the formation of staurolite and kyanite as stress minerals, and that of cordierite and andalusite as anti-stress minerals, whereas Miyashiro (1949, 1951, 1953) attempted to advocate Eskola's ideas by the hypothesis that kyanite was stable at higher pressures than andalusite.

At that time most petrologists accepted Harker's doctrine, and ignored Eskola's, as is seen, for example, in Loewinson-Lessing's (1954) history. Most of those who knew of Eskola's metamorphic facies used it only in the range where it does not conflict with Harker. A most typical example was Turner (1948). He devised an emasculated version of metamorphic facies which does not conflict with Harker's doctrine. Although Turner used most metamorphic facies, he denied the existence of the glaucophane-schist facies, and ascribed the formation of glaucophane to Na-metasomatism (Turner 1948: 99–100). Further, Turner regarded regional-scale metamorphic regions with andalusite and cordierite (including the Orijärvi studied by Eskola) as being of contact-metamorphic origin simply because of the occurrence of "anti-stress minerals" (Turner 1948: 77).

III.2.3.2 General acceptance of the second paradigm. Miyashiro (1949, 1951) claimed that the modes of occurrence of andalusite, kyanite and sillimanite in metamorphic rocks could be explained in terms of temperature and pressure without invoking Harker's stress-mineral hypothesis. He proposed a hypothetical phase diagram which indicated that andalusite was stable at low pressure, kyanite at a higher pressure, and sillimanite at a higher temperature than both the other two, like the one shown in Figure 8.1. In the 1950s, W. P. de Roever (1950, 1955a,b) and Miyashiro & Banno (1958) advocated Eskola's idea of glaucophane-schist facies. At that time, the main significance of these works laid in their attempts to shift the philosophical framework of metamorphic petrology from Harker's to Eskola's, and not in the origin of particular kinds of rocks and minerals.

Miyashiro (1961) proposed the existence of three major types of regional metamorphism, representing distinct ranges of *P–T* conditions. One of the three corresponded to what Harker had called normal regional metamorphism, whereas the other two represented types of regional metamorphism at lower pressure producing andalusite and cordierite, and at higher pressure producing glaucophane and jadeite (Ch. 8). The textbooks by Ramberg (1952) and Barth (1952), both from Norway, also greatly contributed to the widespread acceptance of Eskola's ideas and metamorphic facies.

In the late 1950s kyanite was synthesized in the laboratory at high pressure without shearing stress (Clark et al. 1957), and the fact that pressure exerts large effects on the mineral formation was demonstrated by thermodynamic calculations as well as by synthetic experiments (e.g. Adams 1953, Coes 1953, Robertson et al. 1957). At the end of the 1950s, Eskola's philosophy came to be accepted by almost all relevant workers.

It may be noted that Harker's hypothesis was not disproved at that time. Eskola did not give clear-cut arguments on the question (Eskola 1939: 331–4). Miyashiro (1951) described kyanite–quartz veins in which crystals of kyanite appeared to have grown in fluid-filled cavities without nonhydrostatic stresses. These particular cases, however, fell short of disproving the possible existence of stress minerals in general. The essential point of these arguments was to show the existence of an alternative system of explanation for all observed petrologic relations of metamorphic rocks that did not use Harker's hypotheses. This was in harmony with Kuhn's (1962) view that the rejection of an old paradigm usually occurs without decisive, logical disproof of it.

III.2.3.3 Completion of the second paradigm

Open system models. Eskola's doctrine was based on his crude and incomplete thermodynamic analysis of metamorphic mineral assemblages. It was particularly serious that Eskola did not realize the petrogenetic significance of the fact that volatile components such as H_2O and CO_2 could be much more mobile than other ordinary silicate components during metamorphism. A more rigorous thermodynamic analysis was urgently needed. James B. Thompson (1955, 1957, 1970, 1972) formulated a rigorous theory of thermodynamic equilibrium in metamorphic rocks based on the assumption that metamorphic rock systems were open to H_2O and CO_2. He and A. B. Thompson made great contributions to the rigorous analysis of paragenetic relations of minerals in metamorphic rocks (e.g. J. B. Thompson & A. B. Thompson 1976, A. B. Thompson 1976). D. S. Korzhinskii appears to have formulated a similar theory independently. He proposed a modification of the definition of metamorphic facies so as to adapt Eskola's idea to the open system model (Korzhinskii 1959).

Limitation of open system models and the importance of internal buffering. From about 1970, certain geologists began to doubt or deny the idea of metamorphic rocks as being open to H_2O and CO_2, partly because they found metamorphic regions to which the *AFM* diagram was not rigorously applicable. Rumble (1978), Ferry (1979, 1987), Rice & Ferry (1982) and Rumble et al. (1982) showed that the chemical potentials of H_2O and CO_2 were not uniform in a small metamorphic area, and even in a single outcrop. Thus, it was realized that the applicability of open system models was limited.

Greenwood (1975) advanced an idea of internal buffering of the fluid composition by the mineral assemblage of metamorphic rocks. The existence of such metamorphic rocks was shown by Rice (1977), Ferry (1983a) and others.

III.2.4 Towards the third paradigm: a framework involving temporal changes

III.2.4.1 Introduction. Both the first and second paradigms were based on static thermal models of the crust and static thermodynamic models of metamorphic rocks, and so tended to treat metamorphism as a static state rather than as a process involving temporal changes. Hence, the interest of metamorphic petrologists was focused on the description and theoretical analysis of equilibrium mineral assemblages that formed at the thermal peak of individual rocks.

In the mid-1970s, a new trend of thinking that required the introduction of temporal changes into the theoretical framework developed on two research fronts: one concerning the thermal model of metamorphism in the continental collision zone, and the other concerning the compositional change and flow of fluid during metamorphism. They are now co-operating towards the formulation of the third paradigm.

III.2.4.2 Impact of thermal model studies. Since the advent of plate tectonics in geology in the late 1960s, a number of authors have discussed plate-tectonic thermal models of orogenic belts. In the middle and late 1970s, these studies, particularly on the Cretaceous–Tertiary orogeny in the eastern Alps by Oxburgh, Bickle, Richardson, England and their associates, then at Oxford University, showed the rôle of time to be as important as those of temperature and pressure in controlling metamorphism. It led to the realization of essential defects in the second paradigm.

First, Oxburgh & Turcotte (1974) showed that in the large-scale overthrust region of the eastern Alps, the thermal history of the region was dominated by the thermal effect of thrusting. Such overthrusting in a continental collision zone was expected to have taken place at the common rate of plate movements, such as a few centimers per year. Since thermal conduction could not keep step with such relatively fast movements, tectonic movements must have drastically changed the temperature distribution in the orogenic complex. Prior to the advent of plate tectonics, the geosynclinal theory assumed that geosynclinal sinking and sediment accumulation took place very slowly, say, at a rate of 10 km per 100 million years, or 1 mm yr^{-1}. In movement as slow as this, thermal conduction could keep step with the sinking, maintaining nearly horizontal, stationary isothermal surfaces in the sedimentary pile: a condition supposed by Harker. In plate-tectonic models of collision zones, isothermal surfaces cannot be stationary.

In the eastern Alps, the thrusting was completed in a short time and apparently was not accompanied by large-scale magmatism. This greatly simplifies the thermal models. For further simplification, they used one-dimensional mathematical analysis, that is, assumed horizontal isothermal surfaces, though not stationary. Bickle et al. (1975) examined the effects of variations of parameters involved in the calculation of the thermal history of an overthrust region of this type, and concluded that the Tertiary regional metamorphism that reached a thermal peak about 30 million years after the overthrusting could be explained by radioactive heating in the crust thickened by the overthrusting under normal heat flow from the underlying mantle. This means the possible existence of regional-metamorphic belts with no abnormally high heat introduction from the mantle.

Full realization of the implication of these thermal model studies for the understanding of metamorphism came when the *P–T* history of individual metamorphic rocks was quantitatively traced by England & Richardson (1977) and then by England & A. B. Thompson (1984) and A. B. Thompson & England (1984). The mineral assemblages of rocks record the *P–T* conditions at or near the thermal peak, which is not synchronous, but may differ in age up to several tens of millions of years between different parts within a regional-metamorphic complex. The time of metamorphic crystallization differs commonly in different parts of an isograd. In other words, each isograd is generally polychronous. Different isograds in the same region usually form at different times. So a system of isograds in a region does not represent a distribution of *P–T* conditions at any specific time. Moreover, the pressure at the thermal peak will not usually be the maximum value of the pressure which the rock experiences.

To understand regional metamorphism, the whole course of tectonothermal history including uplift and erosion must be considered. This is because increases of temperature in the metamorphic complex may be extremely slow, continuing over a period of tens of millions of years so that it may be concurrent with uplift and erosion. Advance of erosion on the surface decreases the pressure of, and gives a cooling effect on, the rock mass undergoing metamorphism. Hence, the rate of erosion may have a considerable influence on the values of temperature and pressure of the thermal peak.

In the period when the first and second paradigms were generally accepted, it was considered that the mineralogical variations through successive zones across a progressive metamorphic region were identical to the successive mineralogical changes which a metamorphic rock experienced temporally during increasing temperature, although even at that time the validity of this idea was doubted by a few geologists (e.g. Read 1957: 24–5, 31–2, Kojima 1952: 62–3, Miyashiro 1973: 65–6). The thermal model studies of the eastern Alps have clearly shown for the first time that the idea cannot be valid at least generally, even though one-dimensional thermal models as used in the eastern Alps are far from the real situations in most orogenic belts.

Prior to the mid-1970s, models of regional metamorphism were discussed mainly for island-arc or continental-arc zones (e.g. Miyashiro 1973). However, we did not have reliable knowledge of the deep structure of arc zones. Moreover, the thermal histories of such zones are very complicated by repeated intrusions of magmas from the underlying mantle, as well as by enduring crustal movements over a long period. Therefore, the thermal histories of these zones were not easily amenable to mathematical treatment. This is one reason why studies of arc zones failed to induce critical reconsideration of the second paradigm.

Since then, various attempts have been made empirically to determine the paths of temporal changes of P–T condition that metamorphic rocks experienced. It has been observed that different parts of a metamorphic region experienced different tectonic histories, and consequently different courses of change of P–T conditions (e.g. Harte & Dempster 1987). Common middle-sized folding, faulting and igneous intrusion have noticeable influences on the course of P–T condition of metamorphic rocks (e.g. Sleep 1979, Chamberlain & Karabinos 1987).

III.2.4.3 Fluid flows as agents of metamorphic reactions, and temporal changes in composition of intergranular fluid. The existence of fluids flowing through rocks during regional metamorphism had long been imagined by many authors. It was not until Carmichael's pioneering work that reliable petrologic evidence to support it began to appear (Carmichael 1970).

Ferry (1976a, 1980, 1983a, 1986, 1987), in particular, made an elaborate series of investigationd on this problem in a regional-metamorphic complex in Maine. He showed that externally derived fluids flowed pervasively through rocks at all stages of prograde metamorphism, and that prograde metamorphism was driven not only by increasing temperature, but also by reactions of rocks with intergranular fluid whose composition had changed by mixing with infiltrating fluids. Even at a constant temperature, reactions could be driven by infiltration of externally derived fluids, if by mixing of infiltrating fluid, the composition of intergranular fluid came to be deviated from that which had initially been equilibrated with the minerals. Thus, flow of fluid is an agent of metamorphic reactions (§6.7).

An intergranular fluid in metamorphic rocks is commonly buffered by the mineral assemblage of each rock. It changes in composition with increasing temperature. An external fluid may infiltrate the rock and mix with the pre-existing intergranular fluid, causing reactions anew. Therefore, metamorphic rocks have a history of flow and compositional changes of fluids.

Flows of fluid have been envisaged by many other authors in recent years in studies of metabasites, charnockites and other rocks, and are thought to be a widespread agent of metamorphism, contributing also to heat transfer and chemical migration during metamorphism (e.g. Chamberlain & Rumble 1988).

III.3 Thermodynamic modeling and thermobarometry

III.3.1 Pioneering attempts of Goldschmidt and Ramberg

In their famous study of evaporite deposits, van't Hoff et al. (1903) calculated the effect of the composition of aqueous solution on the equilibrium of the reaction: gypsum ($CaSO_4.2H_2O$) = anhydrite ($CaSO_4$) + 2 H_2O. This has been called the first use of thermodynamic calculation in the geologic problem.

The first important pioneering attempts to apply thermodynamics to metamorphic petrology were made by Victor Moritz Goldschmidt (1912a). He calculated, from thermochemical data, the univariant curve for the wollastonite reaction: calcite + quartz = wollastonite + CO_2. Then, Goldschmidt (1912b) applied the phase rule to the mineral assemblages of metamorphic rocks, and obtained the so-called Goldschmidt's mineralogical phase rule (§ 5.1).

In both cases he applied thermodynamic relations that hold in the chemical laboratory directly to natural metamorphic rocks, and did not consider the possibility that natural rocks may have different thermodynamic constraints from laboratory chemical systems. His calculation of the wollastonite reaction is valid in the presence of pure CO_2 gas, whereas in natural metamorphic rocks, the reaction usually occurs in the presence of an impure fluid with a large mole fraction of H_2O. His mineralogical phase rule is valid in a closed system, whereas natural metamorphic rock systems are usually open with regard to certain components. Valid applications of thermodynamics require a better understanding of the thermodynamic constraints of metamorphic rock systems.

There was no essential progress between 1912 and the Second World War, as seen in Eskola (1939: 315–16), who cited the above attempts of Goldschmidt as valid.

In the late 1940s and early 1950s, Hans Ramberg made various stimulating attempts to apply thermodynamics to metamorphic and orogenic studies. Although many of his conclusions were highly controversial and were rejected by the majority of contemporary geologists, his attempts contributed greatly to arousing interest in thermodynamics among young students of petrology, and were a prelude to the active advance of thermodynamic studies in the 1950s.

Ramberg (1944a, 1952) discussed qualitative subsolidus phase relations between metamorphic minerals. Many of the discussions are still essentially valid. Ramberg & DeVore (1951) formulated the thermodynamic relation of a Fe–Mg exchange reaction with particular reference to coexisting olivine and pyroxene. This opened a new epoch in the rigorous thermodynamic calculation of reaction equilibria among rock-forming minerals.

Prior to the Second World War, virtually all petrologists believed that an aqueous fluid existed in the intergranular space during metamorphism, and that metamorphic reactions and chemical migration took place through this medium. In the 1940s, many of the radical advocates of large-scale granitization, such as Perrin, Doris Reynolds and Hans Ramberg, denied the existence of such a fluid phase. They claimed that long-range chemical migration occurred by diffusion of ions and molecules mainly along grain boundaries, causing granitization (e.g. Perrin & Roubault 1949, Ramberg 1952: 174–82). Thus, Ramberg (1949, 1952) proposed a fantastic grand hypothesis of crustal-scale granitization by the long-range chemical diffusion of almost all rock-forming elements in the gravitational field. It was not widely accepted, and tended to cause widespread antipathy to him among more traditional geologists.

III.3.2 Rigorous thermodynamic models for metamorphic reactions

Danielsson (1950) published a rigorous calculation of the equilibrium curve of the wollastonite reaction under conditions where a fluid phase of pure CO_2 composition may not be present. This was a pioneer step towards the formulation of rigorous thermodynamic models of metamorphic reactions.

It appears that Dmitrii Sergeevich Korzhinskii (1899–1985) began his attempts to formulate a thermodynamic model of the open system in the 1930s (e.g. Korzhinskii 1936), although his work was not known outside the Soviet Union. His mineralogical phase rule for the open system first came to be known in the West through his paper presented to the 18th International Geological Congress (1948) in Britain (Korzhinskii 1950).

James B. Thompson, Jr began his theoretical investigation of chemical reactions in the crust in the late 1940s. His first comprehensive treatise (J. B. Thompson 1955) rigorously dealt with the nature and thermodynamic relations of metamorphic reactions. In addition, it contained a theory of the open system similar to Korzhinskii's as well as a thermodynamic analysis of crustal-scale chemical diffusion in the gravitational field, a problem initiated by Ramberg. For example, a clear formulation for dehydration equilibria in fluid-present and fluid-absent metamorphism was first given in this paper.

To illustrate how Thompson's analyses were fresh at that time, an example is cited below. Bowen (1940) proposed the idea of a petrogenetic grid, which consists of intersecting equilibrium curves with different slopes in the P–T diagram, thus cutting P–T space up into a grid (§10.4). Following an idea predominant at that time Bowen thought that the equilibrium curves of solid–solid reactions had much greater slopes (dP/dT) than those of dehydration and decarbonation reactions. J. B. Thompson (1955) showed, from a theoretical viewpoint as well as from results of early synthetic experiments, that the actual relation was contrary to this idea. In other words, dehydration curves in the presence of a pure water phase usually have a greater slope than solid–solid reaction curves in the ordinary pressure range of regional metamorphism (see Figs 2.2 and 10.7, for example).

The theory of the open system exerted a strong influence on petrographic work, partly because it led to the introduction of a relatively rigorous composition–paragenesis diagram, called the *AFM* diagram (J. B. Thompson 1957). A. B. Thompson contributed to the expansion and refinement of rigorous thermodynamic analysis of this issue (e.g. A. B. Thompson 1976, J. B. Thompson & A. B. Thompson 1976).

J. B. Thompson (1955) accepted the possible effectiveness of diffusion as a mechanism even of long-range migration of H_2O and CO_2. Here, we may detect a moderated influence of Ramberg. In the 1970s, younger authors began to undermine this view from various sides. They presented evidence indicating a very limited effectiveness of diffusion and the importance of internal buffering. This problem was pursued by Greenwood, Rumble, Rice, Ferry and others in the 1970s and 1980s. During the course of this research, Ferry came up with a new idea of fluid flows controlling metamorphic reactions (§6.7).

III.3.3 Thermodynamic calculation of metamorphic equilibria

Since the mid-1970s, a large number of books have been published dealing with the application of thermodynamics to metamorphic studies (Froese 1976, Ferry 1982) as well as to petrology and adjacent fields of geology (e.g. Fraser 1977, Greenwood 1977, Wood

& Fraser 1977, Powell 1978, Nordstrom & Munoz 1985). This shows the recent arousal of great interest in the thermodynamic calculation of equilibrium states in geology. Various equations convenient for the purpose have been derived in these books.

Rumble and Spear developed a powerful new method (named the **Gibbs method**) for the numerical calculation of the change of equilibrium states in metamorphic rocks with temperature and pressure or with the composition of solid-solution minerals (Rumble 1976a, Spear et al. 1982, Spear 1988c).

Concurrently, thermodynamic and thermochemical data of individual minerals have increased in amount and have been improved in quality (e.g. Helgeson et al. 1978, Robie et al. 1978, Holland & Powell 1985, 1990, Powell & Holland 1985, 1988, Berman 1988).

III.3.4 Geothermometers and geobarometers

From the beginning of the 20th century, "determination of the temperatures and pressures of rock-forming processes" became a high-spirited slogan to ornament the physico-chemical investigation of igneous and metamorphic rocks. Now that the advent of plate tectonics has provided us with a better understanding of tectonic processes involving magmatism and metamorphism, and the formulation of thermodynamic models has provided us with a better understanding of the nature of metamorphic reactions, the significance of the determination of *P–T* conditions has come to lie in its contribution to the clarification of the relevant tectonic and chemical processes. It provides us with basic constraints for possible tectonic models.

Comprehensive reviews of thermobarometry have been published by Essene (1982, 1989). The development of the four widely used methods of *P–T* determination is discussed briefly below, largely in chronological order of development.

Use of univariant curves. In the late 1940s and the 1950s, synthetic experiments provided us with a lot of data on univariant curves of simplified metamorphic reactions. They showed possible ranges of *P–T* conditions of mineral assemblages. Because most data at that time were of the breakdown curves of hydrous minerals with end-member compositions, their usefulness was very limited. Moreover, early data tended to give too high temperatures. In the 1960s the reliability of such data was greatly improved.

Particularly important as a broad framework of *P–T* conditions of metamorphic processes, were the stability relations of andalusite, kyanite and sillimanite as well as the equilibrium curve for the reaction: jadeite + quartz = albite (see Fig. 8.1). Francis Birch and his co-workers at Harvard University began experimental determination of them in the late 1950s. In the period 1966–71, at least seven papers were published reporting experimentally determined *P–T* values of the triple point of the Al_2SiO_5 minerals. The values of pressure given were 7.8, 6.5, 5.5, 4.0, 3.76, 2.5 and 2.0 kbar in the decreasing order. More recent, and probably more reliable results are reviewed in §3.4.3.

Solvus geothermometers. Experimental determination of solvi for use as geothermometers began in the 1950s, i.e. earlier than the experimental studies on other types of solid-solution geothermometers. The solvus between calcite and dolomite as was determined by Graf & Goldsmith (1955) and others, was widely used as a geothermometer in metamorphic studies. The solvi between plagioclase and alkali feldspar and between Ca-poor and Ca-rich pyroxenes, have been investigated by many workers since the late 1950s. Some solvi are sensitive to a variation of pressure, whereas others are not.

Fe–Mg exchange geothermometers. Ramberg & DeVore (1951) opened the thermodynamic study of exchange reactions. Exchange reactions cause little change in volume, and so are insensitive to pressure. It was found that Fe–Mg exchange reactions sensitive to temperature are ideal as geothermometers (§ 3.7).

For pelitic metamorphic rocks, the Fe–Mg exchange reaction between garnet and biotite was found to be a good geothermometer (A. B. Thompson 1976, Ferry & Spear 1978). For high-grade metabasites, the Fe–Mg exchange reaction between garnet and clinopyroxene was found to be useful (Ellis & Green 1979, Pattison & Newton 1989).

Solid–solid reaction geobarometers Equilibria of solid–solid reactions are usually sensitive to both temperature and pressure. If temperature is known by some other method, such as an Fe–Mg exchange geothermometer, solid–solid reactions involving solid-solution minerals could be good geobarometers. Active investigation of this type of geobarometer began in the 1970s.

The most successful was Ghent's (1976) geobarometer using the plagioclase–garnet–Al_2SiO_5–quartz assemblage, which was based on the end-member reaction: anorthite = grossular + sillimanite(kyanite) + quartz (§3.4.2). Geobarometers using the garnet–plagioclase–orthopyroxene (clinopyroxene)–quartz assemblage (e.g. Newton & Perkins 1982) were also used widely.

References

Abraham, K. and Schreyer, W. (1976) A talc-phengite assemblage in piemontite schist from Brezovica, Serbia, Yugoslavia. *J. Petrol.* **17**, 421–439.

Adams, L. H. (1953) A note on the stability of jadeite. *Am. J. Sci.* **251**, 299–308.

Albee, A. L. (1965a) Phase equilibria in three assemblages of kyanite-zone pelitic schists, Lincoln Mountain quadrangle, central Vermont. *J. Petrol.* **6**, 246–301.

Albee, A. L. (1965b) A petrogenetic grid for the Fe-Mg silicates of pelitic schists. *Am. J. Sci.* **263**, 512–536.

Albee, A. L. (1968) Metamorphic zones in northern Vermont. In: E-an Zen et al. (eds.) *Studies of Appalachian Geology, Northern and Maritime*, pp. 329–341. Interscience, New York.

Albee, A. L. (1972) Metamorphism of pelitic schists: reaction relations of chloritoid and staurolite. *Geol. Soc. Amer. Bull.* **83**, 3249–3268.

Apted, M. J. and Liou, J. G. (1983) Phase relations among greenschist, epidote-amphibolite, and amphibolite in a basalt system. *Am. J. Sci.* **283-A**, 328–354.

Arima, M. and Gower, C. F. (1991) Osumilite-bearing granulites in the eastern Grenville Province, eastern Labrador, Canada: mineral parageneses and metamorphic conditions. *J. Petrol.* **32**, 29–61.

Armbruster, Th., Schreyer, W. and Hoefs, J. (1982) Very high CO_2 cordierite from Norwegian Lapland: mineralogy, petrology, and carbon isotopes. *Contr. Mineral. Petrol.* **81**, 262–267.

Armstrong, R. L. and Dick, H. J. B. (1974) A model for the development of thin overthrust sheets of crystalline rock. *Geology* **2**, 35–40.

Ashworth, J. R. (ed.) (1985) *Migmatites*. Blackie, Glasgow. 302 pp.

Atherton, M. P. (1968) The variation in garnet, biotite and chlorite composition in medium grade pelitic rocks from the Dalradian, Scotland, with particular reference to the zonation in garnet. *Contr. Mineral. Petrol.* **18**, 347–371.

Atherton, M. P. (1977) The metamorphism of the Dalradian rocks of Scotland. *Scott. J. Geol.* **13**, 331–370.

Atherton, M. P. and Gribble, C. D. (eds.) (1983) *Migmatites, Melting and Metamorphism*. Shiva Nantwich, England, 326 pp.

Baker, A. J. (1985) Pressure and temperature of metamorphism in the eastern Dalradian. *J. Geol. Soc. London* **142**, 137–148.

Baker, A. J. (1987) Models for the tectonothermal evolution of the eastern Dalradian of Scotland. *J. Metamorphic Geol.* **5**, 101–118.

Baker, A. J. and Droop, G. T. R. (1983) Grampian metamorphic conditions deduced from mafic granulites and sillimanite-K-feldspar gneisses in the Dalradian of Glen Muick, Scotland. *J. Geol. Soc. London* **140**, 489–497.

Baker, J., Holland, T. and Powell, R. (1991) Isograds in internally buffered systems without solid solutions: principles and examples. *Contr. Mineral. Petrol.***106**, 170–182.

Banno, S. (1964) Petrologic studies on Sanbagawa crystalline schists in the Bessi-Ino district, central Sikoku, Japan. *Tokyo Univ. Fac. Sci. J.*, Sec. 2, **15**, 203–319.

Banno, S. (1986) The high-pressure metamorphic belts of Japan: a review. *Geol. Soc. Amer. Mem.* **164**, 365–374.

Banno, S., Higashino, T., Otsuki, M., Itaya, T. and Nakajima, T. (1978) Thermal structure of the Sanbagawa metamorphic belt in central Shikoku. In: S. Uyeda, et al. (eds.) *Geodynamics of the Western Pacific (Advances in Earth and Planetary Sciences*, Vol.6), pp.345–356. Japan Scientific Societies Press, Tokyo.

Banno, S. and Sakai, C. (1989) Geology and metamorphic evolution of the Sanbagawa metamorphic belt, Japan. In: J. S. Daly et al. (eds.) *Evolution of Metamorphic Belts* (Geol. Soc. London, Spec. Publ., No.43), pp.519–532. Blackwell Scientific, Oxford.

Barbey, P. and Cuney, M. (1982) K, Rb, Sr, Ba, U and Th geochemistry of the Lapland granulites (Fennoscandia). LILE fractionation controlling factors. *Contr. Mineral. Petrol.* **81**, 304–316.

Barnicoat, A. C. (1983) Metamorphism of the Scourian complex, N. W. Scotland. *J. Metamorphic Geol.* **1**, 163–182.

Barnicoat, A. C. and Treloar, P. J. (1989) Himalayan metamorphism – an introduction. *J. Metamorphic Geol.* **7**, 3–8.

Barrow, G. (1893) On an intrusion of muscovite-biotite gneiss in the southeastern Highlands of Scotland, and its accompanying metamorphism. *Q. J. Geol. Soc. London* **49**, 330–358.

Barrow, G. (1912) On the geology of lower Dee-side and the southern Highland Border. *Proc. Geol. Assoc.* **23**, 274–290.

Barth, Tom. F. W. (1936) Structural and petrologic studies in Dutchess County, New York, II. *Geol. Soc. Amer. Bull.* **47**, 775–850.

Barth, Tom. F. W. (1952, 1962) *Theoretical Petrology*. 1st and 2nd edns. John Wiley, New York.

Barton, M. D. and Hanson, R. B. (1989) Magmatism and the development of low-pressure metamorphism: implications from thermal modelling and the western United States. *Geol. Soc. Amer. Bull.* **101**, 1051–1065.

Bearth, P. (1962) Versuch einer Gliederung alpinmetamorpher Serien der Westalpen. *Schweiz. Mineral. Petrogr. Mitt.* **42**, 127–137.

Becke, F. (1903) Ueber Mineralbestand und Struktur der kristallinischen Schiefer. *IX. Session du Congres geologique internat. (Vienne), Compte rendu*, Part 2, 553–570.

Berg, J. H. and Wheeler, E. P., II (1976) Osumilite of deep-seated origin in the contact aureole of the anorthositic Nain complex, Labrador. *Am. Mineralogist* **61**, 29–37.

Berman, H. (1937) Constitution and classification of the natural silicates. *Am. Mineralogist* **22**, 342–408.

Berman, R. G. (1988) Internally-consistent thermodynamic data for minerals in the system Na_2O-K_2O-CaO-MgO-FeO-Fe_2O_3-Al_2O_3-SiO_2-TiO_2-H_2O-CO_2. *J. Petrol.* **29**, 445–522.

Béthune, S. de (1976) Formation of metamorphic biotite by decarbonation. *Lithos* **9**, 309–318.

Bhattacharya, A. and Sen, S. K. (1985) Energetics of hydration of cordierite and water barometry in cordierite-granulites. *Contr. Mineral. Petrol.* **89**, 370–378.

Bhattacharya, A. and Sen, S. K. (1986) Granulite metamorphism, fluid buffering, and dehydration melting in the Madras charnockite and metapelites. *J. Petrol.* **27**, 1119–1141.

Bickle, M. J., Hawkesworth, C. J., England, P. C. and Athey, D. R. (1975) A preliminary thermal model for regional metamorphism in the Eastern Alps. *Earth Planet. Sci. Letters* **26**, 13–28.

Birch, F. and LeComte, P. (1960) Temperature-pressure plane for albite composition. *Am. J. Sci.* **258**, 209–217.

Bishop, D. G. (1972) Progressive metamorphism from prehnite-pumpellyite to greenschist facies in the Dansey Pass area, Otago, New Zealand. *Geol. Soc. Amer. Bull.* **83**,

3177–3198.

Black, P. M. (1977) Regional high-pressure metamorphism in New Caledonia: phase equilibria in the Ouegoa district. *Tectonophys.* **43**, 89–107.

Black, P. M., Brothers, R. N. and Yokoyama, K. (1988) Mineral paragenesis in eclogite facies meta-acidites in northern New Caledonia. In: D. C. Smith (ed.) *Eclogites and Eclogite-facies Rocks*, pp.271–289. Elsevier, Amsterdam.

Blundy, J. D. and Holland, T. J. B. (1990) Calcic amphibole equilibria and a new amphibole-plagioclase geothermometer. *Contr. Mineral. Petrol.* **104**, 208–224.

Boettcher, A. L. and Wyllie, P. J. (1969) Phase relationships in the system $NaAlSiO_4$-SiO_2-H_2O to 35 kbars pressure. *Am. J. Sci.* **267**, 875–909.

Bohlen, S. R. (1987) Pressure-temperature-time paths and a tectonic model for the evolution of granulites. *J. Geol.* **95**, 617–632.

Bohlen, S. R. (1991) On the formation of granulites. *J. Metamorphic Geol.* **9**, 223–229.

Bohlen, S. R. and Boettcher, A. L. (1982) The quartz-coesite transformation: a precise determination and the effect of other components. *J. Geophys. Res.* **87**, 7073–7078.

Bohlen, S. R. and Mezger, K. (1989) Origin of granulite terranes and the formation of the lowermost continental crust. *Science* **244**, 326–329.

Boles, J. R. (1977) Zeolites in low-grade metamorphic rocks. In: F. A. Mumpton (ed.) Mineralogy and *Geology of Natural Zeolites* (Reviews in Mineralogy, Vol 4.), Ch. 6. Mineral. Soc. America.

Boles, J. R. and Coombs, D. S. (1975) Mineral reactions in zeolitic Triassic tuff, Hokonui Hills, New Zealand. *Geol. Soc. Amer. Bull.* **86**, 163–173.

Boles, J. R. and Coombs, D. S. (1977) Zeolite facies alteration of sandstones in the Southland syncline, New Zealand. *Am. J. Sci.* **277**, 982–1012.

Bowen, N. L. (1940) Progressive metamorphism of siliceous limestone and dolomite. *J. Geol.* **48**, 225–274.

Bowers, T. S., Jackson, K. J. and Helgeson, H. C. (1984) *Equilibrium Activity Diagrams*. Springer-Verlag, Berlin, 397 pp.

Brace, W. F., Ernst, W. G., and Kallberg, R. W. (1970) An experimental study of tectonic overpressure in Franciscan rocks. *Geol. Soc. Amer. Bull.* **81**, 1325–1338.

Brady, J. B. (1974) Coexisting actinolite and hornblende from west-central New Hampshire. *Am. Mineralogist* **59**, 529–535.

Brady, J. B. (1988) The role of volatiles in the thermal history of metamorphic terranes. *J. Petrol.* **29**, 1187–1213.

Brothers, R. N. and Yokoyama, K. (1982) Comparison of the high-pressure schist belts of New Caledonia and Sanbagawa, Japan. *Contr. Mineral. Petrol.* **79**, 219–229.

Brown, E. H. (1971) Phase relations of biotite and stilpnomelane in the greenschist facies. *Contr. Mineral. Petrol.* **31**, 275–299.

Brown, E. H. (1977) Phase equilibria among pumpellyite, lawsonite, epidote and associated minerals in low-grade metamorphic rocks. *Contr. Mineral. Petrol.* **64**, 123–136.

Buerger, M. J. (1948) The role of temperature in mineralogy. *Am. Mineralogist* **33**, 101–121.

Burnell, J. R., Jr. and Rutherford, M. J. (1984) An experimental investigation of the chlorite terminal equilibrium in pelitic rocks. *Am. Mineralogist* **69**, 1015–1024.

Buseck, P. R. and Huang, B.-J. (1985) Conversion of carbonaceous material to graphite during metamorphism. *Geochim. Cosmochim. Acta* **49**, 2003–2016.

Carman, J. H. and Gilbert, M. C. (1983) Experimental studies on glaucophane stability. *Am. J. Sci.* **283-A**, 414–437.

Carmichael, D. M. (1970) Intersecting isograds in the Whetstone Lake area, Ontario. *J. Petrol.* **11**, 147–181.

Carmichael, D. M. (1978) Metamorphic bathozones and bathograds: a measure of the depth of post-metamorphic uplift and erosion on the regional scale. *Am. J. Sci.* **278**,

769–797.

Carpenter, M. A. (1978) Kinetic control of ordering and exsolution in omphacite. *Contr. Mineral. Petrol.* **67**, 17–24.

Carswell, D. A. (ed., 1990) *Eclogite Facies Rocks*. Blackie, Glasgow, 396 pp.

Chamberlain, C. P. and Karabinos, P. (1987) Influence of deformation on pressure-temperature paths of metamorphism. *Geology* **15**, 42–44.

Chamberlain, C. P. and Lyons, J. B. (1983) Pressure, temperature and metamorphic zonation studies of pelitic schists in the Merrimack synclinorium, south-central New Hampshire. *Am. Mineralogist* **68**, 530–540.

Chamberlain, C. P. and Rumble, D. (1988) Thermal anomalies in a regional metamorphic terrane: an isotopic study of the role of fluids. *J. Petrol.* **29**, 1215–1232.

Chatterjee, N. D. (1970) Synthesis and upper stability of paragonite. *Contr. Mineral. Petrol.* **27**, 244–257.

Chatterjee, N. D. (1972) The upper stability limit of the assemblage paragonite + quartz and its natural occurrence. *Contr. Mineral. Petrol.* **34**, 288–303.

Chatterjee, N. D. and Johannes, W. (1974) Thermal stability and standard thermodynamic properties of synthetic $2M_1$-muscovite, $KAl_2[AlSi_3O_{10}(OH)_2]$. *Contr. Mineral. Petrol.* **48**, 89–114.

Chatterjee, N. D., Johannes, W. and Leistner, H. (1984) The system CaO-Al_2O_3-SiO_2-H_2O: new phase equilibria data, some calculated phase relations, and their petrological applications. *Contr. Mineral. Petrol.* **88**, 1–13.

Chinner, G. A. (1960) Pelitic gneisses with varying ferrous/ferric ratios from Glen Clova, Angus, Scotland. *J. Petrol.* **1**, 178–217.

Chinner, G. A. (1961) The origin of sillimanite in Glen Clova, Angus. *J. Petrol.* **2**, 312–323.

Chinner, G. A. (1965) The kyanite isograd in Glen Clova, Angus, Scotland. *Mineral. Mag.* **34**, 132–143.

Chinner, G. A. (1966) The distribution of pressure and temperature during Dalradian metamorphism. *Q. J. Geol. Soc. London* **122**, 159–186.

Chinner, G. A. (1967) Chloritoid, and the isochemical character of Barrow's zones. *J. Petrol.* **8**, 268–282.

Chinner, G. A. (1980) Kyanite isograds of Grampian metamorphism. *J. Geol. Soc. London* **137**, 35–39.

Chinner, G. A. and Heseltine, F. J. (1979) The Grampide andalusite/kyanite isograd. *Scott. J. Geol.* **15**, 117–127.

Cho, M., Maruyama, S., and Liou, J. G. (1987) An experimental investigation of heulandite–laumontite equilibrium at 1000 to 2000 bar P_{fluid}. *Contr. Mineral. Petrol.* **97**, 43–50.

Chopin, C. (1981) Talc-phengite: a widespread assemblage in high-grade pelitic blueschists of the Western Alps. *J. Petrol.* **22**, 628–650.

Chopin, C. (1984) Coesite and pure pyrope in high-grade blueschists of the Western Alps: a first record and some consequence. *Contr. Mineral. Petrol.* **86**, 107–118.

Chopin, C. and Schreyer, W. (1983) Magnesiocarpholite and magnesiochloritoid: two index minerals of pelitic blueschists and their preliminary phase relations in the model system MgO-Al_2O_3-SiO_2-H_2O. *Am. J. Sci.* **283-A**, 72–96.

Clark, S. P., Jr., Robertson, E. C. and Birch, F. (1957) Experimental determination of kyanite-sillimanite equilibrium relations at high temperatures and pressures. *Am. J. Sci.* **255**, 628–640.

Coes, L., Jr. (1953) A new dense crystalline silica. *Science*, Washington, D. C. **118**, 131–132.

Cohen, I. B. (1985) *Revolution in Science*. Harvard Univ. Press, Cambridge, Mass., 711 pp.

Compagnoni, R. (1977) The Sesia-Lanzo zone: high pressure-low temperature metamorphism in the Austroalpine continental margin. *Rend. Soc. Italiana Min. Petr.* **33**, 335–374.

Compagnoni, R., Dal Piaz, G. V., Hunziker, J. C., Gosso, G., Lombardo, B. and Williams, P. F. (1977) The Sesia-Lanzo zone, a slice of continental crust with Alpine high pressure-low temperature assemblages in the Western Italian Alps. *Rend. Soc. Italiana Min. Petr.* **33**, 281–334.

Condie, K. C., Allen, P. and Narayana, B. L. (1982) Geochemistry of the Archean low to high-grade transition zone, southern India. *Contr. Mineral. Petrol.* **81**, 157–167.

Connolly, J. A. D. and Thompson, A. B. (1989) Fluid and enthalpy production during regional metamorphism. *Contr. Mineral. Petrol.* **102**, 347–366.

Coombs, D. S. (1954) The nature and alteration of some Triassic sediments from Southland, New Zealand. *Trans. R. Soc. N. Z.* **82**, 65–109.

Coombs, D. S. (1960) Lower grade mineral facies in New Zealand. *Internatl. Geol. Congr. 21st Sess. (Copenhagen) Rept.*, Part 13, 339–351.

Coombs, D. S. (1961) Some recent work on the lower grade metamorphism. *Australian J. Sci.* **24**, 203–215.

Coombs, D. S. (1971) Present status of the zeolite facies. *Advances in Chemistry Series No.101 (Molecular Sieve Zeolites*, Vol.1), pp.317–327. American Chemical Society.

Coombs, D. S., Ellis, A. J., Fyfe, W. S. and Tayler, A. M. (1959) The zeolite facies, with comments on the interpretation of hydrothermal synthesis. *Geochim. Cosmochim. Acta* **17**, 53–107.

Cooper, A. F. (1972) Progressive metamorphism of metabasic rocks from the Haast Schist Group of southern New Zealand. *J. Petrol.* **13**, 457–592.

Crawford, M. L. (1966) Composition of plagioclase and associated minerals in some schists from Vermont, U.S.A., and South Westland, New Zealand, with inferences about the peristerite solvus. *Contr. Mineral. Petrol.* **13**, 269–294.

Crawford, M. L. and Hollister, L. S. (1986) Metamorphic fluids: the evidence from fluid inclusions. In: J. V. Walther and B. J. Wood (eds.) *Fluid-Rock Interactions during Metamorphism*, pp.36–59. Springer-Verlag, New York.

Cuthbert, S. J. and Carswell, D. A. (1990) Formation and exhumation of medium-temperature eclogites in the Scandinavian Caledonides. In: D. A. Carswell (ed.) *Eclogite Facies Rocks*, pp.180–203. Blackie, Glasgow.

Cuthbert, S. J., Harvey, M. A. and Carswell, D. A. (1983) A tectonic model for the metamorphic evolution of the Basal Gneiss Complex, western South Norway. *J. Metamorphic Geol.* **1**, 63–90.

Daly, R. A. (1914) *Igneous Rocks and their Origin.* McGraw-Hill, New York, 563 pp.

Daly, R. A. (1933) *Igneous Rocks and the Depths of the Earth.* McGraw-Hill, New York, 598 pp.

Danielsson, A. (1950) Das Calcit-Wollastonitgleichgewicht. *Geochim. Cosmochim. Acta* **1**, 55–69.

Dawson, J. B. and Carswell, D. A. (1990) High temperature and ultra-high pressure eclogites. In: D. A. Carswell (ed.) *Eclogite Facies Rocks*, pp.315–349. Blackie, Glasgow.

Dempster, T. J. (1985) Garnet zoning and metamorphism of the Barrovian type area, Scotland. *Contr. Mineral. Petrol.* **89**, 30–38.

de Roever, E. W. F. (1972) Lawsonite-albite facies metamorphism near Fuscaldo, Calabria (southern Italy); its geological significance and petrological aspects. *GUA Papers of Geology,* Ser.1, No.3.

de Roever, W. P. (1950) Preliminary notes on glaucophane-bearing and other crystalline schists from South-East Celebes, and on the origin of glaucophane-bearing rocks. *K. Nederl. Akad. van Wetenschappen (Amsterdam), Proc.* **53**, 1455–1465.

de Roever, W. P. (1955a) Genesis of jadeite by low-grade metamorphism. *Am. J. Sci.*

253, 283–298.

de Roever, W. P. (1955b) Discussion: some remarks concerning the origin of glaucophane in the North Berkeley Hills, California. *Am. J. Sci.* **253**, 240–244.

de Roever, W. P. (1956) Some differences between post-Paleozoic and older regional metamorphism. *Geol. en Mijinbouw* (nw. ser.) **18**, 123–127.

de Roever, W. P., de Roever, E. W. F., Beunk, F. F. and Lahaye, P. H. J. (1967) Preliminary note on ferrocarpholite from a glaucophane- and lawsonite-bearing part of Calabria, southern Italy. *K. Nederl. Akad. van Wetenschappen (Amsterdam) Proc.* B, **70**, 534–537.

de Roever, W. P. and Nijhuis, H. J. (1963) Plurifacial alpine metamorphism in the eastern Betic Cordilleras (SE Spain), with special reference to the genesis of the glaucophane. *Geol. Rundschau* **53**, 324–336.

Dewey, J. F. and Pankhurst, R. J. (1970) The evolution of the Scottish Caledonides in relation to their isotopic age pattern. *Trans. R. Soc. Edinburgh* **68**, 361–389.

De Yoreo, J. J. , Lux, D. R. and Guidotti, C. V. (1991) Thermal modelling in low pressure/high-temperature metamorphic belts. *Tectonophys.* **188**, 209–238.

De Yoreo, J. J., Lux, D. R., Guidotti, C. V., Decker, E. R. and Osberg, P. H. (1989) The Acadian thermal history of western Maine. *J. Metamorphic Geol.* **7**, 169–190.

Dickenson, M. P. (1988) Local and regional differences in the chemical potential of water in amphibolite facies pelitic schists. *J. Metamorphic Geol.* **6**, 365–381.

Dobretsov, N. L. and Sobolev, V. S. (1975) Eclogite-glaucophane schist complexes of the USSR and their bearing on the genesis of blueschist terranes. *Geol. Soc. Amer. Spec. Paper* **151**, 145–155.

Dodge, F. C. (1971) Al_2SiO_5 minerals in rocks of the Sierra Nevada and Inyo Mountains, California. *Am. Mineralogist* **56**, 1443–1451.

Draper, G., Harding, R. R., Horsfield, W. T., Kemp, A. W. and Tresham, A. E. (1976) Low-grade metamorphic belt in Jamaica and its tectonic implications. *Geol. Soc. Amer. Bull.* **87**, 1283–1290.

Droop, G. T. R. (1985) Alpine metamorphism in the south-east Tauern Window, Austria: 1. *P-T* variations in space and time. *J. Metamorphic Geol.* **3**, 371–402.

Eade, K. E. and Fahrig, W. F. (1971) Geochemical evolutionary trends of continental plates: a preliminary study of the Canadian shield. *Geol. Surv. Canada Bull.* **179.**

Edwards, R. L. and Essene, E. J. (1988) Pressure, temperature and C-O-H fluid fugacities across the amphibolite-granulite transition, northwest Adirondack Mountains, New York. *J. Petrol.* **29**, 39–72.

Eitel, W. (1958) Structural conversions in crystalline systems and their importance for geological problems. *Geol. Soc. Amer. Spec. Paper* **66**. 183 pp.

Ellis, D. J. (1980) Osumilite-saphirine-quartz granulites from Enderby Land, Antarctica: *P-T* conditions of metamorphism, implications for garnet-cordierite equilibria and the evolution of the deep crust. *Contr. Mineral. Petrol.* **74**, 201–210.

Ellis, D. J. and Green, D. H. (1979) A experimental study of the effect of Ca upon garnet-clinopyroxene exchange equilibria. *Contr. Mineral. Petrol.* **71**, 13–22.

Ellis, D. J., Sheraton, J. W., England, R. N. and Dallwitz, W. B. (1980) Osumilite-sapphirine-quartz granulites from Enderby Land Antarctica: mineral assemblages and reactions. *Contr. Mineral. Petrol.* **72**, 123–143.

Enami, M. (1983) Petrology of pelitic schists in the oligoclase-biotite zone of the Sanbagawa metamorphic terrain, Japan: phase equilibria in the highest grade zone of a high-pressure intermediate type of metamorphic belt. *J. Metamorphic Geol.* **1**, 141–161.

Engel, A. E. J. and Engel, C. G. (1962) Progressive metamorphism of amphibolite, northwest Adirondack Mountains, New York. In: A. E. J. Engel et al. (eds.) *Petrologic Studies (Buddington Volume)*, pp.37–82. Geol. Soc. America.

England, P. C. (1978) Some thermal considerations of the Alpine metamorphism - past, present and future. *Tectonophys.* **46**, 21–40.

England, P. C. and Richardson, S. W. (1977) The influence of erosion upon the mineral facies of rocks from different metamorphic environments. *J. Geol. Soc. London* **134**, 201–213.

England, P. C. and Thompson, A. B. (1984) Pressure-temperature-time paths of regional metamorphism, I. *J. Petrol.* **25**, 894–928.

Ernst, W. G. (1963) Significance of phengitic micas from low-grade schists. *Am. Mineralogist* **48**, 1357–1373.

Ernst, W. G. (1970) Tectonic contact between the Franciscan melange and the Great Valley Sequence - crustal expression of a late Mesozoic Benioff zone. *J. Geophys. Res.* **75**, 886–901.

Ernst, W. G. (1972a) CO_2-poor composition of the fluid attending Franciscan and Sanbagawa low-grade metamorphism. *Geochim. Cosmochim. Acta* **36**, 497–504.

Ernst, W. G. (1972b) Occurrence and mineralogic evolution of blueschist belts with time. *Am. J. Sci.* **272**, 657–668.

Ernst, W. G. (1975) Systematics of large-scale tectonics and age progressions in Alpine and circum-Pacific blueschist belts. *Tectonophys.* **26**, 229–246.

Ernst, W. G. (1977) Mineral parageneses and plate tectonic settings of relatively high-pressure metamorphic belts. *Fortschr. Miner.* **54**, 192–222.

Ernst, W. G. (1983) Phanerozoic continental accretion and the metamorphic evolution of northern and central California. *Tectonophys.* **100**, 287–320.

Ernst, W. G., Seki, Y., Onuki, H. and Gilbert, M. C. (1970) Comparative study of low-grade metamorphism in the California Coast Ranges and the outer metamorphic belt of Japan. *Geol. Soc. Amer. Mem.* **124**.

Eskola, P. (1914) On the petrology of the Orijärvi region in southwestern Finland. *Bull. Comm. geol. Finlande*, No.40.

Eskola, P. (1915) On the relations between the chemical and mineralogical composition in the metamorphic rocks of the Orijärvi region. *Bull. Comm. geol. Finlande*, No.44.

Eskola, P. (1920) The mineral facies of rocks. *Norsk Geol. Tidsskr.* **6**, 143–194.

Eskola, P. (1921) On the eclogites of Norway. *Videnskaps Skrifter I. Mat.-Naturv. Kl.* **1921**, No.8.

Eskola, P. (1929) Om mineral facies. *Geol. Foren. Stockholm Forh.* **51**, 157–173.

Eskola, P. (1939) Die metamorphen Gesteine. In: Tom. F. W. Barth, C. W. Correns and P. Eskola (1939) *Die Entstehung der Gesteine*, pp.263–407. Julius Springer, Berlin. (Reprinted in 1960 and 1970).

Essene, E. J. (1982) Geologic thermometry and barometry. In: J. M. Ferry (ed.) *Characterization of Metamorphism through Mineral Equilibria* (Reviews in Mineralogy, Vol. 10), pp. 153–206. Mineral. Soc. America.

Essene, E. J. (1989) The current status of thermobarometry in metamorphic rocks. In: J. S. Daly et al. (eds.) *Evolution of Metamorphic Belts* (Geol. Soc. London, Spec. Publ. No.43), pp.1–44.

Essene, E. J., Hensen, B. J. and Green, D. H. (1970) Experimental study of amphibolite and eclogite stability. *Phys. Earth Planet. Interiors* **3**, 378–384.

Eugster, H. P. (1959) Reduction and oxidation in metamorphism. In: P. H. Abelson (ed.) *Researches in Geochemistry*, pp.397–426. John Wiley, New York.

Eugster, H. P. and Skippen, G. B. (1967) Igneous and metamorphic reactions involving gas equilibria. In: P. H. Abelson (ed.) *Researches in Geochemistry*, Vol.2, pp. 492–520. John Wiley, New York.

Eugster, H. P. and Wones, D. R. (1962) Stability relations of the ferruginous biotite, annite. *J. Petrol.* **3**, 82–125.

Evans, B. W. (1964) Coexisting albite and oligoclase in some schists from New Zealand. *Am. Mineralogist* **49**, 173–179.

Evans, B. W. and Guidotti, C. V. (1966) The sillimanite-potash feldspar isograd in western Maine, U.S.A. *Contr. Mineral. Petrol.* **12**, 25–62.

Ferry, J. M. (1976a) Metamorphism of calcareous sediments in the Waterville-Vassalboro area, south-central Maine: mineral reactions and graphical analysis. *Am. J. Sci.* **276**, 841–882.

Ferry, J. M. (1976b) P, T, fCO_2, and fH_2O during metamorphism of calcareous sediments in the Waterville-Vassalboro area, south-central Maine. *Contr. Mineral. Petrol.* **57**, 119–143.

Ferry, J. M. (1980) A case study of the amount and distribution of heat and fluid during metamorphism. *Contr. Mineral. Petrol.* **71**, 373–385.

Ferry, J. M. (ed.), (1982) *Characterization of Metamorphism through Mineral Equilibria* (Reviews in Mineralogy, Vol. 10). Mineral. Soc. America. 397 pp.

Ferry, J. M. (1983a) Regional metamorphism of the Vassalboro Formation, south-central Maine, USA: a case study of the role of fluid in metamorphic petrogenesis. *J. Geol. Soc. London* **140**, 551–576.

Ferry, J. M. (1983b) Application of the reaction progress variable in metamorphic petrology. *J. Petrol.* **24**, 343–376.

Ferry, J. M. (1984) A biotite isograd in south-central Maine, U.S.A.: mineral reactions, fluid transfer, and heat transfer. *J. Petrol.* **25**, 871–893.

Ferry, J. M. (1986) Infiltration of aqueous fluid and high fluid: rock ratios during greenschist facies metamorphism: a reply. *J. Petrol.* **27**, 695–714.

Ferry, J. M. (1987) Metamorphic hydrology at 13-km depth and 400–550°C. *Am. Mineralogist* **72**, 39–58.

Ferry, J. M. (1988) Infiltration-driven metamorphism in northern New England, USA. *J. Petrol.* **29**, 1121–1159.

Ferry, J. M. and Spear, F. S. (1978) Experimental calibration of the partitioning of Fe and Mg between biotite and garnet. *Contr. Mineral. Petrol.* **66**, 113–117.

Fettes, D. J. (1979) The metamorphic map of the British and Irish Caledonides. In: A. L. Harris et al. (eds.) *The Caledonides of the British Isles – Reviewed* (Geol. Soc. London, Spec. Publ. No.8), pp.307–321. Scottish Academic Press, Edinburgh.

Fowler, C. M. R. and Nisbet, E. G. (1988) Geotherm in the continental crust and metamorphism. In: E. G. Nisbet and C. M. R. Fowler (eds.) *Heat, Metamorphism and Tectonics*. (Short Course Handbook, Vol.14), pp. 34–50. Mineral. Assoc. Canada.

Fraser, D. G. (ed., 1977) *Thermodynamics in Geology*. D. Reidel Publ. Dordrecht, 410pp.

French, B M. (1966) Some geological implications of equilibrium between graphite and a C-H-O gas phase at high temperatures and pressures. *Reviews Geophys.* **4**, 223–253.

Frey, M., Bucher, K., Frank, E. and Millis, J. (1980) Alpine metamorphism along the geotraverse Basel–Chiasso – a review. *Eclogae geol. Helv.* **73**, 527–546.

Frey, M., Hunziker, J. C., Frank, W., Bocquet, J., Dal Piaz, G. V., Jaeger, E. and Niggli, E. (1974) Alpine metamorphism of the Alps. A review. *Schweiz. Mineral. Petrogr. Mitt.* **54**, 247–290.

Froese, E. (1976) Application of thermodynamics in metamorphic petrology. *Geol. Surv. Canada Paper* 75–43.

Froese, E. (1977) Oxidation and sulphidation reactions. In: H. J. Greenwood (ed.) *Short Course in Application of Thermodynamics to Petrology and Ore Deposits*, pp. 84–98. Min. Assoc. Canada.

Froese, E. (1978) The graphical representation of mineral assemblages in biotite-bearing granulites. *Geol. Surv. Canada Paper* 78–1A, 323–325.

Froese, E. (1981) Application of thermodynamics in the study of mineral deposits. *Geol. Surv. Canada Paper* 80–28.

Frost, B. R. (1979) Mineral equilibria involving mixed-volatiles in a C-O-H fluid phase:

the stabilities of graphite and siderite. *Am. J. Sci.* **279**, 1033–1059.

Frost, B. R. and Chacko, T. (1989) The granulite uncertainty principle: limitation on thermobarometry in granulites. *J. Geol.* **97**, 435–450.

Fry, N. and Fyfe, W. S. (1969) Eclogites and water pressure. *Contr. Mineral. Petrol.* **24**, 1–6.

Fyfe, W. S., Price, N. J. and Thompson, A. B. (1978) *Fluids in the Earth's Crust.* Elsevier, Amsterdam, 383 pp.

Fyfe, W. S., Turner, F. J. and Verhoogen, J. (1958) Metamorphic reactions and metamorphic facies. *Geol. Soc. Amer. Mem.* **73**.

Ganguly, J. and Saxena, S. K. (1984) Mixing properties of aluminosilicate garnets: constraints from natural and experimental data, and applications to geothermometry. *Am. Mineralogist* **69**, 88–97.

Gasparik, T. (1985) Experimentally determined compositions of diopside-jadeite pyroxene in equilibrium with albite and quartz at 1200–1350 °C and 15–34 kbar. *Geochim. Cosmochim. Acta* **49**, 865–870.

Ghent, E. D. (1976) Plagioclase-garnet-Al_2SiO_5-quartz: a potential geobarometer-geothermometer. *Am. Mineralogist* **61**, 710–714.

Ghent, E. D. (1988) A review of chemical zoning in eclogite garnets. In: D. C. Smith (ed.) *Eclogites and Eclogite-facies Rocks*, pp.207–236. Elsevier, Amsterdam.

Ghent, E. D. and Stout, M. Z. (1988) Determination of metamorphic pressure-temperature-time (*P-T-t*) paths. In: E. G. Nisbet and C. M. R. Fowler (eds) *Heat, Metamorphism and Tectonics* (Short Course Handbook, Vol.14), chapter 6. Mineral. Assoc. Canada.

Gillet, Ph., Choukroune, P., Balle'vre, M. and Davy, Ph. (1986) Thickening history of the Western Alps. *Earth Planet. Sci. Letters* **78**, 44–52.

Glassley, W. E. (1983) The role of CO_2 in the chemical modification of deep continental crust. *Geochim. Cosmochim. Acta* **47**, 597–616.

Goffé, B. and Chopin, C. (1986) High-pressure metamorphism in the Western Alps: zoneography of metapelites, chronology and consequences. *Schweiz. Mineral. Petrogr. Mitt.* **66**, 41–52.

Goldschmidt, V. M. (1911) Die Kontaktmetamorphose im Kristianiagebiet. *Vidensk. Skrifter. I. Mat.-Naturv. Kl.* 1911, No.11.

Goldschmidt, V. M. (1912a) Die Gesetze der Gesteinsmetamorphose, mit Beispielen aus der Geologie des suedlichen Norwegens. *Vidensk. Skrifter. I. Mat.-Naturv. Kl.* 1912, No.22.

Goldschmidt, V. M. (1912b) Ueber die Anwendung der Phasnregel auf die Gesetze der Mineralassoziation. *Centralblatt Mineral. Geol. Palaeont.* (1912) 574–576.

Goldschmidt, V. M. (1915) Geologisch-petrographische Studien im Hochgebirge des suedlichen Norwegens. III. Die Kalksilikatgneise und Kalksilikatglimmerschiefer im Trondhjem-Gebiete. *Vidensk. Skrifter. I. Mat.-Naturv. Kl.* 1915, No.10.

Gonzalez-Bonorino, F. (1971) Metamorphism of the crystalline basement of Central Chile. *J. Petrol.* **12**, 149–175.

Graf, D. F. and Goldsmith, J. R. (1955) Dolomite-magnesian calcite relations at elevated temperatures and CO_2 pressures. *Geochim. Cosmochim. Acta* **7**, 109–128.

Graham, C. M. and England, P. C. (1976) Thermal regimes and regional metamorphism in the vicinity of overthrust faults. *Earth Planet. Sci. Letters* **31**, 142–152.

Graham, C. M., Greig, K. M., Sheppard, S. M. F. and Turi, B. (1983) Genesis and mobility of the H_2O-CO_2 fluid phase during regional greenschist and epidote amphibolite facies metamorphism: a petrological and stable isotope study in the Scottish Dalradian. *J. Geol. Soc. London* **140**, 577–599.

Grambling, J. A. (1983) Reversal in Fe-Mg partitioning between chloritoid and staurolite. *Am. Mineralogist* **68**, 373–388.

Grambling, J. A. (1986) A regional gradient in the composition of metamorphic fluids in pelitic schists, Pecos Baldy, New Mexico. *Contr. Mineral. Petrol.* **94**, 149–164.

Grambling, J. A. (1990) Internally-consistent geothermometry and H_2O barometry in metamorphic rocks: the example garnet-chlorite-quartz. *Contr. Mineral. Petrol.* **105**, 617–628.

Grambling, J. A. and Williams, M. L. (1985) The effect of Fe^{3+} and Mn^{3+} on aluminum silicate phase relations in north-central New Mexico, U.S.A. *J. Petrol.* **26**, 324–354.

Grapes, R. H. and Graham, C. M. (1978) The actinolite-hornblende series in metabasites and the so-called miscibility gap: a review. *Lithos* **11**, 85–97.

Green, D. H. and Ringwood, A. E. (1967a) An experimental investigation of the gabbro to eclogite transformation and its petrological application. *Geochim. Cosmochim. Acta* **31**, 767–833.

Green, D. H. and Ringwood, A. E. (1967b) The genesis of basaltic magmas. *Contr. Mineral. Petrol.* **15**, 103–190.

Green, T. E. (1982) Anatexis of mafic crust and high pressure crystallization of andesite. In: R. S. Thorpe (ed.) *Andesites*, pp.465–488. John Wiley, New York.

Greenwood, H. J. (1967) Mineral equilibria in the system MgO-SiO_2-H_2O-CO_2. In: P. H. Abelson (ed.) *Researches in Geochemistry*, Vol.2, pp. 542–567. John Wiley, New York.

Greenwood, H. J. (1975) Buffering of pore fluids by metamorphic reactions. *Am. J. Sci.* **275**, 573–593.

Greenwood, H. J. (1976) Metamorphism at moderate temperatures and pressures. In: D. K. Bailey and R. Macdonald (eds.) *The Evolution of the Crystalline Rocks*, pp. 187–259. Academic Press, London.

Greenwood, H. J. (ed.) (1977) *Short Course in Application of Thermodynamics to Petrology and Ore Deposits*. Mineral. Assoc. Canada, 231 pp.

Grew, E. S. (1982) Osumilite in the sapphirine-quartz terrane of Erderby Land, Antarctica: implications for osumilite petrogenesis in the granulite facies. *Am. Mineralogist* **67**, 762–787.

Griffin, W. L. (1987) 'On the eclogite of Norway'--65 year later. *Mineral. Mag.* **51**, 333–343.

Grove, T. L., Ferry, J. M. and Spear, F. S. (1983) Phase transitions and decomposition relations in calcic plagioclase. *Am. Mineralogist* **68**, 41–59.

Grover, J. E. and Orville, P. M. (1969) The partitioning of cations between coexisting single and multi-site phases with application to the assemblages: orthopyroxene-clinopyroxene and orthopyroxene-olivine. *Geochim. Cosmochim. Acta* **33**, 205–226.

Grubenmann, U. (1904–6) *Die kristallinen Schiefer*. 1st ed. Gebrueder Borntraeger, Berlin.

Grubenmann, U. and Niggli, P. (1924) *Die Gesteinsmetamorphose*. I. Gebrueder Borntraeger, Berlin.

Guidotti, C. V. (1970) The mineralogy and petrology of the transition from the lower to upper sillimanite zone in the Oquossoc area, Maine. *J. Petrol.* **11**, 277–336.

Guidotti, C. V. (1974) Transition from staurolite to sillimanite zone, Rangeley quadrangle, Maine. *Geol. Soc. Amer. Bull.* **85**, 475–490.

Guiraud, M., Holland, Tim and Powell, R. (1990) Calculated mineral equilibria in the greenschist-blueschist-eclogite facies in Na_2O-FeO-MgO-Al_2O_3-SiO_2-H_2O. Methods, results and geologic applications. *Contr. Mineral. Petrol.* **104**, 85–98.

Hallimond, A. F. (1943) On the graphical representation of the calciferous amphiboles. *Am. Mineralogist* **28**, 65–89.

Hamilton, W. (1969) Mesozoic California and the underflow of Pacific mantle. *Geol. Soc. Amer. Bull.* **80**, 2409–2430.

Hamilton, W. and Myers, W. B. (1967) The nature of batholiths. *U. S. Geol. Surv.*

Prof. Paper 554-C.

Hanson, R. B. (1992) Effects of fluid production on fluid flow during regional and contact metamorphism. *J. Metamorphic Geol.* **10**, 87–97.

Hanson, R. B. and Barton, M. D. (1989) Thermal development of low-pressure metamorphic belts: results from two dimensional numerical models. *J. Geophys. Res.* **94**, 10363–10377.

Hara, I., Shiota, T, Hida, K., Okamoto, K., Takeda, K., Hayasaka, Y. and Sakura, Y. (1990) Nappe structure of the Sanbagawa belt. *J. Metamorphic Geol.* **8**, 441–456.

Harker, A. (1909) *The Natural History of Igneous Rocks*. Macmillan, London, 384 pp. (Reprinted in 1965 by Hafner, New York.)

Harker, A. (1918) The present position and outlook of the study of metamorphism in rock masses (presidential address). *Q. J. Geol. Soc. London* **74**, li-lxxx.

Harker, A. (1932, 1939) *Metamorphism. A Study of the Transformations of Rock-Masses*. 1st and 2nd edns. Methuen, London, 360 and 362 pp.

Harley, S. L. (1989) The origin of granulites: a metamorphic perspective. *Geol. Mag.* **126**, 215–247.

Harte, B. and Dempster, T. J. (1987) Regional metamorphic zones: tectonic controls. *Phil. Trans. R. Soc. London* **A321**, 105–127.

Harte, B. and Hudson, N. F. C. (1979) Pelitic facies series and the temperatures and pressures of Dalradian metamorphism in E Scotland. In: A. J. Harris et al. (eds.) *The Caledonides of the British Isles - Reviewed* (Geol. Soc. London, Spec. Publ. No.8), pp.323–337. Scottish Academic Press, Edinburgh.

Harte, B. and Johnson, M. R. W. (1969) Metamorphic history of Dalradian rocks in Glens Clova, Esk and Lethnot, Angus, Scotland. *Scott. J. Geol.* **5**, 54–80.

Hashimoto, M. (1966) On the prehnite-pumpellyite metagreywacke facies (in Japanese with English abstract). *J. Geol. Soc. Japan* **72**, 253–265.

Hashimoto, M. (1991) Ferric-ferrous ratios of glaucophane schists and actinolite greenschists of the Sangun and Sanbagawa belts, Japan (in Japanese). *J. Jap. Assoc. Mineral. Petrol. Econ. Geol.* **86**, 45–48.

Hashimoto, M., Tagiri, M., Kusakabe, K., Masuda, K. and Yano, T. (1992) Geologic structure formed by tectonic stacking of sliced layers in the Sambagawa metamorphic terrain, Kodama-Nagatoro area, Kanto Mountains (in Japanese with English abstract) *J. Geol. Soc. Japan* **98**, 953–965.

Hawkesworth, C. J., Waters, D. J. and Bickle, M. J. (1975) Plate tectonics in the Eastern Alps. *Earth Planet. Sci. Letters* **24**, 405–413.

Hay, R. S. and Evans, B. (1988) Intergranular distribution of pore fluid and the nature of high-angle grain boundaries in limestone and marble. *J. Geophys. Res.* **93**, 8959–8974.

Heier, K. S. (1957) Phase relations of potash feldspar in metamorphism. *J. Geol.* **65**, 468–479.

Heier, K. S. (1960) Petrology and geochemistry of high-grade metamorphic and igneous rocks on Langoey, northern Norway. *Norges Geol. Unders.* No.207.

Heinrich, C. A. (1986) Eclogite facies regional metamorphism of hydrous mafic rocks in the Cetral Alpine Adula nappe. *J. Petrol.* **27**, 123–154.

Helgeson, H. C., Delany, J. M., Nesbitt, H. W. and Bird, D. K. (1978) Summary and critique of the thermodynamic properties of rock-forming minerals. *Am. J. Sci.* **278-A**, 229 pp.

Helms, T. S. and Labotka, T. C. (1991) Petrogenesis of Early Proterozoic pelitic schists of the southern Black Hills, South Dakota: Constraints on regional low-pressure metamorphism. *Geol. Soc. Amer. Bull.* **103**, 1324–1334.

Hemingway, B. S., Bobie, R. A., Evans, H. T. Jr., and Kerrick, D. M. (1991) Heat capacities and entropies of sillimanite, fibrolite, andalusite, kyanite, and quartz and the Al_2SiO_5 phase diagram. *Am. Mineralogist* **76**, 1597–1613.

Hensen, B. J. and Green, D. H. (1973) Experimental study of the stability of cordierite and garnet in pelitic compositions at high pressures and temperatures, part 3. *Contr. Mineral. Petrol.* **38**, 151–166.

Hess, P. C. (1969) The metamorphic paragenesis of cordierite in pelitic rocks. *Contr. Mineral. Petrol.* **24**, 191–207.

Hess, P. C. (1971) Prograde and retrograde equilibria in garnet-cordierite gneisses in south-central Massachusetts. *Contr. Mineral. Petrol.* **30**, 177–195.

Hewitt, D. A. (1973) The metamorphism of micaceous limestones from south-central Connecticut. *Am. J. Sci.* **273-A**, 444–469.

Higashino, T. (1990) The higher grade metamorphic zonation of the Sambagawa metamorphic belt in central Shikoku, Japan. *J. Metamorphic Geol.* **8**, 413–423.

Higashino, T., Sakai, C., Otsuki, M., Itaya, T. and Banno, S. (1981) Electron microprobe analyses of rock-forming minerals from the Sanbagawa metamorphic rocks, Shikoku, Part I. *Kanazawa Univ. Science Reports* **26**, 73–123.

Hlabse, T. and Kleppa, O. J. (1968) The thermochemistry of jadeite. *Am. Mineralogist* **53**, 1281–1292.

Hochella, M. F., Jr. and White, A. F. (eds.) (1990) *Mineral-Water Interface Geochemistry*. (Reviews in Mineralogy, Vol.23). Mineral. Soc. America, 603 pp.

Hodges, K. V. and Spear, F. S. (1982) Geothermometry, geobarometry and the Al_2SiO_5 triple point at Mt. Moosilauke, New Hampshire. *Am. Mineralogist* **67**, 1118–1134.

Hoisch, T. D. (1990) Empirical calibration of six geobarometers for the mineral assemblage quartz + muscovite + biotite + plagioclase + garnet. *Contr. Mineral. Petrol.* **104**, 225-234.

Holdaway, M. J. (1971) Stability of andalusite and aluminum silicate phase diagram. *Am. J. Sci.* **271**, 97–131.

Holdaway, M. J., Dutrow, B. L. and Hinton, R. W. (1988) Devonian and Carboniferous metamorphism in west-central Maine: the muscovite-almandine geobarometer and the staurolite problem revisited. *Am. Mineralogist* **73**, 20–47.

Holdaway, M. J., Guidotti, C. V., Novak, J. M. and Henry, W. E. (1982) Polymetamorphism in medium- to high-grade pelitic metamorphic rocks, west-central Maine. *Geol. Soc. Amer. Bull.* **93**, 572–584.

Holdaway, M. J. and Lee, S. M. (1977) Fe-Mg cordierite stability in high-grade pelitic rocks based on experimental, theoretical, and natural observations. *Contr. Mineral. Petrol.* **63**, 175–198.

Holland, T. J. B. (1979a) Experimental determination of the reaction paragonite = jadeite + kyanite + H_2O, and internally consistent thermodynamic data for part of the system Na_2O-Al_2O_3-SiO_2-H_2O, with applications to eclogites and blueschists. *Contr. Mineral. Petrol.* **68**, 293–301.

Holland, T. J. B. (1979b) High water activities in the generation of high pressure kyanite eclogites of the Tauern Window, Austria. *J. Geol.* **87**, 1–27.

Holland, T. J. B. (1980) The reaction albite = jadeite + quartz determined experimentally in the range 600–1200°C. *Am. Mineralogist* **65**, 129–134.

Holland, T. J. B. and Powell, R. (1985) An internally consistent thermodynamic dataset with uncertainties and corrections: 2. Data and results. *J. Metamorphic Geol.* **3**, 343–370.

Holland, T. J. B. and Powell, R. (1990) An enlarged and updated internally consistent thermodynamic dataset with uncertainties and corrections: the system K_2O-Na_2O-CaO-MgO-MnO-FeO-Fe_2O_3-Al_2O_3-TiO_2-SiO_2-C-H_2-O_2. *J. Metamorphic Geol.* **8**, 89–124.

Holland, T. J. B. and Richardson, S. W. (1979) Amphibole zonation in metabasites as a guide to the evolution of metamorphic conditions. *Contr. Mineral. Petrol.* **70**, 143–148.

Hollister, L. S. (1966) Garnet zoning: an interpretation based on the Rayleigh fractionation model. *Science* **154**, 1147–1150.

Hollister, L. S. and Crawford, M. L. (eds.) (1981) *Short Course in Fluid Inclusions: Applications to Petrology*. Mineral. Assoc. Canada, 305 pp.

Huang, W. L. and Wyllie, P. J. (1973) Melting relations of muscovite-granite to 35kbar as a model for fusion of metamorphosed subducted ocenic sediments. *Contr. Mineral. Petrol.* **42**, 1–14.

Huang, W. L. and Wyllie, P. J. (1974) Melting relations of muscovite with quartz and sanidine in the K_2O-Al_2O_3-SiO_2-H_2O system to 30 kilobars and an outline of paragonite melting relations. *Am. J. Sci.* **274**, 378-395.

Indares, A. and Martignole, J. (1985) Biotite-garnet geothermometry in the granulite facies : the influence of Ti and Al in biotite. *Am. Mineralogist* **70**, 272–278.

Irifune, T., Seki, T, Ringwood, A. E. and Hibberson, W. O. (1986) The eclogite-garnetite transformation at high pressure and some geophysical implications. *Earth Planet. Sci. Letters* **77**, 245–256.

Isozaki, Y. and Maruyama, S. (1991) Studies on orogeny based on plate tectonics in Japan and new geotectonic sudivision of the Japanese Islands (in Japanese with English abstract) *Chigaku-Zasshi* **100**, 697–761.

Isozaki, Y. and Maruyama, S. (1992) New geotectonic subdivision of the Francsican-Klamath region redefined by formation age of accretionary complexes. *Amer. Assoc. Petrol. Geol. Bull* **76**, 423.

Itaya, T. (1981) Carbonaceous material in pelitic schists of the Sanbagawa metamorphic belt in central Shikoku, Japan. *Lithos* **14**, 215–224.

Jacobs, G. K. and Kerrick, D. M. (1981) Devolatilization equilibria in H_2O-CO_2 and H_2O-CO_2-NaCl fluids: an experimental and thermodynamic evaluation at elevated pressures and temperatures. *Am. Mineralogist* **66**, 1135–1153.

Jaeger, J. C. (1968) Cooling and solidification of igneous rocks. In: H. H. Hess and A. Poldervaart (eds.) *Basalts*, Vol.2, pp.503–536. Interscience, New York.

Jaffe, H. W., Robinson, P., Tracy, R. J. and Ross, M. (1975) Orientation of pigeonite exsolution lamellae in metamorphic augite: correlation with composition and calculated optimal phase boundary. *Am. Mineralogist* **60**, 9–28.

James, H. L. and Howland, A. L. (1955) Mineral facies in iron and silica rich rocks (abstr.). *Geol. Soc. Amer. Bull.* **66**, 1580–1581.

Janardhan, A. S., Newton, R. C. and Hansen, E. C. (1982) The transformation of amphibolite facies gneiss to charnockite in southern Karnataka and northern Tamil Nadu, India. *Contr. Mineral. Petrol.* **79**, 130–149.

Johannes, W. and Puhan, D. (1971) The calcite-aragonite transition, reinvestigated. *Contr. Mineral. Petrol.* **31**, 28–38.

Jolly, W. T. (1974) Regional metamorphic zonation as an aid in study of Archean terrains: Abitibi region, Ontario. *Canad. Mineralogist* **12**, 499–508.

Kars, H., Jansen, B. H., Tobi, A. C. and Poorter, R. P. E. (1980) Metapelitic rocks of the polymetamorphic Precambrian of Rogaland, SW Norway. II. *Contr. Mineral. Petrol.* **74**, 235–244.

Kawachi, Y. (1974) Geology and petrochemistry of weakly metamorphosed rocks in the Upper Wakatipu district, southern New Zealand. *New Zealand J. Geol. Geophys.* **17**, 169–208.

Kennedy, W. Q. (1948) On the significance of thermal structure in the Scottish Highlands. *Geol. Mag.* **85**, 229–234.

Kerrick, D. M. (1968) Experiments on the upper stability limit of pyrophyllite at 1.8 and 3.9 kb water pressure. *Am. J. Sci.* **266**, 204–214.

Kerrick, D. M. (1972) Experimental determination of muscovite + quartz stability with $P\,H_2O < P$ total. *Am. J. Sci.* **272**, 946–958.

Kerrick, D. M. (1974) Review of metamorphic mixed-volatile (H_2O-CO_2) equilibria. *Am. Mineralogist* **59**, 729–762.

Kerrick, D. M. (1990) *The Al_2SiO_5 Polymorphs* (Reviews in Mineralogy, Vol. 22), Mineral. Soc. America, 406 pp.

Kerrick, D. M. and Speer, J. A. (1988) The role of minor element solid solution on the andalusite-sillimanite equilibrium in metapelites and peraluminous granitoids. *Am. J. Sci.* **288**, 152–192.

Kienast, J. R., Lombardo, B., Biino, G. and Pinardon, J. L. (1991) Petrology of very-high-pressure eclogitic rocks from the Brossasco-Isasca complex, Dora-Maria massif, Italian Western Alps. *J. Metamorphic Geol.* **9**, 19–34.

Klein, C., Jr. (1969) Two-amphibole assemblages in the system actinolite-hornblende-glaucophane. *Am. Mineralogist* **54**, 212–237.

Koehler, A. (1941) Die Abhängigkeit der Plagioklasoptik vom vorangegangenen Wärmeverhalten. *Min. Petrogr. Mitt.* **53**, 24–49.

Koehler, A. (1949) Recent results of investigations on the feldspars. *J. Geol.* **57**, 592–599.

Kohn, M. J. and Spear, F. S. (1990) Two new geobarometers from garnet amphibolites, with applications to southeastern Vermont. *Am. Mineralogist* **75**, 89–96.

Kojima, G. (1952) *Henseigan* (in Japanese). Minka-Chidanken, Tokyo.

Koons, P. O. and Thompson, A. B. (1985) Non-mafic rocks in the greenschist, blueschist and eclogite facies. *Chem. Geol.* **50**, 3–30.

Korsman, K. (1977) Progressive metamorphism of the metapelites in the Rantasalmi-Sulkava area, southeastern Finland. *Geol. Surv. Finland Bull.* **290**.

Korzhinskii, D. S. (1936) Mobility and inertness of components in metasomatosis. *Acad. Nauk. SSSR. Bull.*, Ser. Geol. 1936, No.1, 35–60.

Korzhinskii, D. S. (1950) Phase rule and geochemical mobility of elements. *Internatl. Geol. Congr. 18th Sess. (London, 1948) Rept.*, Part 2, 50–65.

Korzhinskii, D. S. (1959) *Physicochemical Basis of the Analysis of the Paragenesis of Minerals*. Consultants Bureau, New York. 142 pp.

Korzhinskii, D. S. (1965) The theory of systems with perfectly mobile components and processes of mineral formation. *Am. J. Sci.* **263**, 193–205.

Koziol, A. M. and Newton, R. C. (1988) Redetermination of the anorthite breakdown reaction and improvement of the plagioclase-garnet-Al_2SiO_5-quartz geobarometer. *Am. Mineralogist* **73**, 216–223.

Kretz, R. (1981) Site-occupancy interpretation of the distribution of Mg and Fe between orthopyroxene and clinopyroxene in metamorphic rocks. *Canad. Mineralogist* **19**, 483–500.

Kretz, R. (1983) Symbols for rock-forming minerals. *Am. Mineralogist* **68**, 277–279.

Kreulen, R. (1988) High integrated fluid/rock ratios during metamorphism at Naxos. *Contr. Mineral. Petrol.* **98**, 28–32.

Krogh, E. J. (1977) Evidence of Precambrian continent-continent collision in western Norway. *Nature,* London **267**, 17–19.

Kuhn, T. S. (1962, 1970) *The Structure of Scientific Revolutions*. 1st and 2nd edns. Univ. of Chicago Press, Chicago.

Laird, J. (1980) Phase equilibria in mafic schist from Vermont. *J. Petrol.* **21**, 1–37.

Laird, J. and Albee, A. L. (1981a) High-pressure metamorphism in mafic schist from northern Vermont. *Am. J. Sci.* **281**, 97–126.

Laird, J. and Albee, A. L. (1981b) Pressure, temperature and time indicators in mafic schist: their application to reconstructing the polymetamorphic history of Vermont. *Am. J. Sci.* **281**, 127–175.

Lamb, W. M., Brown, P. E. and Valley, J. W. (1991) Fluid inclusions in Adirondack granulites: implications for the retrograde *P-T* path. *Contr. Mineral. Petrol.* **107**,

472–483.

Lamb, W. M. and Valley, J. W. (1984) Metamorphism of reduced granulites in low-CO_2 vapor-free environment. *Nature,* London **321**, 56–58.

Lamb, W. M. and Valley, J. W. (1988) Granulite facies amphibole and biotite equilibria, and calculated peak-metamorphic water activities. *Contr. Mineral. Petrol.* **100**, 349–360.

Lambert, R. St. J. (1959) The mineralogy and metamorphism of the Moine schists of the Morar and Knoydart districts of Inverness-shire. *Trans. R. Soc. Edinburgh* **63**, 553–588.

Lambert, R. St. J. and McKerrow, W. S. (1976) The Grampian orogeny. *Scott. J. Geol.* **12**, 271–292.

Landis, C. A. and Coombs, D. S. (1967) Metamorphic belts and orogenesis in southern New Zealand. *Tectonophys.* **4**, 501–517.

Lattard, D. and Schreyer, W. (1981) Experimental results bearing on the stability of the blueschist-facies minerals deerite, howieite, and zussmanite, and their petrological significance. *Bull. Soc. francaise Minéral. Cristal.* **104**, 431–440.

Leake, B. E. (1978) Nomenclature of amphiboles. *Am. Mineralogist* **63**, 1023–1052.

Le Breton, N. and Thompson, A. B. (1988) Fluid-absent (dehydration) melting of biotite in metapelites in the early stages of crustal anatexis. *Contr. Mineral. Petrol.* **99**, 226–237.

Lee, S. M. (ed.) (1984) *Metamorphic Map of South and East Asia.* IUGS, Commission for the Geological Map of the World (Subcommission for Metamorphic Maps). Korea Institute of Energy and Resources, Seoul.

Le Fort, P. (1975) Himalayas: the collided range. The present knowledge of the continental arc. *Am. J. Sci.* 275A, 1–44.

Liou, J. G. (1971a) *P-T* stabilities of laumontite, wairakite, lawsonite, and related minerals in the system $CaAl_2Si_2O_8$-SiO_2-H_2O. *J. Petrol.* **12**, 397–411.

Liou, J. G. (1971b) Analcime equilibria. *Lithos* **4**, 389–402.

Liou, J. G., Kuniyoshi, S. and Ito, K. (1974) Experimental studies of the phase relations between greenschist and amphibolite in a basalt system. *Am. J. Sci.* **274**, 613–632.

Liou, J. G. and Maruyama, S. (1987) Parageneses and compositions of amphiboles from Franciscan jadeite-glaucophane type facies series metabasites at Cazadero, California. *J. Metamorphic Geol.* **5**, 371–395.

Loewinson-Lessing, F. Y. (1954) *A Historical Survey of Petrology* (translated by S. I. Tomkeieff). Oliver and Boyd, Edinburgh, 112 pp.

Lonker, S. (1980) Conditions of metamorphism in high-grade pelitic gneisses from the Frontenac Axis, Ontario, Canada. *Canad. J. Earth Sci.* **17**, 1666–1684.

Lonker, S. W. (1981) The *P-T-X* relations of the cordierite-garnet-sillimanite-quartz equilibrium. *Am. J. Sci.* **281**, 1056–1090.

Luth, W. C., Jahns, R. H. and Tuttle, O. F. (1964) The granite system at pressures of 4 to 10 kilobars. *J. Geophys. Res.* **69**, 759–773.

Lux, D. R., De Yoreo, J. J., Guidotti, C. V. and Decker, E. R. (1986) Role of plutonism in low-pressure metamorphic belt formation. *Nature,* London **323**, 794–797.

Lyons, J. B. (1955) Geology of the Hanover quadrangle, New Hampshire-Vermont. *Geol. Soc. Amer. Bull.* **66**, 106–146.

Martignole, J. and Sisi, J.-C. (1981) Cordierite-garnet-H_2O equilibrium: a geological thermometer, barometer and water fugacity indicator. *Contr. Mineral. Petrol.* **77**, 38–46.

Maruyama, S., Cho, M. and Liou, J. G. (1986) Experimental investigations of blueschist-greenschist transition equilibria: pressure dependence of Al_2O_3 contents in sodic amphiboles – a new geobarometer. *Geol. Soc. Amer. Mem.* **164**, 1–16.

Maruyama, S., Coleman, R. G. and Liou, J. G. (1989) Blueschists in the world.

Internatl. Geol. Congr. 28th Sess. (Washington, D. C.) Rept., Vol.2, pp.380–381.

Maruyama, S. and Isozaki, Y. (1992) Tectonics of the early Earth (in Japanese). *Kagaku* **62**, 175–184.

Maruyama. S. Isozaki, Y., Takeuchi, R., Tominaga, N., Takeshita, H. and Itaya, T. (1992) Application of K-Ar mapping on the Franciscan complex in northern California and redefined geotectonic subdivision and boundaries. *Amer. Assoc. Petrol. Geol. Bull.* **76**, 425.

Maruyama, S. and Liou, J. G. (1987) Clinopyroxene – a mineral telescoped through the process of blueschist facies metamorphism. *J. Metamorphic Geol.* **5**, 529–552.

Maruyama, S. and Liou, J. G. (1988) Petrology of Franciscan metabasites along the jadeite-glaucophane type facies series, Cazadero, California. *J. Petrol.* **29**, 1–37.

Maruyama, S., Liou, J. G. and Sasakura, Y. (1985) Low-temperature recrystallization of Franciscan greywackes from Pacheco Pass, California. *Mineral. Mag.* **49**, 345–355.

Maruyama, S., Liou, J. G. and Suzuki, K. (1982) The peristerite gap in low-grade metamorphic rocks. *Contr. Mineral. Petrol.* **81**, 268–276.

Maruyama, S., Suzuki, K. and Liou, J. G. (1983) Greenschist-amphibolite transition equilibria at low pressure. *J. Petrol.* **24**, 583–604.

Mason, R. (1984) Inverted isograds at Sulitjelma, Norway: the result of shear-zone deformation. *J. Metamorphic Geol.* **2**, 77–82.

Massonne, H.-J. and Schreyer, W. (1989) Stability field of the high-pressure assemblage talc + phengite and two new phengite barometers. *Eur. J. Mineral.* **1**, 391–410.

Mather, J. D. (1970) The biotite isograd and the lower greenschist facies in the Dalradian rocks of Scotland. *J. Petrol.* **11**, 253–275.

Matsuda, T. and Kuriyagawa, S. (1965) Lower grade metamorphism in the eastern Akaishi Mountains, central Japan (in Japanese with English abstract). *Tokyo Univ. Earthquake Res. Institute Bull.* **43**, 209–235.

Matsui, T. and Nishizawa, O. (1974) Iron(II)-magnesium exchange equilibrium between olivine and calcium-free pyroxene over a temperature range 800–1300°C. *Bull. Soc. francaise minéral. cristal.* **97**, 122–130.

McLellan, E. (1985) Metamorphic reactions in the kyanite and sillimanite zones of the Barrovian type area. *J. Petrol.* **26**, 789–818.

Mielke, H. and Schreyer, W. (1972) Magnetite-rutile assemblages in metapelites of the Fichtelgebirge, Germany. *Earth Planet. Sci. Letters* **16**, 423–428.

Misch, P. (1966) Tectonic evolution of the Northern Cascades of Washington State. In: H. C. Cunning (ed.) *Tectonic History and Mineral Deposits of the Western Cordillera.* (Canad. Inst. Min. Metal. Spec. Vol. 8), pp.101–148. Canad. Inst. Min. Metal.

Miyashiro, A. (1949) The stability relation of kyanite, sillimanite and andalusite, and the physical conditions of the metamorphic processes (in Japanese with English abstract). *J. Geol. Soc. Japan* **55**, 218–223.

Miyashiro, A. (1951) Kyanites in druses in kyanite-quartz veins from Saiho-ri in the Fukushinzan district, Korea. *J. Geol. Soc. Japan* **57**, 59–63.

Miyashiro, A. (1953) Calcium-poor garnet in relation to metamorphism. *Geochim. Cosmochim. Acta* **4**, 179–208.

Miyashiro, A. (1956) Osumilite, a new silicate mineral, and its crystal structure. *Am. Mineralogist* **41**, 104–116.

Miyashiro, A. (1957) Cordierite-indialite relations. *Am. J. Sci.* **255**, 43–62.

Miyashiro, A. (1958) Regional metamorphism of the Gosaisyo-Takanuki district in the central Abukuma Plateau. *Tokyo Univ. Fac. Sci. J.*, Sec.2, **11**, 219–272.

Miyashiro, A. (1961) Evolution of metamorphic belts. *J. Petrol.* **2**, 277–311.

Miyashiro, A. (1964) Oxidation and reduction in the earth's crust with special reference to the role of graphite. *Geochim. Cosmochim. Acta* **28**, 717–729.

Miyashiro, A. (1967a) Aspects of metamorphism in the circum-Pacific region. *Tectonophys.* **4**, 519–521.

Miyashiro, A. (1967b) Orogeny, regional metamorphism, and magmatism in the Japanese Islands. *Medd. fra Dansk Geol. Forening* **17**, 390–446.

Miyashiro, A. (1972) Metamorphism and related magmatism in plate tectonics. *Am. J. Sci.* **272**, 629–656.

Miyashiro, A. (1973) *Metamorphism and Metamorphic Belts*. George Allen & Unwin, London, 492 pp.

Miyashiro, A. (1991) Reorganization of geological sciences and particularly of metamorphic geology by the advent of plate tectonics: a personal view. *Tectonophys.* **187**, 51–60.

Miyashiro, A., Aki, K. and Sengor, A. M. C (1982) *Orogeny*. John Wiley, New York, 242 pp.

Miyashiro, A. and Banno, S. (1958) Nature of glaucophanitic metamorphism. *Am. J. Sci.* **256**, 97–110.

Miyashiro, A. and Shido, F. (1970) Progressive metamorphism in zeolite assemblages. *Lithos* **3**, 251–260.

Miyashiro, A. and Shido, F. (1985) Tschermak substitution in low- and middle-grade pelitic schists. *J. Petrol.* **26**, 449–487.

Moorbath, S. and Taylor, P. N. (1986) Geochronology and related isotope geochemistry of high-grade metamorphic rocks from the lower continental crust. In: J. B. Dawson et al. (eds.) *The Nature of the Lower Continental Crust* (Geol. Soc. London, Spec. Publ. No.24), pp.211–220. Blackwell Scientific, Oxford.

Morand, V. J. (1990) Low-pressure regional metamorphism in the Omeo metamorphic complex, Victoria, Australia. *J. Metamorphic Geol.* **8**, 1–12.

Mueller, R. F. (1962) Energetics of certain silicate solid solutions. *Geochim. Cosmochim. Acta* **26**, 581–598.

Mysen, B. O. (1977) The solubility of H_2O and CO_2 under predicted magma genesis conditions and some petrological and geophysical implications. *Reviews of Geophys. Space Phys.* **15**, 351–361.

Nagle, F. (1974) Blueschist, eclogite, paired metamorphic belts and the early tectonic history of Hispaniola. *Geol. Soc. Amer. Bull.* **85**, 1461–1466.

Nathenson, M. and Guffanti, M. (1988) Geothermal gradients in the conterminous United States. *J. Geophys. Res.* **93**, 6437–6480.

Newton, R. C. (1966) Kyanite-sillimanite equilibrium at 750°C. *Science* **151**, 1222–1225.

Newton, R. C. (1983) Geobarometry of high grade metamorphic rocks. *Am. J. Sci.* **283-A**, 1–28.

Newton, R. C. (1986a) Fluids of granulite facies metamorphism. In: J. V. Walther & B. J. Wood (eds.) *Fluid-Rock Interactions during Metamorphism*, pp.36–59. Springer-Verlag, New York.

Newton, R. C. (1986b) Metamorphic temperatures and pressures of group B and C eclogites. *Geol. Soc. Amer. Mem.* **164**, 17–30.

Newton, R. C. (1989) Metamorphic fluids in the deep crust. *Ann. Review of Earth Planet. Sci.* **17**, 385–412.

Newton, R. C. and Kennedy, G. C. (1963) Some equilibrium relations in the join $CaAl_2Si_2O_8$-H_2O. *J. Geophys. Res.* **68**, 2967–2983.

Newton, R. C. and Perkins, D., III (1982) Thermodynamic calibration of geobarometers based on the assemblages garnet-plagioclase-orthopyroxene (clinopyroxene)-quartz. *Am. Mineralogist* **67**, 203–222.

Newton, R. C. and Smith, J. V. (1967) Investigations concerning the breakdown of albite at depth in the earth. *J. Geol.* **75**, 268–286.

Newton, R. C., Smith, J. V. and Windley, B. F. (1980) Carbonic metamorphism, granulites and crustal growth. *Nature*, London **288**, 45–50.

Newton, R. C. and Wood, B. J. (1979) Thermodynamics of water in cordierite and some petrologic consequences of cordierite as a hydrous phase. *Contr. Mineral. Petrol.* **68**, 391–405.

Niggli, E. (1960) Mineral-Zonen der alpinen Metamorphose in der Schweizer Alpen. *Internatl. Geol. Congr. 21st Sess. (Copenhagen) Rept.*, Part 13, pp.132–138.

Nordstrom, D. K. and Munoz, J. L. (1985) *Geochemical Thermodynamics*. Benjamin/Cummings Publ., Menlo Park, Calif., 477 pp.

Oba, T. (1980) Phase relations in the tremolite-pargasite join. *Contr. Mineral. Petrol.* **71**, 247–256.

Oba, T. and Yagi, K. (1987) Phase relations on the actinolite-pargasite join. *J. Petrol.* **28**, 23–36.

O'Hara, M. J. (1961) Zoned ultrabasic and basic gneiss masses in the early Lewisian metamorphic complex at Scourie, Sutherland. *J. Petrol.* **2**, 248–276.

Ohmoto, H. and Kerrick, D. (1977) Devolatilization equilibria in graphitic systems. *Am. J. Sci.* **277**, 1013–1044.

Okay, A. I. (1989) Alpine-Himalayan blueschists. *Ann. Rev. Earth Planet. Sci.* **17**, 55–87.

Orville, P. M. (1974) The < <peristerite gap> > as an equilibrium between ordered albite and disordered plagioclase solid solution. Bull. Soc. francaise Minéral. Cristal. 97, 386–392.

Oxburgh, E. R. and England, P. C. (1980) Heat flow and the metamorphic evolution of the Eastern Alps. *Eclogae geol. Helv.* **73**, 379–398

Oxburgh, E. R. and Turcotte, D. L. (1974) Thermal gradients and regional metamorphism in overthrust terrains with special reference to the Eastern Alps. *Schweiz. Mineral. Petrogr. Mitt.* **54**, 641–662.

Parsons, I. (1981) Effect of pressure on the alkali feldspar solvus. *Progress in Experimental Petrology: Fifth Progress Report on Research Supported by Natural Environment Research Council, 1978–1980*, pp.222–224. Natural Environ. Res. Council, U. K.

Pattison, D. R. M. and Newton, R. C. (1989) Reversed experimental calibration of the garnet-clinopyroxene Fe-Mg exchange thermometer. *Contr. Mineral. Petrol.* **101**, 87–103.

Peacock, S. M. (1987a) Thermal effects of metamorphic fluids in subduction zones. *Geology* **15**, 1057–1060.

Peacock, S. M. (1987b) Creation and preservation of subduction-related inverted metamorphic gradients. *J. Geophys. Res.* **92**, 12763–12781.

Perrin, R. and Roubault, M. (1949) On the granite problem. *J. Geol.* **57**, 357–379.

Phillips, G. N. (1980) Water activity changes across an amphibolite-granulite facies transition, Broken Hill, Australia. *Contr. Mineral. Petrol.* **75**, 377–386.

Pin, C. and Vielzeuf, D. (1983) Granulites and related rocks in Variscan median Europe: a dualistic interpretation. *Tectonophys.* **93**, 47–74.

Platt, J. P. (1987) The uplift of high-pressure--low-temperature metamorphic rocks. *Phil. Trans. R. Soc. London* **A321**, 87–103.

Pognante, U. (1991) Petrological constraints on the eclogite- and blueschist-facies metamorphism and *P-T-t* paths in the Western Alps. *J. Metamorphic Geol.* **9**, 5–17.

Poldervaart, A. (1955) Chemistry of the Earth's crust. *Geol. Soc. Amer. Spec. Paper* **62**, 119–144.

Popp, P. K. and Gilbert, M. C. (1972) Stability of acmite-jadeite pyroxenes at low pressure. *Am. Mineralogist* **57**, 1210–1231.

Popp, R. K., Gilbert, M. C. and Craig, J. R. (1977) Stability of Fe-Mg amphiboles with respect to oxygen fugacity. *Am. Mineralogist* **62**, 1–12.

Poulson, S. R. and Ohmoto, H. (1989) Devolatilization equilibria in graphite-pyrite-pyrrhotite bearing pelites with application to magma-pelite interaction. *Contr. Mineral. Petrol.* **101**, 418–425.

Powell, R. (1978) *Equilibrium Thermodynamics in Petrology*. Harper and Row, London, 284 pp.

Powell, R. and Holland, T. J. B. (1985) An internally consistent thermodynamic dataset with uncertainties and corrections: 1. Methods and a worked example. *J. Metamorphic Geol.* **3**, 327–342.

Powell, R. and Holland, T. J. B. (1988) An internally consistent dataset with uncertainties and correlations: 2. Applications to geobarometry, worked examples and a computer program. *J. Metamorphic Geol.* **6**, 173–204.

Powell, R. and Holland, T. (1990) Calculated mineral equilibria in the pelite system KFMASH (K_2O-FeO-MgO-Al_2O_3-SiO_2-H_2O). *Am. Mineralogist* **75**, 367–380.

Prigogine, I. and Defay, R. (1954) *Chemical Thermodynamics*. Longmans, Green and Co., London.

Putnis, A. and Holland, T. J. B. (1986) Sector trilling in cordierite and equilibrium overstepping in metamorphism. *Contr. Mineral. Petrol.* **93**, 265–272.

Raase, P., Raith, M., Ackermand, D. and Lal, R. K. (1986) Progressive metamorphism of mafic rocks from greenschist to granulite facies in the Dharwar craton of southern India. *J. Geol.* **94**, 261–282.

Ramberg, H. (1944a) Petrological significance of subsolidus phase transitions in mixed crystals. *Norsk Geol. Tidsskr.* **24**, 42–73.

Ramberg, H. (1944b, 1945) The thermodynamics of the earth's crust. I, II. *Norsk Geologisk Tidsskr.* **24**, 98–111; **25**, 307–326.

Ramberg, H. (1948) Radial diffusion and chemical stability in the gravitational field. *J. Geol.* **56**, 448–458.

Ramberg, H. (1949) The facies classification of rocks: a clue to the origin of quartzo-feldspathic massifs and veins. *J. Geol.* **57**, 18–54.

Ramberg, H. (1951) Remarks on the average chemical composition of granulite and amphibolite-to-epidote amphibolite facies gneisses in west Greenland. *Meddr. Dansk Geol. Foren.* **12**, 27–34.

Ramberg, H. (1952) *The Origin of Metamorphic and Metasomatic Rocks*. University of Chicago Press, Chicago, Illinois, 317 pp.

Ramberg, H. and DeVore, G. (1951) The distribution of Fe^{++} and Mg^{++} in coexisting olivines and pyroxenes. *J. Geol.* **59**, 193–210.

Read, H. H. (1957) *The Granite Controversy*. Thomas Murby, London, 431 pp.

Rice, J. M. (1977) Progressive metamorphism of impure dolomitic limestone in the Marysville aureole, Montana. *Am. J. Sci.* **277**, 1–24.

Rice, J. M. and Ferry, J. M. (1982) Buffering, infiltration, and the control of intensive variables during metamorphism. In: J. M. Ferry (ed.) *Characterization of Metamorphism through Mineral Equilibria* (Reviews in Mineralogy, Vol. 10), pp.263–326. Mineral. Soc. America.

Richardson, S. W., Gilbert, M. C. and Bell, P. M. (1969) Experimental determination of kyanite-andalusite and andalusite-sillimanite equilibria; the aluminum silicate triple point. *Am. J. Sci.* **267**, 259–272.

Richardson, S. W. and Powell, R. (1976) Thermal causes of the Dalradian metamorphism in the central Highlands of Scotland. *Scott. J. Geol.* **12**, 237–268.

Richter, D. A. and Roy, D. C. (1974) Sub-greenschist metamorphic assemblages in northern Maine. *Canad. Mineralogist* **12**, 469–474.

Ringwood, A. E. (1975) *Composition and Petrology of the Earth's Mantle*. McGraw-Hill, New York, 618 pp.

Robertson, E. C., Birch, F. and MacDonald, G. J. F. (1957) Experimental determination of jadeite stability relations to 25 000 bars. *Am. J. Sci.* **255**, 115–135.

Robertson, J. K. and Wyllie, P. J. (1971) Rock-water system, with special reference to the water-deficient region. *Am. J. Sci.* **271**, 252–277.

Robie, R. A., Hemingway, B. S. and Fisher, J. R. (1978) Thermodynamic properties of minerals and related substances at 298.15 K and 1 bar (10^5 pascals) pressure and at higher temperatures. *U. S. Geol.* Surv. Bull. 1452.

Roedder, E. (1965) Liquid CO_2 inclusions in olivine-bearing nodules and phenocrysts from basalts. *Am. Mineralogist* **50**, 1746–1782.

Roedder, E. (1984) *Fluid Inclusions* (Reviews in Mineralogy, Vol.12). Mineral. Soc. America, 644 pp.

Rosenfeld, J. L. (1968) Garnet rotations due to the major Paleozoic deformations in southeast Vermont. In: E-an Zen et al. (eds) *Studies of Appalachian Geology: Northern and Maritime*, pp.185–202. Interscience, New York.

Rossi, G. (1988) A review of the crystal chemistry of clinopyroxenes in eclogites and other high-pressure rocks. In: D. C. Smith (ed.) *Eclogites and Eclogite-facies Rocks*, pp.237–270. Elsevier, Amsterdam.

Rumble, D., III (1976a) The use of mineral solid solutions to measure chemical potential gradients in rocks. *Am. Mineralogist* **61**, 1167–1174.

Rumble, D., III (1976b) Oxide minerals in metamorphic rocks. In: D. Rumble, III (ed.) *Oxide Minerals* (Reviews in Mineralogy, Vol. 3), Ch. 3. Mineral. Soc. America.

Rumble D., III (1978) Mineralogy, petrology, and oxygen isotopic geochemistry of the Clough Formation, Black Mountain, western New Hampshire, U.S.A. *J. Petrol.* **19**, 317–340.

Rumble, D., III, Chamberlain, C. P., Zeitler, P. K. and Barreiro, B. (1989) Hydrothermal graphite veins and Acadian granulite facies metamorphism, New Hampshire, U.S.A. In: D. Bridgwater (ed.) *Fluid Movements-Element Transport and the Composition of the Deep Crust*, pp.117–119. Kluwer Academic, Dordrecht.

Rumble, D., III, Ferry, J. M., Hoering, T. C. and Boucot, A. J. (1982) Fluid flow during metamorphism at the Beaver Brook fossil locality, New Hampshire. *Am. J. Sci.* **282**, 886–919.

Rutland, R. W. R. (1965) Tectonic overpressures. In: W. S. Pitcher and G. W. Flinn (eds.) *Controls of Metamorphism*, pp.119–139. Oliver & Boyd, Edinburgh.

Salje, E. (1986) Heat capacities and entropies of andalusite and sillimanite: the influence of fibrolitization on the phase diagram of the Al_2SiO_5 polymorphs. *Am. Mineralogist* 71, 1366–1371.

Sandiford, M. (1985) The metamorphic evolution of granulites at Fyfe Hills; implications for Archaean crustal thickness in Enderby Land, Antarctica. *J. Metamorphic Geol.* **3**, 155–178.

Sandiford, M. (1989) Horizontal structures in granulite terrains: a record of mountain building or mountain collapse ? *Geology* **17**, 449–452.

Sandiford, M., Neall, F. B. and Powell, R. (1987) Metamorphic evolution of aluminous granulites from Labwor Hills, Uganda. *Contr. Mineral. Petrol.* **95**, 217–225.

Sandiford, M. and Powell, R. (1986) Deep crustal metamorphism during continental extension: modern and ancient examples. *Earth Planet. Sci. Letters* **79**, 151–158.

Schiffman, P. and Liou, J. G. (1980) Synthesis and stability relations of Mg-Al pumpellyite, $Ca_4Al_5MgSi_6O_{21}(OH)_7$. *J. Petrol.* **21**, 441–474.

Schliestedt, M. (1986) Eclogite-blueschist relations as evidenced by mineral equilibria in the high-pressure metabasic rocks of Sifnos (Cyclades Islands), Greece. *J. Petrol.* **27**, 1437–1459.

Schliestedt, M. (1990) Occurrence and stability conditions of low-temperature eclogites. In: D. A. Carswell (ed.) *Eclogite Facies Rocks*.pp.160–179. Blackie, Glasgow.

Schliestedt, M. and Okrusch, M. (1988) Meta-acidites and silicic meta-sediments related to eclogites and glaucophanites in northern Sifnos, Cycladic Archipelago, Greece. In: D. C. Smith (ed.) *Eclogites and Eclogite-facies Rocks*, pp.291–334. Elsevier, Amsterdam.

Schreurs, J. (1984) The amphibolite-granulite facies transition in west Uusimaa, S. W. Finland. A fluid inclusion study. *J. Metamorphic Geol.* **2**, 327–341.

Schreurs, J. (1985) Prograde metamorphism of metapelites, garnet-biotite thermometry and prograde changes of biotite chemistry in high-grade rocks of West Uusimaa, southwest Finland. *Lithos* **18**, 69–80.

Schreurs, J. and Westra, L. (1986) The thermotectonic evolution of a Proterozoic, low pressure, granulite dome, west Uusimaa, S W Finland. *Contr. Mineral. Petrol.* **93**, 236–250.

Schreyer, W. (1976) Experimental metamorphic petrology at low pressures and high temperatures. In: D. K. Bailey and R. Macdonald (eds.) *The Evolution of the Crystalline Rocks*, pp. 261–331. Academic Press, London.

Schreyer, W. (1988) Experiemntal studies on metamorphism of crustal rocks under mantle pressures. *Mineral. Mag.* **52**, 1–26.

Schreyer, W. and Baller, T. (1977) Talc-muscovite: synthesis of a new high-pressure phyllosilicate assemblage. *Neues Jb. Mineral. Mh.* **1977**, 421–425.

Schreyer, W., Maresch, W. V., Daniels, P. and Wolfsdorff, P. (1990) Potassic cordierites: characteristic minerals for high-temperature, very low-pressure environments. *Contr. Mineral. Petrol.* **105**, 162–172.

Schreyer, W., Massonne, H.-J., and Chopin, C. (1987) Continental crust subducted to depths near 100 km: implications for magma and fluid genesis in collision zones. In: B. O. Mysen (ed.) Magmatic Processes: Physicochemical Principles (Geochem. Soc. Spec. Publ. No.1), pp.155–163. Geochem. Soc.

Schuiling, R. D. and Kreulen, R. (1979) Are thermal domes heated by CO_2-rich fluids from the mantle. *Earth Planet. Sci. Letters* **43**, 298–302.

Schulze, D. J. and Helmstaedt, H. (1988) Coesite-sanidine eclogites from kimberlite: products of mantle fractionation or subduction ? *J. Geol.* **96**, 435–443.

Sederholm, J. J. (1923, 1926) On migmatites and associated Pre-Cambrian rocks of south-western Finland. I, II. *Bull. Comm. geol. Finlande*, Nos. 58 and 77.

Seifert, F. (1970) Low-temperature compatibility relations of cordierite in haplopelites of the system K_2O-MgO-Al_2O_3-SiO_2-H_2O. *J. Petrol.* **11**, 73–99.

Seki, Y. (1961a) Calcareous hornfelses in the Arisu district of the Kitakami Mountains, northeastern Japan. *Jap. Jour. Geol. Geogr.* **32**, 55–78.

Seki, Y. (1961b) Pumpellyite in low-grade metamorphism. *J. Petrol.* **2**, 407–423.

Seki, Y. (1966) Wairakite in Japan. *J. Jap. Assoc. Mineral. Petrol. Econ. Geol.* **55**, 254–261; **56**, 30–39.

Seki, Y. (1969) Facies series in low-grade metamorphism. *J. Geol. Soc. Japan* **75**, 255–266.

Seki, Y., Oba, T., Mori, R. and Kuriyagawa, S. (1964) Sanbagawa metamorphism in the central part of Kii Peninsula (in Japanese with English abstract). *J. Jap. Assoc. Mineral. Petrol. Econ. Geol.* **52**, 73–89.

Seki, Y., Oki, Y., Matsuda, T., Mikami, K. and Okumura, K. (1969) Metamorphism in the Tanzawa Mountains, central Japan. *J. Jap. Assoc. Mineral. Petrol. Econ. Geol.* **61**, 1–29, 50–75.

Selverstone, J. and Chamberlain, C. P. (1990) Apparent isobaric cooling paths from granulites: two counterexamples from British Columbia and New Hampshire. *Geology* **18**, 307–310.

Shaw, D. M. (1956) Geochemistry of pelitic rocks, III: major elements and general geochemistry. *Geol. Soc. Amer. Bull.* **67**, 919–934.

Shiba, M. (1988) Metamorphic evolution of the southern part of the Hidaka belt, Hokkaido, Japan. *J. Metamorphic Geol.* **6**, 273–296.

Shido, F. (1958) plutonic and metamorphic rocks of the Nakoso and Iritono districts in the central Abukuma Plateau. *Tokyo Univ. Fac. Sci. J.*, Sec.2, **11**, 131–217.

Shido, F. and Miyashiro, A. (1959) Hornblendes of basic metamorphic rocks. *Tokyo Univ. Fac. Sci. J.*, Sec.2, **12**, 85–102.

Skippen, G. (1974) An experimental model for low pressure metamorphism of siliceous dolomitic marble. *Am. J. Sci.* **274**, 487–509.

Sleep, N. H. (1979) A thermal constraint on the duration of folding with reference to Acadian geology, New England (USA). *J. Geol.* **87**, 583–589.

Smelik, E. A., Nyman, M. W., and Veblen, D. R. (1991) Pervasive exsolution within the calcic amphibole series: TEM evidence for a miscibility gap between actinolite and hornblende in natural samples. *Am. Mineralogist* **76**, 1184–1204.

Smith, D. C. (1984) Coesite in clinopyroxene in the Caledonides and its implications for geodynamics. *Nature* **310**, 641–644.

Smith, J. V. (1975) Phase equilibria of plagioclase. In: P. H. Ribbe (ed.) *Feldspar Mineralogy* (Reviews in Mineralogy, Vol.2), Ch. 9. Min. Soc. America.

Smyth, J. R. and Hatton, C. J. (1977) A coesite-sanidine grospydite from the Roberts Victor kimberlite. *Earth Planet. Sci. Letters* **34**, 284–290.

Sobolev, N. V. and Shatsky, V. S. (1990) Diamond inclusions in garnets from metamorphic rocks: a new environment for diamond formation. *Nature,* London **343**, 742–746.

Spear, F. S. (1980a) NaSi $\rightleftharpoons$ CaAl exchange equilibrium between plagioclase and amphibole. An empirical model. *Contr. Mineral. Petrol.* **72**, 33–41.

Spear, F. S. (1980b) The gedrite-anthophyllite solvus and the composition limits of orthoamphibole from the Post Pond Volcanics, Vermont. *Am. Mineralogist* **65**, 1103–1118.

Spear, F. S. (1981) Amphibole-plagioclase equilibria: an empirical model for the relation albite + tremolite = edenite + 4 quartz. *Contr. Mineral. Petrol.* **77**, 355–364.

Spear, F. S. (1982) Phase equilibria of amphibolites from the Post Pond Volcanics, Mt. Cube Quadrangle, Vermont. *J. Petrol.* **23**, 383–426.

Spear, F. S. (1988a) Thermodynamic projection and extrapolation of high-variance mineral assemblages. *Contr. Mineral. Petrol.* **98**, 346–351.

Spear, F. S. (1988b) Metamorphic fractional crystallization and internal metasomatism by diffusional homogenization of zoned garnets. *Contr. Mineral. Petrol.* **99**, 507–517.

Spear, F. S. (1988c) The Gibbs method and Duhem's theorem. *Contr. Mineral. Petrol.* **99**, 249–256.

Spear, F. S. and Cheney, J. T. (1989) A petrogenetic grid for pelitic schists in the system SiO_2-Al_2O_3-FeO-MgO-K_2O-H_2O. *Contr. Mineral. Petrol.* **101**, 149–164.

Spear, F. S., Ferry, M. J. and Rumble, D., III (1982) Analytical formulation of phase equilibria: the Gibbs' method. In: J. M. Ferry (ed.) *Characterization of Metamorphism through Mineral Equilibria* (Reviews in Mineralogy, Vol. 10), pp.105–152. Mineral. Soc. America.

Spear, F. S. and Selverstone, J. (1983) Quantitative *P-T* paths from zoned minerals: theory and tectonic applications. *Contr. Mineral. Petrol.* **83**, 348–357.

Stern, C. R. and Wyllie, P. J. (1973) Water-saturated and undersaturated melting relations of a granite to 35 kilobars. *Earth Planet. Sci. Letters* **18**, 163–167.

Stockwell, C. H. (1968) Geochronology of stratified rocks on the Canadian shield. *Can. J. Earth Sci.* **5**, 693–698.

St-Onge, M. R. (1981) "Normal" and "inverted" metamorphic isograds and their relation to syntectonic Proterozoic batholiths in the Wopmay orogen, Northwest Territories, Canada. *Tectonophys.* **76**, 295–316.

Strunz, H. (1941, 1966) *Mineralogische Tabellen.* 1st and 4th eds. Akademische Verlag, Leipzig.

Symmes, G. H. and Ferry, J. M. (1991) Evidence from mineral assemblages for infiltration of pelitic schists by aqueous fluids during metamorphism. *Contr. Mineral. Petrol.* **108**, 419–438.

Symmes, G. H. and Ferry, J. M. (1992) The effect of whole-rock MnO content on the stability of garnet in pelitic schists during metamorphism. *J. Metamorphic Geol.* **10**, 221–237.

Tagiri, M. (1977) Fe-Mg partition and miscibility gap between coexisting calcic amphiboles from the southern Abukuma Plateau (Japan). *Contr. Mineral. Petrol.* **62**, 271–281.

Tagiri, M. (1981) A measurement of the graphitizing-degree by the X-ray powder diffractometer. *J. Jap. Assoc. Mineral. Petrol. Econ. Geol.* **76**, 345–352.

Takasu, A. (1984) Prograde and retrograde eclogites in the Sambagawa metamorphic belt, Besshi district, Japan. *J. Petrol.* **25**, 619–643.

Thompson, A. B. (1971a) Analcite-albite equilibria at low temperatures. *Am. J. Sci.* **271**, 79–92.

Thompson, A. B. (1971b) PCO_2 in low-grade metamorphism; zeolite, carbonate, clay mineral, prehnite relations in the system CaO-Al_2O_3-SiO_2-CO_2-H_2O. *Contr. Mineral. Petrol.* **33**, 145–161.

Thompson, A. B. (1976) Mineral reactions in pelitic rocks, I and II. *Am. J. Sci.* **276**, 401–424, 425–454.

Thompson, A. B. (1983) Fluid-absent metamorphism. *J. Geol. Soc. London* **140**, 533–547.

Thompson, A. B. and Algor, J. R. (1977) Model systems for anatexis of pelitic rocks. I. Theory of melting reactions in the system $KAlO_2$-$NaAlO_2$-Al_2O_3-SiO_2-H_2O. *Contr. Mineral. Petrol.* **63**, 247–269.

Thompson, A. B. and England, P. C. (1984) Pressure-temperature-time paths of regional metamorphism, II. *J. Petrol.* **25**, 929–955.

Thompson, A. B. and Ridley, J. R. (1987) Pressure-temperature-time (*P-T-t*) histories of orogenic belts. *Phil. Trans. R. Soc. London* **A321**, 27–45.

Thompson, A. B., Tracy, R. J., Lyttle, P. T. and Thompson, J. B., Jr. (1977) Prograde reaction histories deduced from compositional zonation and mineral inclusions in garnet from the Gassetts schist, Vermont. *Am. J. Sci.* **277**, 1152–1167.

Thompson, J. B., Jr. (1955) Thermodynamic basis for the mineral facies concept. *Am. J. Sci.* **253**, 65–103.

Thompson, J. B. Jr. (1957) The graphical analysis of mineral assemblages in pelitic schists. *Am. Mineralogist* **42**, 842–858.

Thompson, J. B., Jr. (1959) Local equilibrium in metasomatic processes. In: P. H. Abelson (ed.) *Researches in Geochemistry*, pp.427–457. John Wiley, New York.

Thompson, J. B., Jr. (1970) Geochemical reaction and open system. *Geochim. Cosmochim. Acta* **34**, 529–551.

Thompson, J. B. Jr. (1972) Oxides and sulfides in regional metamorphism of pelitic schists. *Internatl. Geol. Congr. 24th Sess. (Montreal) Rept.* Sec. 10, 27–35.

Thompson, J. B., Jr. (1979) The Tschermak substitution and reactions in pelitic schists (in Russian). In: V. A. Zharikov et al. (eds.) *Problems in Physicochemical Petrology*, Vol.1, pp.146–159. Acad. Sciences USSR. Moscow.

Thompson, J. B. Jr. (1982a) Composition space: an algebraic and geometric approach. In: J. M. Ferry (ed.) *Characterization of Metamorphism through Mineral Equilibria* (Reviews in Mineralogy, Vol. 10), pp. 1–31. Mineral. Soc. America.

Thompson, J. B. Jr. (1982b) Reaction space: an algebraic and geometric approach. In: J. M. Ferry (ed.) *Characterization of Metamorphism through Mineral Equilibria* (Reviews in Mineralogy, Vol. 10), pp. 33–52. Mineral. Soc. America.

Thompson, J. B., Jr. (1987) A simple thermodynamic model for grain interfaces: some insights on nucleation, rock textures, and metamorphic differentiation. In: H. C. Helgeson (ed.) *Chemical Transport in Metasomatic Processes*, pp.169–188. D. Reidel, Dordrecht.

Thompson, J. B. Jr., Laird, Jo and Thompson, A. B. (1982) Reactions in amphibolite, greenschist and blueschist. *J. Petrol.* **23**, 1–27.

Thompson, J. B. Jr. and Norton, S. A. (1968) Paleozoic regional metamorphism in New England and adjacent areas. In: E-an Zen et al. (eds.) *Studies of Appalachian Geology, Northern and Maritime*, pp. 319–327. Interscience Publishers, New York.

Thompson, J. B. Jr. and Thompson, A. B. (1976) A model system for mineral facies in pelitic schists. *Contr. Mineral. Petrol.* **58**, 243–277.

Thompson, P. H. (1977) Metamorphic *P-T* distributions and the geothermal gradients calculated from geophysical data. *Geology* **5**, 520–522.

Tilley, C. E. (1924) The facies classification of rocks. *Geol. Mag.* **61**, 167–171.

Tilley, C. E. (1925) A preliminary survey of metamorphic zones in the southern Highlands of Scotland. *Q. J. Geol. Soc. London* **81**, 100–112.

Tobi, A. C., Hermans, G. A. E. M., Maijer, C. and Jensen, J. B. H. (1985) Metamorphic zoning in the high-grade Proterozoic of Rogaland-Vest Agar, S. W. Norway. In: A. C. Tobi and L. R. Touret (eds.) *The Deep Proterozoic Crust in the North Atlantic Provinces*. NATO Advanced Science Institutes, Ser. C, **158**, pp.477–497. D. Reidel, Dordrecht.

Tobschall, H. J. (1969) A sequence of subfacies of the greenschist facies in the Cevennes Medianes (Dep. Ardeche, France) with pyrophyllite-bearing parageneses. *Contr. Mineral. Petrol.* **24**, 76–91.

Touret, J. (1971) Le facies granulite en Norvege meridionale, II. Les inclusions fluides. *Lithos* **4**, 423–436.

Touret, J. (1981) Fluid inclusions in high grade metamorphic rocks. In: L. S. Hollister and M. L. Crawford (eds.) *Short Course in Fluid Inclusions: Applications to Petrology*, pp.182–204. Mineral. Assoc. Canada.

Tracy, R. J. (1978) High grade metamorphic reactions and partial melting in pelitic schist, west-central Massachusetts. *Am. J. Sci.* **278**, 150–178.

Turner, F. J. (1948) Mineralogical and structural evolution of the metamorphic rocks. *Geol. Soc. Amer. Mem.* **30**.

Turner, F. J. (1968, 1981) *Metamorphic Petrology*. 1st and 2nd edns. McGraw-Hill, New York.

Turner, F. J. and Verhoogen, J. (1951, 1960) *Igneous and Metamorphic Petrology*. 1st and 2nd edns. McGraw-Hill, New York.

van't Hoff, J. H., Armstrong, E. F., Hinrichsen, W., Weigert, F. and Just, G. (1903) Gips und Anhydrit. *Zs. phys. Chem.* **45**, 257–306.

Veblen, D. R. (ed.) (1981) *Amphiboles and Other Hydrous Pyriboles – Mineralogy* (Reviews in Mineralogy, Vol. 9A), Mineral. Soc. America, 372 pp.

Veblen, D. R. and Ribbe, P. H. (eds.) (1982) *Amphiboles: Petrology and Experimental Phase Relations (Reviews in Mineralogy*, Vol. 9B). Mineral. Soc. America. 390 pp.

Vielzeuf, D. and Holloway, J. R. (1988) Experimental determination of the fluid-absent melting relations in the pelitic system. Consequence of crustal differentiation. *Contr. Mineral. Petrol.* **98**, 257–276.

Vogt, Th. (1927) Sulitelmafeltets geologi og petrografi. *Norges Geol. Unders*. No.121.

Wallace, J. H. and Wenk, H. R. (1980) Structure variation in low cordierite. *Am. Mineralogist* **65**, 96–111.

Walther, J. V. and Orville, P. M. (1982) Volatile production and transport in regional metamorphism. *Contr. Mineral. Petrol.* **79**, 252–257.

Warren, E. B. (1930a) The structure of tremolite $H_2Ca_2Mg_5(SiO_3)_8$. *Zs. Krist.* **72**,

42–57.

Warren, E. B. (1930b) The crystal structure and chemical composition of the monoclinic amphiboles. *Zs. Krist.* **72**, 493–517.

Waters, D. J. and Whales, C. J. (1985) Dehydration melting and the granulite transition in metapelites from southern Namaqualand, S. Africa. *Contr. Mineral. Petrol.* **88**, 269–275.

Watkins, K. P. (1985) Geothermometry and geobarometry of inverted metamorphic zones in the W central Scottish Dalradian. *J. Geol. Soc. London* **142**, 157–165.

Weber, K. (1984) Variscan events: early Paleozoic continental rift metamorphism and late Paleozoic crustal shortening. In: D. H. W. Hutton, and D. J. Sanderson (eds.) *Variscan Tectonics of the North Atlantic Region* (Geol. Soc. London, Spec. Publ. No.14), pp.3–22. Blackwell Scientific, Oxford.

Wegmann, C. E. (1935) Zur Deutung der Migmatite. *Geol. Rundschau* **26**, 305–350.

Wells, P. R. A. (1979a) *P-T* conditions in the Moines of the central Highlands, Scotland. *J. Geol. Soc. London* **136**, 663–671.

Wells, P. R. A. (1979b) Chemical and thermal evolution of Archaean sialic crust, southern west Greenland. *J. Petrol.* **20**, 187–226.

Wendlandt, R. F. (1981) Influence of CO_2 on melting of model granulite facies assemblages: a model for the genesis of charnockites. *Am. Mineralogist* **66**, 1164–1174.

Wenk, E. and Wenk, H. R. (1977) An-variation and intergrowths of plagioclases in banded metamorphic rocks from Val Carecchio (Central Alps). *Schweiz. Mineral. Petrogr. Mitt.* **57**, 41–57.

Wenk, H. R. (1979) An albite-anorthite assemblage in low-grade amphibolite facies rocks. *Am. Mineralogist* **64**, 1294–1299.

White, A. J. R. (1964) Clinopyroxenes from eclogites and basic granulites. *Am. Mineralogist* **49**, 883–888.

White, J. C. and White, S. H. (1981) On the structure of grain boundaries in tectonites. *Tectonophys.* **78**, 613–628.

Wickham, S. M. and Oxburgh, E. R. (1985) Continental rifts as a setting for regional metamorphism. *Nature,* London **318**, 330–333.

Wickham, S. M. and Oxburgh, E. R. (1987) Low-pressure regional metamorphism in the Pyrenees and its implications for the thermal evolution of rifted continental crust. *Phil. Trans. R. Soc. London* **A321**, 219–242.

Will, T. M., Powell, R., Holland, Tim and Guiraud, M. (1990) Calculated greenschist facies mineral equilibria in the system CaO-FeO-MgO-Al_2O_3-SiO_2-CO_2-H_2O. *Contr. Mineral. Petrol.* **104**, 353–368.

Wilson, J. T. (1968a) Static or mobile earth: the current scientific revolution. *Amer. Phil. Soc. (Philadelphia) Proc.* **112**, 309–320.

Wilson, J. T. (1968b) A revolution in earth science. *GeoTimes* **13**, 10–16.

Wilson, J. T. (1968c) Reply to V. V. Beloussov. *GeoTimes* **13**, 20–22.

Wilson, J. T. (1971) Presentation of the 1968 Penrose Medal to John Tuzo Wilson – response. *Geol. Soc. Amer. Proc. Vol. for 1968*, 90–101.

Winchell, A. N. (1927, 1933) *Elements of Optical Mineralogy. Part II: Descriptions of Minerals.* 2nd and 3rd edns. John Wiley, New York.

Winchester, J. A. (1974) The zonal pattern of regional metamorphism in the Scottish Caledonides. *J. Geol. Soc. London* **130**, 509–524.

Windley, B. F. (1981) Phanerozoic granulites. *J. Geol. Soc. London* **138**, 745–751.

Winkler, H. G. F. (1965, 1979) *Petrogenesis of Metamorphic Rocks.* 1st and 5th edns. Springer-Verlag, New York.

Wiseman, J. D. H. (1934) The central and south-west Highland epidiorites: a study in progressive metamorphism. *Q. J. Geol. Soc. London* **90**, 354–417.

Wones, D. R. and Eugster, H. P. (1965) Stability of biotite: experiment, theory, and application. *Am. Mineralogist* **50**, 1228–1272.

Wood, B. J. (1975) The influence of pressure, temperature and bulk composition on the appearance of garnet in orthogneisses – an example from South Harris, Scotland. *Earth Planet. Sci. Letters* **26**, 299–311.

Wood, B. J. and Fraser, D. G. (1977) *Elementary Thermodynamics for Geologists*. Oxford Univ. Press, Oxford, 303 pp.

Wood, B. J. and Graham, C. M. (1986) Infiltration of aqueous fluid and high fluid: rock ratios during greenschist facies metamorphism: a discussion. *J. Petrol.* **27**, 751–761.

Wyllie, P. J. (1962) The effect of 'impure' pore fluids on metamorphic dissociation reactions. *Mineral. Mag.* **33**, 9–25.

Wyllie, P. J. (1971) *The Dynamic Earth*. John Wiley, New York, 416 pp.

Yardley, B. W. D. (1989) *An Introduction to Metamorphic Petrology*. Longman, Harlow, England, 248 pp.

Yoder, H. S. and Tilley, C. E. (1962) Origin of basalt magmas: an experimental study of natural and synthetic rock systems. *J. Petrol.* **3**, 342–532.

Yokoyama, K., Brothers, R. N. and Black, P. M. (1986) Regional eclogite facies in the high-pressure metamorphic belt of New Caledonia. *Geol. Soc. Amer. Mem.* **164**, 407–423.

Zen, E-an (1960) Metamorphism of Lower Paleozoic rocks in the vicinity of the Taconic Range in west-central Vermont. *Am. Mineralogist* **45**, 129–175.

Zen, E-an (1961) The zeolite facies: an interpretation. *Am. J. Sci.* **259**, 401–409.

Zen, E-an (1966) Construction of pressure-temperature diagrams for multicomponent systems after the method of Schreinemakers – a geometric approach. *U. S. Geol. Surv. Bull.* 1225.

Zen, E-an (1974a) Burial metamorphism. *Canad. Mineralogist* **12**, 445–455.

Zen, E-an (1974b) Prehnite- and pumpellyite-bearing mineral assemblages, west side of the Appalachian metamorphic belt, Pennsylvania to Newfoundland. *J. Petrol.* **15**, 197–242.

Zen, E-an (1981) Metamorphic mineral assemblages of slightly calcic pelitic rocks in and around the Taconic allochthon, southwestern Massachusetts and adjacent Connecticut and New York. *U. S. Geol. Surv. Prof. Paper* 1113.

Zen, E-an and Thompson, A. B. (1974) Low grade regional metamorphism: mineral equilibrium relations. *Annual Review of Earth and Planetary Sciences*, Vol.2, 179–212.

Index